Mathematics Curriculum

AF510701

Table of Contents[1]

Congruence, Proof, and Constructions

Module Overview 3

Topic A: Basic Constructions (**G-CO.A.1**, **G-CO.D.12**, **G-CO.D.13**) 11

Lessons 1–2: Construct an Equilateral Triangle 12

Lesson 3: Copy and Bisect an Angle 25

Lesson 4: Construct a Perpendicular Bisector 35

Lesson 5: Points of Concurrencies 43

Topic B: Unknown Angles (**G-CO.C.9**) 48

Lesson 6: Solve for Unknown Angles—Angles and Lines at a Point 49

Lesson 7: Solve for Unknown Angles—Transversals 61

Lesson 8: Solve for Unknown Angles—Angles in a Triangle 70

Lesson 9: Unknown Angle Proofs—Writing Proofs 76

Lesson 10: Unknown Angle Proofs—Proofs with Constructions 83

Lesson 11: Unknown Angle Proofs—Proofs of Known Facts 90

Topic C: Transformations/Rigid Motions (**G-CO.A.2**, **G-CO.A.3**, **G-CO.A.4**, **G-CO.A.5**, **G-CO.B.6**, **G-CO.B.7**, **G-CO.D.12**) 98

Lesson 12: Transformations—The Next Level 100

Lesson 13: Rotations 109

Lesson 14: Reflections 119

Lesson 15: Rotations, Reflections, and Symmetry 127

Lesson 16: Translations 134

Lesson 17: Characterize Points on a Perpendicular Bisector 142

Lesson 18: Looking More Carefully at Parallel Lines 149

Lesson 19: Construct and Apply a Sequence of Rigid Motions 159

Lesson 20: Applications of Congruence in Terms of Rigid Motions 164

[1]Each lesson is ONE day, and ONE day is considered a 45-minute period.

Lesson 21: Correspondence and Transformations 170

Mid-Module Assessment and Rubric 175
Topics A through C (assessment 1 day, return 1 day, remediation or further applications 2 days)

Topic D: Congruence (**G-CO.B.7**, **G-CO.B.8**) 189

Lesson 22: Congruence Criteria for Triangles—SAS 190

Lesson 23: Base Angles of Isosceles Triangles 200

Lesson 24: Congruence Criteria for Triangles—ASA and SSS 208

Lesson 25: Congruence Criteria for Triangles—AAS and HL 215

Lessons 26–27: Triangle Congruency Proofs 222

Topic E: Proving Properties of Geometric Figures (**G-CO.C.9**, **G-CO.C.10**, **G-CO.C.11**) 232

Lesson 28: Properties of Parallelograms 233

Lessons 29–30: Special Lines in Triangles 243

Topic F: Advanced Constructions (**G-CO.D.13**) 255

Lesson 31: Construct a Square and a Nine-Point Circle 256

Lesson 32: Construct a Nine-Point Circle 261

Topic G: Axiomatic Systems (**G-CO.A.1**, **G-CO.A.2**, **G-CO.A.3**, **G-CO.A.4**, **G-CO.A.5**, **G-CO.B.6**, **G-CO.B.7**, **G-CO.B.8**, **G-CO.C.9**, **G-CO.C.10**, **G-CO.C.11**, **G-CO.C.12**, **G-CO.C.13**) 265

Lessons 33–34: Review of the Assumptions 267

End-of-Module Assessment and Rubric 280
Topics A through G (assessment 1 day, return 2 days, remediation or further applications 3 days)

Geometry • Module 1

Congruence, Proof, and Constructions

OVERVIEW

Module 1 embodies critical changes in Geometry as outlined by the Common Core. The heart of the module is the study of transformations and the role transformations play in defining congruence.

Students begin this module with Topic A, Basic Constructions. Major constructions include an equilateral triangle, an angle bisector, and a perpendicular bisector. Students synthesize their knowledge of geometric terms with the use of new tools and simultaneously practice precise use of language and efficient communication when they write the steps that accompany each construction (**G.CO.A.1**).

Constructions segue into Topic B, Unknown Angles, which consists of unknown angle problems and proofs. These exercises consolidate students' prior body of geometric facts and prime students' reasoning abilities as they begin to justify each step for a solution to a problem. Students began the proof writing process in Grade 8 when they developed informal arguments to establish select geometric facts (**8.G.A.5**).

Topics C and D, Transformations/Rigid Motions and Congruence, build on students' intuitive understanding developed in Grade 8. With the help of manipulatives, students observed how reflections, translations, and rotations behave individually and in sequence (**8.G.A.1**, **8.G.A.2**). In high school Geometry, this experience is formalized by clear definitions (**G.CO.A.4**) and more in-depth exploration (**G.CO.A.3**, **G.CO.A.5**). The concrete establishment of rigid motions also allows proofs of facts formerly accepted to be true (**G.CO.C.9**). Similarly, students' Grade 8 concept of congruence transitions from a hands-on understanding (**8.G.A.2**) to a precise, formally notated understanding of congruence (**G.CO.B.6**). With a solid understanding of how transformations form the basis of congruence, students next examine triangle congruence criteria. Part of this examination includes the use of rigid motions to prove how triangle congruence criteria such as SAS actually work (**G.CO.B.7**, **G.CO.B.8**).

In Topic E, Proving Properties of Geometric Figures, students use what they have learned in Topics A through D to prove properties—those that have been accepted as true and those that are new—of parallelograms and triangles (**G.CO.C.10**, **G.CO.C.11**). The module closes with a return to constructions in Topic F (**G.CO.D.13**), followed by a review of the module that highlights how geometric assumptions underpin the facts established thereafter (Topic G).

Focus Standards

Experiment with transformations in the plane.

G-CO.A.1 Know precise definitions of angle, circle, perpendicular line, parallel line, and line segment, based on the undefined notions of point, line, distance along a line, and distance around a circular arc.

G-CO.A.2 Represent transformations in the plane using, e.g., transparencies and geometry software; describe transformations as functions that take points in the plane as inputs and give other points as outputs. Compare transformations that preserve distance and angle to those that do not (e.g., translation versus horizontal stretch).

G-CO.A.3 Given a rectangle, parallelogram, trapezoid, or regular polygon, describe the rotations and reflections that carry it onto itself.

G-CO.A.4 Develop definitions of rotations, reflections, and translations in terms of angles, circles, perpendicular lines, parallel lines, and line segments.

G-CO.A.5 Given a geometric figure and a rotation, reflection, or translation, draw the transformed figure using, e.g., graph paper, tracing paper, or geometry software. Specify a sequence of transformations that will carry a given figure onto another.

Understand congruence in terms of rigid motions.

G-CO.B.6 Use geometric descriptions of rigid motions to transform figures and to predict the effect of a given rigid motion on a given figure; given two figures, use the definition of congruence in terms of rigid motions to decide if they are congruent.

G-CO.B.7 Use the definition of congruence in terms of rigid motions to show that two triangles are congruent if and only if corresponding pairs of sides and corresponding pairs of angles are congruent.

G-CO.B.8 Explain how the criteria for triangle congruence (ASA, SAS, and SSS) follow from the definition of congruence in terms of rigid motions.

Prove geometric theorems.

G-CO.C.9 Prove theorems about lines and angles. *Theorems include: vertical angles are congruent; when a transversal crosses parallel lines, alternate interior angles are congruent and corresponding angles are congruent; points on a perpendicular bisector of a line segment are exactly those equidistant from the segment's endpoints.*

G-CO.C.10 Prove theorems about triangles. *Theorems include: measures of interior angles of a triangle sum to* $180°$*; base angles of isosceles triangles are congruent; the segment joining midpoints of two sides of a triangle is parallel to the third side and half the length; the medians of a triangle meet at a point.*

G-CO.C.11 Prove theorems about parallelograms. *Theorems include: opposite sides are congruent, opposite angles are congruent, the diagonals of a parallelogram bisect each other, and conversely, rectangles are parallelograms with congruent diagonals.*

Make geometric constructions.

G-CO.D.12 Make formal geometric constructions with a variety of tools and methods (compass and straightedge, string, reflective devices, paper folding, dynamic geometric software, etc.). *Copying a segment; copying an angle; bisecting a segment; bisecting an angle; constructing perpendicular lines, including the perpendicular bisector of a line segment; and constructing a line parallel to a given line through a point not on the line.*

G-CO.D.13 Construct an equilateral triangle, a square, and a regular hexagon inscribed in a circle.

Foundational Standards

Understand congruence and similarity using physical models, transparencies, or geometry software.

8.G.A.1 Verify experimentally the properties of rotations, reflections, and translations:

a. Lines are taken to lines, and line segments to line segments of the same length.

b. Angles are taken to angles of the same measure.

c. Parallel lines are taken to parallel lines.

8.G.A.2 Understand that a two-dimensional figure is congruent to another if the second can be obtained from the first by a sequence of rotations, reflections, and translations; given two congruent figures, describe a sequence that exhibits the congruence between them.

8.G.A.3 Describe the effect of dilations, translations, rotations, and reflections on two-dimensional figures using coordinates.

8.G.A.5 Use informal arguments to establish facts about the angle sum and exterior angle of triangles, about the angles created when parallel lines are cut by a transversal, and the angle-angle criterion for similarity of triangles. *For example, arrange three copies of the same triangle so that the sum of the three angles appears to form a line, and give an argument in terms of transversals why this is so.*

Focus Standards for Mathematical Practice

MP.3 **Construct viable arguments and critique the reasoning of others**. Students articulate the steps needed to construct geometric figures, using relevant vocabulary. Students develop and justify conclusions about unknown angles and defend their arguments with geometric reasons.

MP.4 **Model with mathematics**. Students apply geometric constructions and knowledge of rigid motions to solve problems arising with issues of design or location of facilities.

MP.5 **Use appropriate tools strategically**. Students consider and select from a variety of tools in constructing geometric diagrams, including (but not limited to) technological tools.

MP.6 **Attend to precision.** Students precisely define the various rigid motions. Students demonstrate polygon congruence, parallel status, and perpendicular status via formal and informal proofs. In addition, students clearly and precisely articulate steps in proofs and constructions throughout the module.

Terminology

New or Recently Introduced Terms

- **Isometry** (An *isometry* of the plane is a transformation of the plane that is distance-preserving.)

Familiar Terms and Symbols[2]

- Congruence
- Reflection
- Rotation
- Transformation
- Translation

Suggested Tools and Representations

- Compass and straightedge
- Geometer's Sketchpad or Geogebra Software
- Patty paper

[2]These are terms and symbols students have seen previously.

Preparing to Teach a Module

Preparation of lessons will be more effective and efficient if there has been an adequate analysis of the module first. Each module in *A Story of Functions* can be compared to a chapter in a book. How is the module moving the plot, the mathematics, forward? What new learning is taking place? How are the topics and objectives building on one another? The following is a suggested process for preparing to teach a module.

Step 1: Get a preview of the plot.

A: Read the Table of Contents. At a high level, what is the plot of the module? How does the story develop across the topics?

B: Preview the module's Exit Tickets to see the trajectory of the module's mathematics and the nature of the work students are expected to be able to do.

Note: When studying a PDF file, enter "Exit Ticket" into the search feature to navigate from one Exit Ticket to the next.

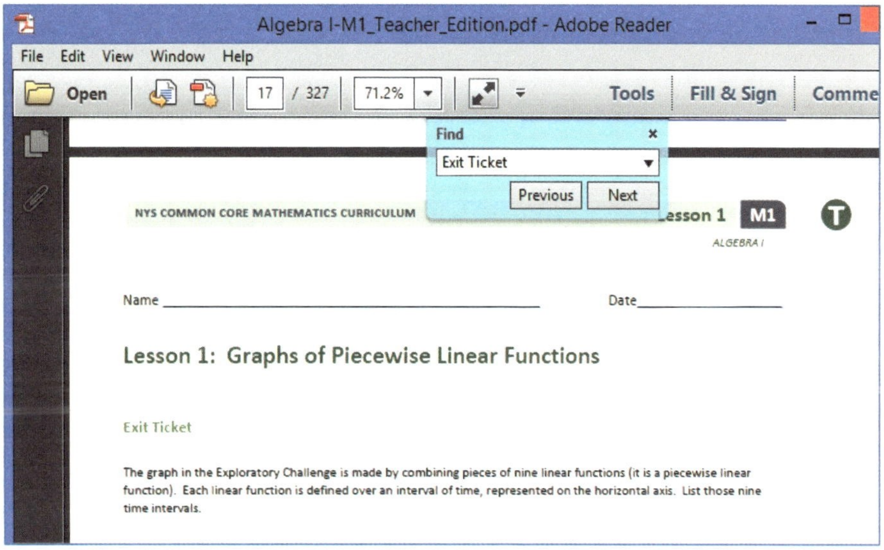

Step 2: Dig into the details.

A: Dig into a careful reading of the Module Overview. While reading the narrative, liberally reference the lessons and Topic Overviews to clarify the meaning of the text—the lessons demonstrate the strategies, show how to use the models, clarify vocabulary, and build understanding of concepts.

B: Having thoroughly investigated the Module Overview, read through the Student Outcomes of each lesson (in order) to further discern the plot of the module. How do the topics flow and tell a coherent story? How do the outcomes move students to new understandings?

Step 3: Summarize the story.

Complete the Mid- and End-of-Module Assessments. Use the strategies and models presented in the module to explain the thinking involved. Again, liberally reference the lessons to anticipate how students who are learning with the curriculum might respond.

Preparing to Teach a Lesson

A three-step process is suggested to prepare a lesson. It is understood that at times teachers may need to make adjustments (customizations) to lessons to fit the time constraints and unique needs of their students. The recommended planning process is outlined below. Note: The ladder of Step 2 is a metaphor for the teaching sequence. The sequence can be seen not only at the macro level in the role that this lesson plays in the overall story, but also at the lesson level, where each rung in the ladder represents the next step in understanding or the next skill needed to reach the objective. To reach the objective, or the top of the ladder, all students must be able to access the first rung and each successive rung.

Step 1: Discern the plot.

- A: Briefly review the module's Table of Contents, recalling the overall story of the module and analyzing the role of this lesson in the module.
- B: Read the Topic Overview related to the lesson, and then review the Student Outcome(s) and Exit Ticket of each lesson in the topic.
- C: Review the assessment following the topic, keeping in mind that assessments can be found midway through the module and at the end of the module.

Step 2: Find the ladder.

- A: Work through the lesson, answering and completing each question, example, exercise, and challenge.
- B: Analyze and write notes on the new complexities or new concepts introduced with each question or problem posed; these notes on the sequence of new complexities and concepts are the rungs of the ladder.
- C: Anticipate where students might struggle, and write a note about the potential cause of the struggle.
- D: Answer the Closing questions, always anticipating how students will respond.

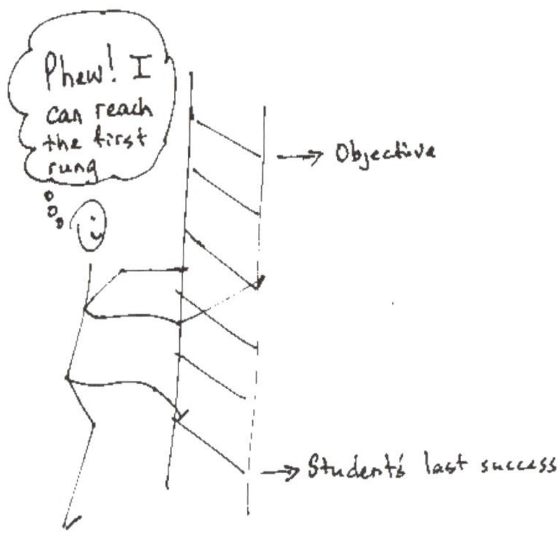

Step 3: Hone the lesson.

Lessons may need to be customized if the class period is not long enough to do all of what is presented and/or if students lack prerequisite skills and understanding to move through the entire lesson in the time allotted. A suggestion for customizing the lesson is to first decide upon and designate each question, example, exercise, or challenge as either "Must Do" or "Could Do."

- A: Select "Must Do" dialogue, questions, and problems that meet the Student Outcome(s) while still providing a coherent experience for students; reference the ladder. The expectation should be that the majority of the class will be able to complete the "Must Do" portions of the lesson within the allocated time. While choosing the "Must Do" portions of the lesson, keep in mind the need for a balance of dialogue and conceptual questioning, application problems, and abstract problems, and a balance between students using pictorial/graphical representations and abstract representations. Highlight dialogue to be included in the delivery of instruction so that students have a chance to articulate and consolidate understanding as they move through the lesson.

B: "Must Do" portions might also include remedial work as necessary for the whole class, a small group, or individual students. Depending on the anticipated difficulties, the remedial work might take on different forms as suggested in the chart below.

Anticipated Difficulty	**"Must Do" Remedial Problem Suggestion**
The first problem of the lesson is too challenging.	Write a short sequence of problems on the board that provides a ladder to Problem 1. Direct students to complete those first problems to empower them to begin the lesson.
There is too big of a jump in complexity between two problems.	Provide a problem or set of problems that bridge student understanding from one problem to the next.
Students lack fluency or foundational skills necessary for the lesson.	Before beginning the lesson, do a quick, engaging fluency exercise[3]. Before beginning any fluency activity for the first time, assess that students have conceptual understanding of the problems in the set and that they are poised for success with the easiest problem in the set.
More work is needed at the concrete or pictorial level.	Provide manipulatives or the opportunity to draw solution strategies.
More work is needed at the abstract level.	Add a set of abstract problems to be completed toward the end of the lesson.

C: "Could Do" problems are for students who work with greater fluency and understanding and can, therefore, complete more work within a given time frame.

D: At times, a particularly complex problem might be designated as a "Challenge!" problem to provide to advanced students. Consider creating the opportunity for students to share their "Challenge!" solutions with the class at a weekly session or on video.

E: If the lesson is customized, be sure to carefully select Closing questions that reflect such decisions, and adjust the Exit Ticket if necessary.

[3]Look for fluency suggestions at www.eureka-math.org.

Assessment Summary

Assessment Type	Administered	Format	Standards Addressed
Mid-Module Assessment Task	After Topic C	Constructed response with rubric	G-CO.A.1, G-CO.A.2, G-CO.A.4, G-CO.A.5, G-CO.B.6, G-CO.C.9, G-CO.D.12
End-of-Module Assessment Task	After Topic G	Constructed response with rubric	G-CO.A.2, G-CO.A.3, G-CO.B.7, G-CO.B.8, G-CO.C.10, G-CO.C.11, G-CO.D.13

Topic A

Basic Constructions

G-CO.A.1, G-CO.D.12, G-CO.D.13

Focus Standards:	G-CO.A.1	Know precise definitions of angle, circle, perpendicular line, parallel line, and line segment, based on the undefined notions of point, line, distance along a line, and distance around a circular arc.
	G-CO.D.12	Make formal geometric constructions with a variety of tools and methods (compass and straightedge, string, reflective devices, paper folding, dynamic geometric software, etc.). *Copying a segment; copying an angle; bisecting a segment; bisecting an angle; constructing perpendicular lines, including the perpendicular bisector of a line segment; and constructing a line parallel to a given line through a point not on the line.*
	G-CO.D.13	Construct an equilateral triangle, a square, and a regular hexagon inscribed in a circle.
Instructional Days:	5	
Lessons 1–2:	Construct an Equilateral Triangle (M, E)[1]	
Lesson 3:	Copy and Bisect an Angle (M)	
Lesson 4:	Construct a Perpendicular Bisector (M)	
Lesson 5:	Points of Concurrencies (E)	

The first module of Geometry incorporates and formalizes geometric concepts presented in all the different grade levels up to high school geometry. Topic A brings the relatively unfamiliar concept of construction to life by building upon ideas with which students are familiar, such as the constant length of the radius within a circle. While the figures that are being constructed may not be novel, the process of using tools to create the figures is certainly new. Students use construction tools, such as a compass, straightedge, and patty paper to create constructions of varying difficulty, including equilateral triangles, perpendicular bisectors, and angle bisectors. The constructions are embedded in models that require students to make sense of their space and to understand how to find an appropriate solution with their tools. Students also discover the critical need for precise language when they articulate the steps necessary for each construction. The figures covered throughout the topic provide a bridge to solving and then proving unknown angle problems.

[1]Lesson Structure Key: **P**-Problem Set Lesson, **M**-Modeling Cycle Lesson, **E**-Exploration Lesson, **S**-Socratic Lesson

Lesson 1: Construct an Equilateral Triangle

Student Outcomes

- Students learn to construct an equilateral triangle.
- Students communicate mathematic ideas effectively and efficiently.

Lesson Notes

Most students have done little more than draw a circle with a compass upon entering high school Geometry. Students truly acquire a whole new set of skills during the first few lessons on constructions.

This lesson begins with a brief Opening Exercise, which requires peer-to-peer conversation and attention to vocabulary. Ensure students understand that, even though the vocabulary terms may be familiar, they should pay careful attention to the *precision of each definition*. For students to develop logical reasoning in geometry, they have to manipulate very exact language, beginning with definitions. Students explore various phrasings of definitions. The teacher guides the discussion until students arrive at a formulation of the standard definition. The purpose of the discussion is to understand why the definition has the form that it does. As part of the discussion, students should be able to test the strength of any definition by looking for possible counterexamples.

Sitting Cats, the main exercise, provides a backdrop to constructing the equilateral triangle. Though students may visually understand where the position of the third cat should be, they spend time discovering how to use their compass to establish the exact location. (The cat, obviously, is in a position that approximates the third vertex. The point constructed is the optimal position of the cat—if cats were points and were perfect in their choice of place to sleep.) Students should work without assistance for some portion of the 10 minutes allotted. As students begin to successfully complete the task, elicit discussion about the use of the compass that makes this construction possible.

In the last segment of class, lead students through Euclid's Proposition 1 of Book 1 (Elements 1:1). Have students annotate the text as they read, noting how labeling is used to direct instructions. After reading through the document, direct students to write in their own words the steps they took to construct an equilateral triangle. As part of the broader goal of teaching students to communicate precisely and effectively in geometry, emphasize the need for clear instruction, for labeling in their diagram and reference to labeling in the steps, and for coherent use of relevant vocabulary. Students should begin the process in class together but should complete the assignment for the Problem Set.

Classwork

Opening Exercise (10 minutes)

Students should brainstorm ideas in pairs. Students may think of the use of counting footsteps, rope, or measuring tape to make the distances between friends precise. The fill-in-the-blanks activity is provided as scaffolding; students may also discuss the terms with a neighbor or as a class and write their own definitions based on discussion.

Opening Exercise

Joe and Marty are in the park playing catch. Tony joins them, and the boys want to stand so that the distance between any two of them is the same. Where do they stand?

How do they figure this out precisely? What tool or tools could they use?

Fill in the blanks below as each term is discussed:

a. *Segment* The _______ between points A and B is the set consisting of A, B, and all points on the line AB between A and B.

b. *Radius* A segment from the center of a circle to a point on the circle

c. *Circle* Given a point C in the plane and a number $r > 0$, the _______ with center C and radius r is the set of all points in the plane that are distance r from point C.

Note that because a circle is defined in terms of a distance, r, we often use a distance when naming the radius (e.g., "radius AB"). However, we may also refer to the specific segment, as in "radius $\overline{AB}$."

Example 1 (10 minutes): Sitting Cats

Students explore how to construct an equilateral triangle using a compass.

MP.5

Example 1: Sitting Cats

You need a compass and a straightedge.

Margie has three cats. She has heard that cats in a room position themselves at equal distances from one another and wants to test that theory. Margie notices that Simon, her tabby cat, is in the center of her bed (at S), while JoJo, her Siamese, is lying on her desk chair (at J). If the theory is true, where will she find Mack, her calico cat? Use the scale drawing of Margie's room shown below, together with (only) a compass and straightedge. Place an M where Mack will be if the theory is true.

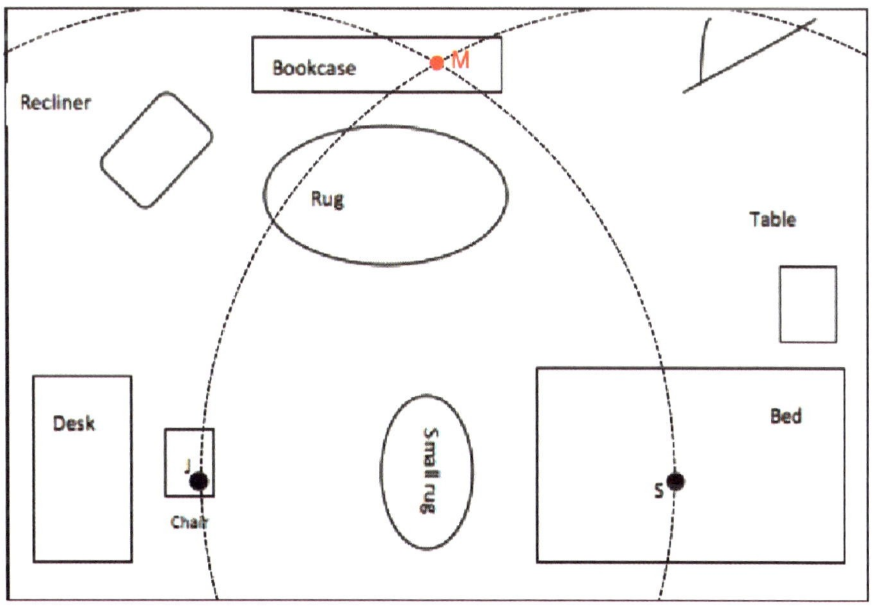

Mathematical Modeling Exercise (12 minutes): Euclid, Proposition 1

Students examine Euclid's solution of how to construct an equilateral triangle.

Lead students through this excerpt, and have them annotate the text as they read it. The goal is for students to form a rough set of steps that outlines the construction of the equilateral triangle. Once a first attempt of the steps is made, review them as if they are a step-by-step guide. Ask the class if the steps need refinement. This is to build to the Problem Set question, which asks students to write a clear and succinct set of instructions for the construction of the equilateral triangle.

Mathematical Modeling Exercise: Euclid, Proposition 1

Let's see how Euclid approached this problem. Look at his first proposition, and compare his steps with yours.

Proposition 1

To construct an equilateral triangle on a given finite straight-line.

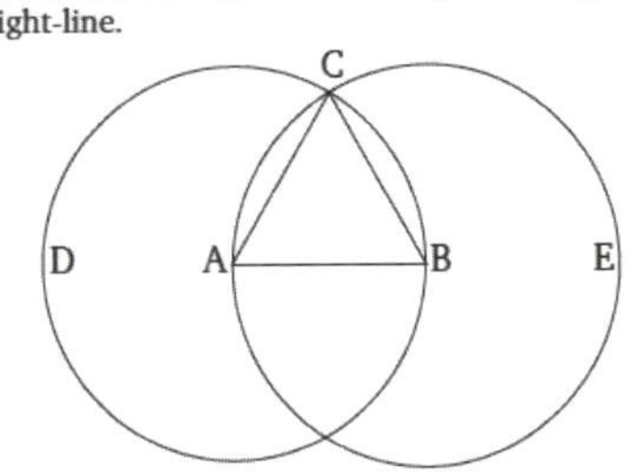

Let AB be the given finite straight-line.

So it is required to construct an equilateral triangle on the straight-line AB.

Let the circle BCD with center A and radius AB have been drawn [Post. 3], and again let the circle ACE with center B and radius BA have been drawn [Post. 3]. And let the straight-lines CA and CB have been joined from the point C, where the circles cut one another,† to the points A and B (respectively) [Post. 1].

And since the point A is the center of the circle CDB, AC is equal to AB [Def. 1.15]. Again, since the point B is the center of the circle CAE, BC is equal to BA [Def. 1.15]. But CA was also shown (to be) equal to AB. Thus, CA and CB are each equal to AB. But things equal to the same thing are also equal to one another [C.N. 1]. Thus, CA is also equal to CB. Thus, the three (straight-lines) CA, AB, and BC are equal to one another.

Thus, the triangle ABC is equilateral, and has been constructed on the given finite straight-line AB. (Which is) the very thing it was required to do.

In this margin, compare your steps with Euclid's.

Geometry Assumptions (7 minutes)

Geometry Assumptions

In geometry, as in most fields, there are specific facts and definitions that we assume to be true. In any logical system, it helps to identify these assumptions as early as possible since the correctness of any proof hinges upon the truth of our assumptions. For example, in Proposition 1, when Euclid says, "Let AB be the given finite straight line," he assumed that, *given any two distinct points, there is exactly one line that contains them*. Of course, that assumes we have two points! It is best if we assume there are points in the plane as well: *Every plane contains at least three noncollinear points.*

Euclid continued on to show that the measures of each of the three sides of his triangle are equal. It makes sense to discuss the measure of a segment in terms of distance. *To every pair of points A and B, there corresponds a real number $\text{dist}(A, B) \geq 0$, called the distance from A to B.* Since the distance from A to B is equal to the distance from B to A, we can interchange A and B: $\text{dist}(A, B) = \text{dist}(B, A)$. Also, *$A$ and B coincide if and only if* $\text{dist}(A, B) = 0$.

Using distance, we can also assume that *every line has a coordinate system*, which just means that we can think of any line in the plane as a number line. Here's how: Given a line, l, pick a point A on l to be "0," and find the two points B and C such that $\text{dist}(A, B) = \text{dist}(A, C) = 1$. Label one of these points to be 1 (say point B), which means the other point C corresponds to -1. Every other point on the line then corresponds to a real number determined by the (positive or negative) distance between 0 and the point. In particular, if after placing a coordinate system on a line, if a point R corresponds to the number r, and a point S corresponds to the number s, then the distance from R to S is $\text{dist}(R, S) = |r - s|$.

History of Geometry: Examine the site http://geomhistory.com/home.html to see how geometry developed over time.

Relevant Vocabulary (3 minutes)

The terms *point*, *line*, *plane*, *distance along a line*, *betweenness*, *space*, and *distance around a circular arc* are all left as undefined terms; that is, they are only given intuitive descriptions. For example, a point can be described as a location in the plane, and a straight line can be said to extend in two opposite directions forever. It should be emphasized that, while we give these terms pictorial representations (like drawing a dot on the board to represent a point), they are concepts, and they only exist in the sense that other geometric ideas depend on them. Spend time discussing these terms with students.

Relevant Vocabulary

GEOMETRIC CONSTRUCTION: A *geometric construction* is a set of instructions for drawing points, lines, circles, and figures in the plane.

The two most basic types of instructions are the following:

1. Given any two points A and B, a straightedge can be used to draw the line AB or segment AB.
2. Given any two points C and B, use a compass to draw the circle that has its center at C that passes through B. (Abbreviation: Draw circle C: center C, radius CB.)

Constructions also include steps in which the points where lines or circles intersect are selected and labeled. (Abbreviation: Mark the point of intersection of the line AB and line PQ by X, etc.)

FIGURE: A (two-dimensional) *figure* is a set of points in a plane.

Usually the term figure refers to certain common shapes such as triangle, square, rectangle, etc. However, the definition is broad enough to include any set of points, so a triangle with a line segment sticking out of it is also a figure.

EQUILATERAL TRIANGLE: An *equilateral triangle* is a triangle with all sides of equal length.

COLLINEAR: Three or more points are *collinear* if there is a line containing all of the points; otherwise, the points are noncollinear.

LENGTH OF A SEGMENT: The *length of* $\overline{AB}$ is the distance from A to B and is denoted AB. Thus, $AB = \text{dist}(A, B)$.

In this course, you have to write about distances between points and lengths of segments in many, if not most, Problem Sets. Instead of writing $\text{dist}(A, B)$ all of the time, which is a rather long and awkward notation, we instead use the much simpler notation AB for both distance and length of segments. Even though the notation always makes the meaning of each statement clear, it is worthwhile to consider the context of the statement to ensure correct usage. Here are some examples:

- $\overleftrightarrow{AB}$ intersects... — $\overleftrightarrow{AB}$ refers to a line.
- $AB + BC = AC$ — Only numbers can be added, and AB is a length or distance.
- Find $\overline{AB}$ so that $\overline{AB} \parallel \overline{CD}$. — Only figures can be parallel, and $\overline{AB}$ is a segment.
- $AB = 6$ — AB refers to the length of $\overline{AB}$ or the distance from A to B.

Here are the standard notations for segments, lines, rays, distances, and lengths:

- A ray with vertex A that contains the point B: $\overrightarrow{AB}$ or "ray AB"
- A line that contains points A and B: $\overleftrightarrow{AB}$ or "line AB"
- A segment with endpoints A and B: $\overline{AB}$ or "segment AB"
- The length of $\overline{AB}$: AB
- The distance from A to B: $\text{dist}(A, B)$ or AB

COORDINATE SYSTEM ON A LINE: Given a line l, a *coordinate system on l* is a correspondence between the points on the line and the real numbers such that: (i) to every point on l, there corresponds exactly one real number; (ii) to every real number, there corresponds exactly one point of l; (iii) the distance between two distinct points on l is equal to the absolute value of the difference of the corresponding numbers.

Closing (1 minute)

- How does a compass aid in the construction of a circle?
  - *Using a compass allows us to draw circles, which we need to determine the last vertex of the equilateral triangle. The first two vertices are determined by the endpoints of the segment selected to be one side of the triangle. The third vertex is determined by drawing circles with radii equal to the length of that segment around each endpoint; either of the two intersections formed by the circles can serve as the third vertex.*

Exit Ticket (3 minutes)

Name ______________________________ Date ______________

Lesson 1: Construct an Equilateral Triangle

Exit Ticket

We saw two different scenarios where we used the construction of an equilateral triangle to help determine a needed location (i.e., the friends playing catch in the park and the sitting cats). Can you think of another scenario where the construction of an equilateral triangle might be useful? Articulate how you would find the needed location using an equilateral triangle.

Exit Ticket Sample Solution

We saw two different scenarios where we used the construction of an equilateral triangle to help determine a needed location (i.e., the friends playing catch in the park and the sitting cats). Can you think of another scenario where the construction of an equilateral triangle might be useful? Articulate how you would find the needed location using an equilateral triangle.

Students might describe a need to determine the locations of fire hydrants, friends meeting at a restaurant, or parking lots for a stadium, etc.

Problem Set Sample Solutions

MP.5

1. **Write a clear set of steps for the construction of an equilateral triangle. Use Euclid's Proposition 1 as a guide.**
   1. *Draw circle J: center J, radius JS.*
   2. *Draw circle S: center S, radius SJ.*
   3. *Label one intersection as M.*
   4. *Join S, J, M.*

2. **Suppose two circles are constructed using the following instructions:**

   Draw circle: center A, radius $\overline{AB}$.

   Draw circle: center C, radius $\overline{CD}$.

   Under what conditions (in terms of distances AB, CD, AC) do the circles have

   a. **One point in common?**

   If $AB + CD = AC$ or $AC + AB = CD$ or $AC + CD = AB$. *Ex.* 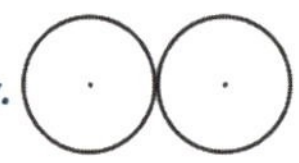

   b. **No points in common?**

   If $AB + CD < AC$ or $AB + AC < CD$ or $CD + AC < AB$. *Ex.* 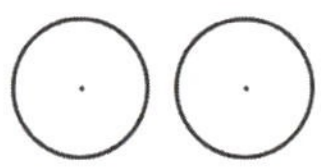

   c. **Two points in common?**

   If $AC < AB + CD$ and $CD < AB + AC$ and $AB < CD + AC$. *Ex.* 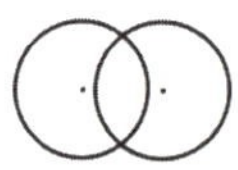

   d. **More than two points in common? Why?**

   If $A = C$ (same points) and $AB = CD$. All the points of the circle coincide since the circles themselves coincide. *Ex.* 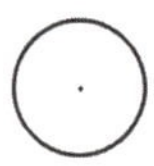

3. You need a compass and a straightedge.

Cedar City boasts two city parks and is in the process of designing a third. The planning committee would like all three parks to be equidistant from one another to better serve the community. A sketch of the city appears below, with the centers of the existing parks labeled as P_1 and P_2. Identify two possible locations for the third park, and label them as P_{3a} and P_{3b} on the map. Clearly and precisely list the mathematical steps used to determine each of the two potential locations.

1. *Draw a circle P_1: center P_1, radius $\overline{P_1P_2}$.*
2. *Draw a circle P_2: center P_2 radius $\overline{P_2P_1}$.*
3. *Label the two intersections of the circles as P_{3a} and P_{3b}.*
4. *Join P_1, P_2, P_{3a} and P_1, P_2, P_{3b}.*

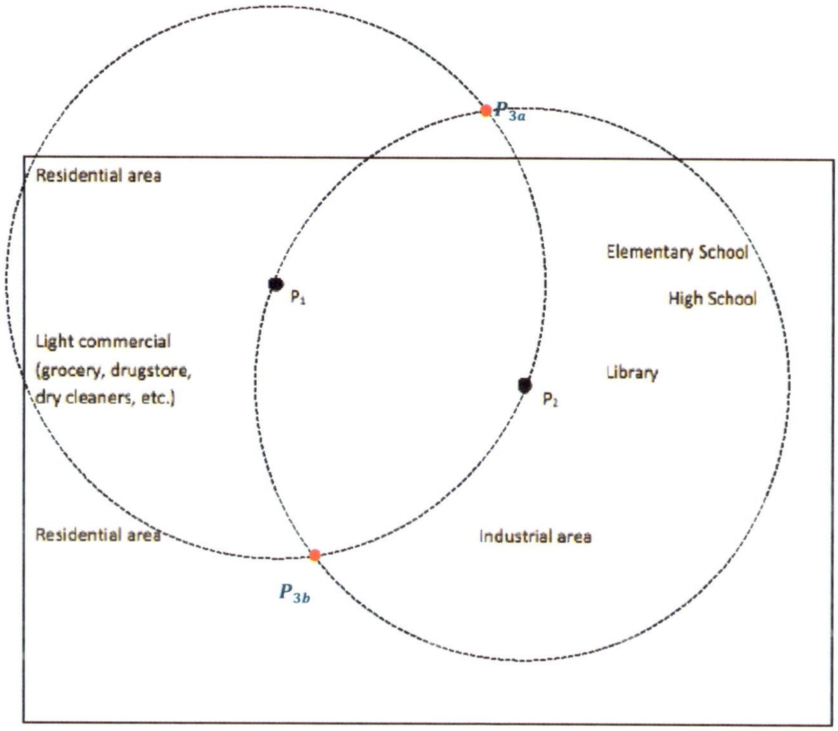

Lesson 2: Construct an Equilateral Triangle

Student Outcomes

- Students apply the equilateral triangle construction to more challenging problems.
- Students communicate mathematical concepts clearly and concisely.

Lesson Notes

Lesson 2 directly builds on the notes and exercises from Lesson 1. The continued lesson allows the class to review and assess understanding from Lesson 1. By the end of this lesson, students should be able to apply their knowledge of how to construct an equilateral triangle to more difficult constructions and to write clear and precise steps for these constructions.

Students critique each other's construction steps in the Opening Exercise; this is an opportunity to highlight Mathematical Practice 3. Through the critique, students experience how a lack of precision affects the outcome of a construction. Be prepared to guide the conversation to overcome student challenges, perhaps by referring back to the Euclid piece from Lesson 1 or by sharing personal writing examples. Remind students to focus on the vocabulary they are using in the directions because it becomes the basis of writing proofs as the year progresses.

In the Exploratory Challenges, students construct three equilateral triangles, two of which share a common side. Allow students to investigate independently before offering guidance. As students attempt the task, ask them to reflect on the significance of the use of circles for the problem.

Classwork

Opening Exercise (5 minutes)

Students should test each other's instructions for the construction of an equilateral triangle. The goal is to identify errors in the instructions or opportunities to make the instructions more concise.

Opening Exercise

You need a compass, a straightedge, and another student's Problem Set.

Directions:

Follow the directions from another student's Problem Set write-up to construct an equilateral triangle.

- **What kinds of problems did you have as you followed your classmate's directions?**
- **Think about ways to avoid these problems. What criteria or expectations for writing steps in constructions should be included in a rubric for evaluating your writing? List at least three criteria.**

Discussion (5 minutes)

MP.5

- What are common errors? What are concrete suggestions to help improve the instruction-writing process?
  - *Correct use of vocabulary, simple and concise steps (making sure each step involves just one instruction), and clear use of labels.*

It is important for students to describe objects using correct terminology instead of pronouns. Instead of "it" and "they," perhaps "the center" and "the sides" should be used.

Exploratory Challenge 1 (15 minutes)

Exploratory Challenge 1

You need a compass and a straightedge.

Using the skills you have practiced, construct three equilateral triangles, where the first and second triangles share a common side and the second and third triangles share a common side. Clearly and precisely list the steps needed to accomplish this construction.

Switch your list of steps with a partner, and complete the construction according to your partner's steps. Revise your drawing and list of steps as needed.

Construct three equilateral triangles here:

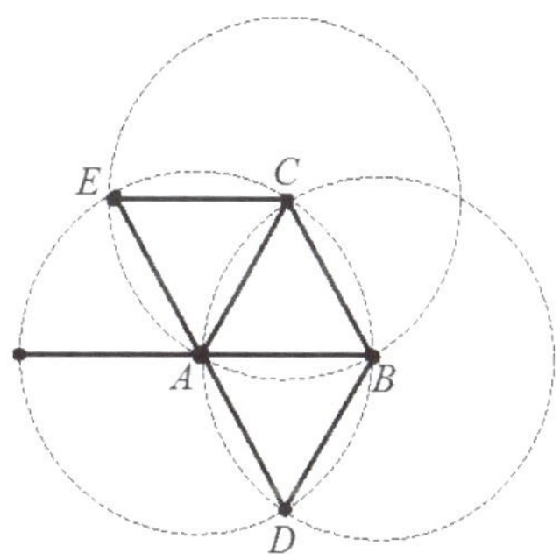

1. *Draw a segment AB.*
2. *Draw circle A: center A, radius AB.*
3. *Draw circle B: center B, radius BA.*
4. *Label one intersection as C; label the other intersection as D.*
5. *Draw circle C: center C, radius CA.*
6. *Label the intersection of circle C with circle A (or the intersection of circle C with circle B) as E.*
7. *Draw all segments that are congruent to $\overline{AB}$ between the labeled points.*

There are many ways to address Step 7; students should be careful to avoid making a blanket statement that would allow segment BE or segment CD.

Exploratory Challenge 2 (15 minutes)

Exploratory Challenge 2

On a separate piece of paper, use the skills you have developed in this lesson to construct a regular hexagon. Clearly and precisely list the steps needed to accomplish this construction. Compare your results with a partner, and revise your drawing and list of steps as needed.

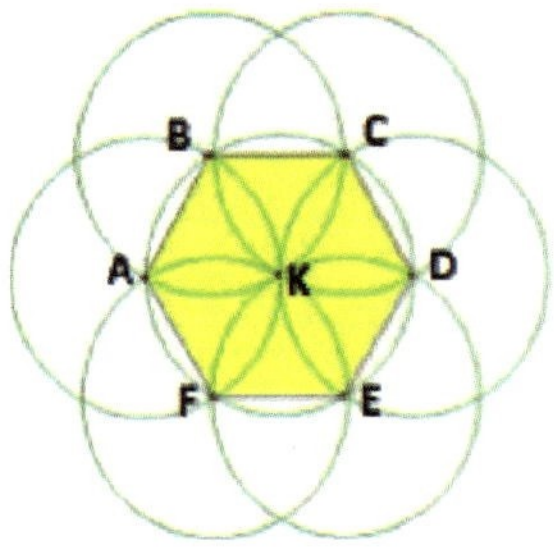

1. ***Draw circle K: center K, any radius.***
2. ***Pick a point on the circle; label this point A.***
3. ***Draw circle A: center A, radius AK.***
4. ***Label the intersections of circle A with circle K as B and F.***
5. ***Draw circle B: center B, radius BK.***
6. ***Label the intersection of circle B with circle K as C.***
7. ***Continue to treat the intersection of each new circle with circle K as the center of a new circle until the next circle to be drawn is circle A.***
8. ***Draw $\overline{AB}$, $\overline{BC}$, $\overline{CD}$, $\overline{DE}$, $\overline{EF}$, $\overline{FA}$.***

Can you repeat the construction of a hexagon until the entire sheet is covered in hexagons (except the edges, which are partial hexagons)?

Yes, this result resembles wallpaper, tile patterns, etc.

Closing (1 minute)

- What should be kept in mind during the instruction-writing process of constructions?
  - *Use specific language and vocabulary instead of pronouns, concise steps that involve one instruction, and articulated labels.*

Exit Ticket (5 minutes)

Name ______________________________ Date________________

Lesson 2: Construct an Equilateral Triangle

Exit Ticket

$\triangle ABC$ is shown below. Is it an equilateral triangle? Justify your response.

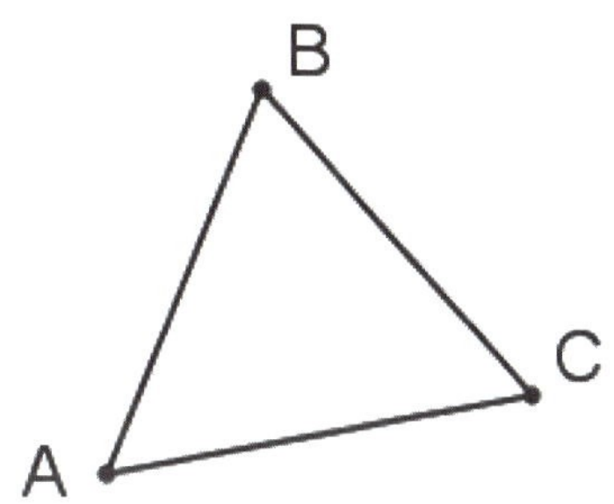

Exit Ticket Sample Solution

$\triangle ABC$ is shown below. Is it an equilateral triangle? Justify your response.

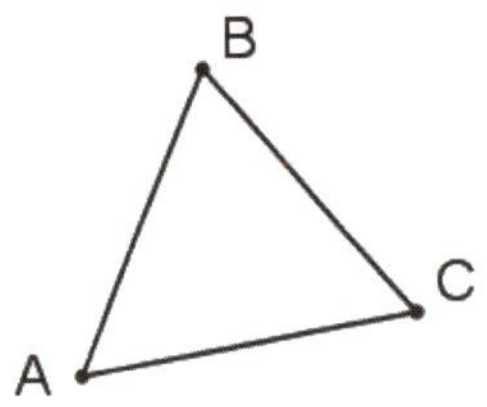

The triangle is not equilateral. Students may prove this by constructing two intersecting circles using any two vertices as the given starting segment. The third vertex will not be one of the two intersection points of the circles.

Problem Set Sample Solution

Why are *circles* so important to these constructions? Write out a concise explanation of the importance of circles in creating equilateral triangles. Why did Euclid use *circles* to create his equilateral triangles in Proposition 1? How does construction of a circle ensure that all relevant segments are of equal length?

The radius of equal-sized circles, which must be used in construction of an equilateral triangle, does not change. This consistent length guarantees that all three side lengths of the triangle are equal.

Lesson 3: Copy and Bisect an Angle

Student Outcomes

- Students learn how to bisect an angle as well as how to copy an angle.

  Note: These more advanced constructions require much more consideration in the communication of the students' steps.

Lesson Notes

In Lesson 3, students learn to copy and bisect an angle. As with Lessons 1 and 2, vocabulary and precision in language are essential to these next constructions.

Of the two constructions, the angle bisection is the simpler of the two and is the first construction in the lesson. Students watch a brief video clip to set the stage for the construction problem. Review the term *bisect*; ask if angles are the only figures that can be bisected. Discuss a method to test whether an angle bisector is really dividing an angle into two equal, adjacent angles. Help students connect the use of circles for this construction as they did for an equilateral triangle.

Next, students decide the correct order of provided steps to copy an angle. Teachers may choose to demonstrate the construction once before students attempt to rearrange the given steps (and after if needed). Encourage students to test their arrangement before making a final decision on the order of the steps.

Classwork

Opening Exercise (5 minutes)

Opening Exercise

In the following figure, circles have been constructed so that the endpoints of the diameter of each circle coincide with the endpoints of each segment of the equilateral triangle.

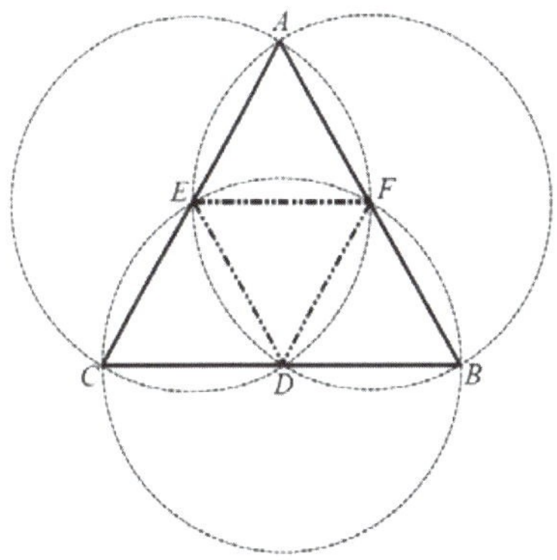

a. **What is special about points D, E, and F? Explain how this can be confirmed with the use of a compass.**

   D, E, and F are midpoints. The compass can be adjusted to the length of CE and used to compare to the lengths AE, AF, BF, CD, and BD.

b. **Draw $\overline{DE}$, $\overline{EF}$, and $\overline{FD}$. What kind of triangle must $\triangle DEF$ be?**

   $\triangle DEF$ is an equilateral triangle.

c. **What is special about the four triangles within $\triangle ABC$?**

   All four triangles are equilateral triangles of equal side lengths; they are congruent.

d. **How many times greater is the area of $\triangle ABC$ than the area of $\triangle CDE$?**

The area of $\triangle ABC$ is four times greater than the area of $\triangle CDE$.

Discussion (5 minutes)

Note that an angle is defined as the union of two *noncollinear* rays with the same endpoint to make the interior of the angle unambiguous; many definitions that follow depend on this clarity. Zero and straight angles are defined at the end of the lesson.

Discussion

Define the terms angle, interior of an angle, and angle bisector.

ANGLE: An *angle* is *the union of two noncollinear rays with the same endpoint.*

INTERIOR: The *interior of $\angle BAC$* is the set of points in the intersection of the half-plane of $\overleftrightarrow{AC}$ that contains B and the half-plane of $\overleftrightarrow{AB}$ that contains C. The interior is easy to identify because it is always the "smaller" region of the two regions defined by the angle (the region that is convex). The other region is called the *exterior* of the angle.

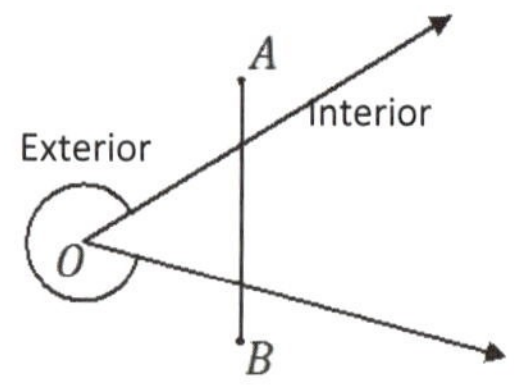

Note that every angle has two angle measurements corresponding to the interior and exterior regions of the angle (e.g., the angle measurement that corresponds to the number of degrees between $0°$ and $180°$ and the angle measurement that corresponds to the number of degrees between $180°$ and $360°$). To ensure there is absolutely no ambiguity about which angle measurement is being referred to in proofs, the angle measurement of an angle is always taken to be the number of degrees between $0°$ and $180°$. This deliberate choice is analogous to how the square root of a number is defined. Every positive number x has two square roots: $\sqrt{x}$ and $-\sqrt{x}$, so while $-\sqrt{x}$ is a square root of x, the square root of x is always taken to be $\sqrt{x}$.

For the most part, there is very little need to measure the number of degrees of an exterior region of an angle in this course. Virtually (if not all) of the angles measured in this course are either angles of triangles or angles formed by two lines (both measurements guaranteed to be less than $180°$). The degree measure of an arc is discussed in Module 5 and can be as large as $360°$, but an arc does not have any ambiguity like an angle does. Likewise, rotations can be specified by any positive *or negative* number of degrees, a point that becomes increasingly important in Algebra II. The main thing to keep straight and to make clear to students is that degree measurements do not automatically correspond to angles; rather, a degree measurement may be referring to an angle, an arc, or a rotation in this curriculum. For example, a degree measurement of $54°$ might be referring to the measurement of an angle, but it might also be referring to the degree measure of an arc or the number of degrees of a rotation. A degree measurement of $-734°$, however, is definitely referring to the number of degrees of a rotation.

ANGLE BISECTOR: If C is in the interior of $\angle AOB$, *and $m\angle AOC = m\angle COB$, then $\overrightarrow{OC}$ bisects $\angle AOB$, and $\overrightarrow{OC}$ is called the bisector of $\angle AOB$.* **When we say $m\angle AOC = m\angle COB$, we mean that the angle measures are equal.**

Geometry Assumptions (8 minutes)

Consider accompanying this discussion with drawn visuals to illustrate the assumptions.

> Geometry Assumptions
>
> In working with lines and angles, we again make specific assumptions that need to be identified. For example, in the definition of interior of an angle above, we assumed that an angle separated the plane into two disjoint sets. This follows from the assumption: *Given a line, the points of the plane that do not lie on the line form two sets called half-planes, such that (1) each of the sets is convex, and (2) if P is a point in one of the sets, and Q is a point in the other, then the segment PQ intersects the line.*
>
> From this assumption, another obvious fact follows about a segment that intersects the sides of an angle: *Given an $\angle AOB$, then for any point C in the interior of $\angle AOB$, the ray OC always intersects the segment AB.*
>
> In this lesson, we move from working with line segments to working with angles, specifically with bisecting angles. Before we do this, we need to clarify our assumptions about measuring angles. These assumptions are based upon what we know about a protractor that measures up to $180°$ angles:
>
> 1. To every $\angle AOB$ there corresponds a quantity $m\angle AOB$ called the degree or measure of the angle so that $0° < m\angle AOB < 180°$.
>
> This number, of course, can be thought of as the angle measurement (in degrees) of the interior part of the angle, which is what we read off of a protractor when measuring an angle. In particular, we have also seen that we can use protractors to "add angles":
>
> 2. If C is a point in the interior of $\angle AOB$, then $m\angle AOC + m\angle COB = m\angle AOB$.
>
> Two angles $\angle BAC$ and $\angle CAD$ form a *linear pair* if $\overrightarrow{AB}$ and $\overrightarrow{AD}$ are opposite rays on a line, and $\overrightarrow{AC}$ is any other ray. In earlier grades, we abbreviated this situation and the fact that the measures of the angles on a line add up to $180°$ as "$\angle$'s on a line." Now, we state it formally as one of our assumptions:
>
> 3. If two angles $\angle BAC$ and $\angle CAD$ form a linear pair, then they are supplementary (i.e., $m\angle BAC + m\angle CAD = 180$).
>
> Protractors also help us to draw angles of a specified measure:
>
> 4. Let $\overrightarrow{OB}$ be a ray on the edge of the half-plane H. For every r such that $0° < r° < 180°$, there is exactly one ray OA with A in H such that $m\angle AOB = r°$.

Mathematical Modeling Exercise 1 (12 minutes): Investigate How to Bisect an Angle

Watch the video, *Angles and Trim* (http://youtu.be/EBP3I8O9gIM).

Ask students to keep the steps in the video in mind as they read the scenarios following the video and attempt the angle bisector construction on their own. (The video actually demonstrates a possible construction.)

> *Note to Teacher:*
>
> The speaker in the clip misspeaks by using the word *protractor* instead of *compass*. The video can even be paused, providing an opportunity to ask if the class can identify the speaker's error.

Consider the following ideas:

- Are angles the only geometric figures that can be bisected?
  - *No. Segments can also be bisected.*
- What determines whether a figure can be bisected? What kinds of figures cannot be bisected?
  - *A line of reflection must exist so that when the figure is folded along this line, each point on one side of the line maps to a corresponding point on the other side of the line. A ray cannot be bisected.*

Mathematical Modeling Exercise 1: Investigate How to Bisect an Angle

You need a compass and a straightedge.

Joey and his brother, Jimmy, are working on making a picture frame as a birthday gift for their mother. Although they have the wooden pieces for the frame, they need to find the angle bisector to accurately fit the edges of the pieces together. Using your compass and straightedge, show how the boys bisected the corner angles of the wooden pieces below to create the finished frame on the right.

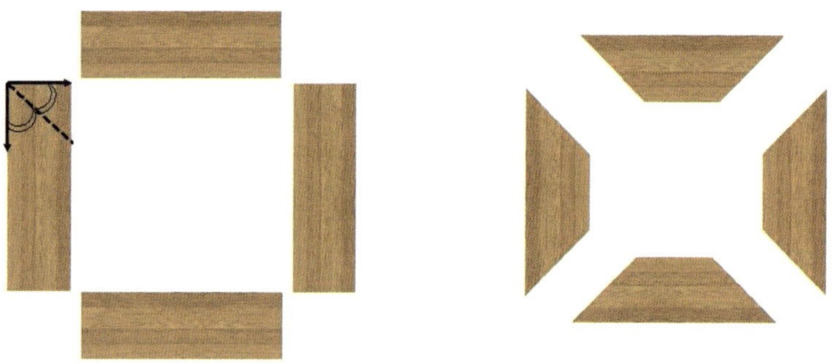

Consider how the use of circles aids the construction of an angle bisector. Be sure to label the construction as it progresses and to include the labels in your steps. Experiment with the angles below to determine the correct steps for the construction.

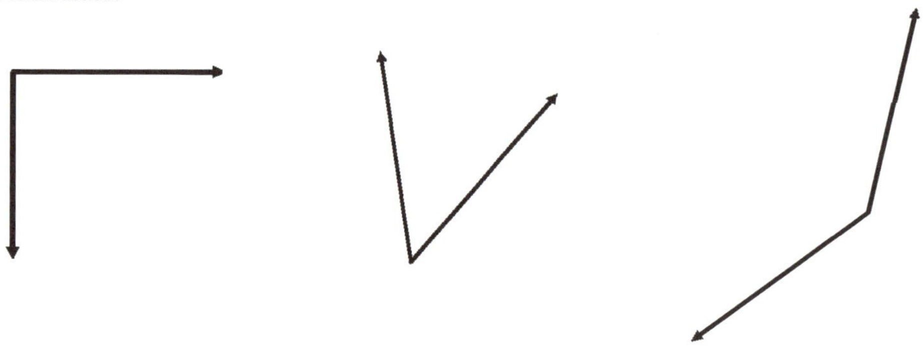

What steps did you take to bisect an angle? List the steps below:

MP.5 & MP.6

Steps to construct an angle bisector are as follows:

1. *Label vertex of angle as A.*
2. *Draw circle A: center A, any size radius.*
3. *Label intersections of circle A with rays of angle as B and C.*
4. *Draw circle B: center B, radius BC.*
5. *Draw circle C: center C, radius CB.*
6. *At least one of the two intersection points of circle B and circle C lie in the angle. Label that intersection point D.*
7. *Draw $\overrightarrow{AD}$.*

- How does the video's method of the angle bisector construction differ from the class's method? Are there fundamental differences, or is the video's method simply an expedited form of the class's method?
  - *Yes, the video's method is an expedited version with no fundamental difference from the class's method.*

After students have completed the angle bisector construction, direct their attention to the symmetry in the construction. Note that the same procedure is done to both sides of the angle, so the line constructed bears the same relationships to each side. This foreshadows the idea of reflections and connects this exercise to the deep themes coming later. (In fact, a reflection along the bisector ray takes the angle to itself.)

Mathematical Modeling Exercise 2 (12 minutes): Investigate How to Copy an Angle

For Mathematical Modeling Exercise 2, provide students with the Lesson 3 Supplement (Sorting Exercise) and scissors. They cut apart the steps listed in the Supplement and arrange them until they yield the steps in correct order.

Mathematical Modeling Exercise 2: Investigate How to Copy an Angle

You need a compass and a straightedge.

You and your partner have been provided with a list of steps (in random order) needed to copy an angle using a compass and straightedge. Your task is to place the steps in the correct order, and then follow the steps to copy the angle below.

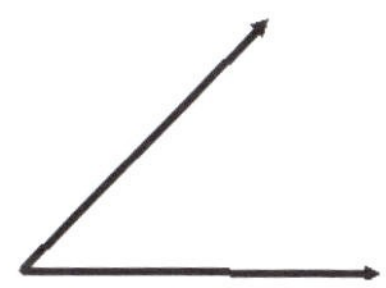

Steps needed (in correct order):

MP.5 & MP.6

Steps to copy an angle are as follows:

1. *Label the vertex of the original angle as B.*
2. *Draw $\overrightarrow{EG}$ as one side of the angle to be drawn.*
3. *Draw circle B: center B, any radius.*
4. *Label the intersections of circle B with the sides of the angle as A and C.*
5. *Draw circle E: center E, radius BA.*
6. *Label intersection of circle E with $\overrightarrow{EG}$ as F.*
7. *Draw circle F: center F, radius CA.*
8. *Label either intersection of circle E and circle F as D.*
9. *Draw $\overrightarrow{ED}$.*

Relevant Vocabulary

Relevant Vocabulary

MIDPOINT: A point B is called a *midpoint* of $\overline{AC}$ if B is between A and C, and $AB = BC$.

DEGREE: Subdivide the length around a circle into 360 arcs of equal length. A central angle for any of these arcs is called a one-degree angle and is said to have angle measure 1 degree. An angle that turns through n one-degree angles is said to have an angle measure of n degrees.

ZERO AND STRAIGHT ANGLE: A *zero angle* is just a ray and measures $0°$. A *straight angle* is a line and measures $180°$ (the ° is a symbol for *degree*).

Closing (1 minute)

- How can we use the expected symmetry of a bisected angle to determine whether the angle bisector construction has been done correctly?
    - *Fold the angle along the bisector: if the rays of the angle coincide, the construction has been performed correctly.*

Exit Ticket (3 minutes)

Name ______________________________ Date ______________

Lesson 3: Copy and Bisect an Angle

Exit Ticket

Later that day, Jimmy and Joey were working together to build a kite with sticks, newspapers, tape, and string. After they fastened the sticks together in the overall shape of the kite, Jimmy looked at the position of the sticks and said that each of the four corners of the kite is bisected; Joey said that they would only be able to bisect the top and bottom angles of the kite. Who is correct? Explain.

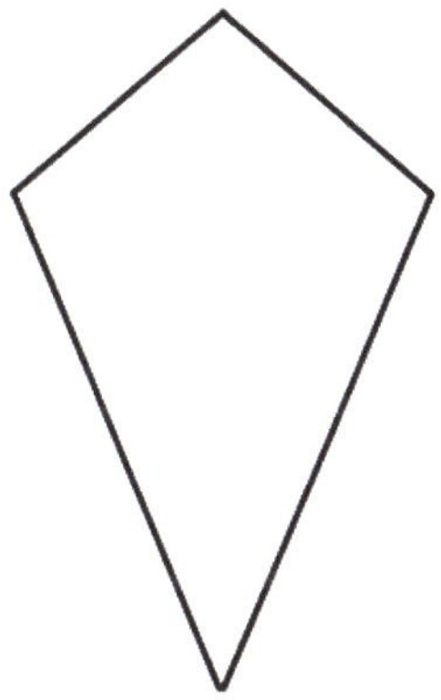

Exit Ticket Sample Solution

Later that day, Jimmy and Joey were working together to build a kite with sticks, newspapers, tape, and string. After they fastened the sticks together in the overall shape of the kite, Jimmy looked at the position of the sticks and said that each of the four corners of the kite is bisected; Joey said that they would only be able to bisect the top and bottom angles of the kite. Who is correct? Explain.

Joey is correct. The diagonal that joins the vertices of the angles between the two pairs of congruent sides of a kite also bisects those angles. The diagonal that joins the vertices of the angles created by a pair of the sides of uneven lengths does not bisect those angles.

Problem Set Sample Solutions

Bisect each angle below.

1.

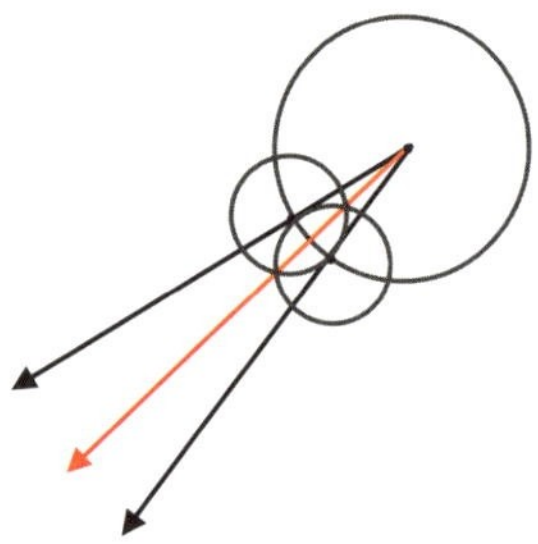

2.

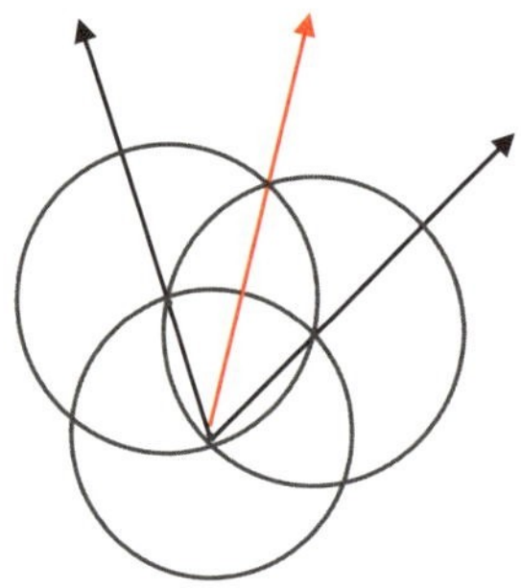

3.

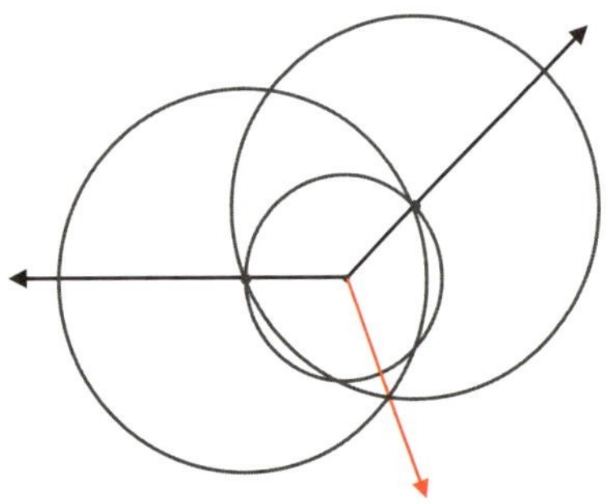

4.

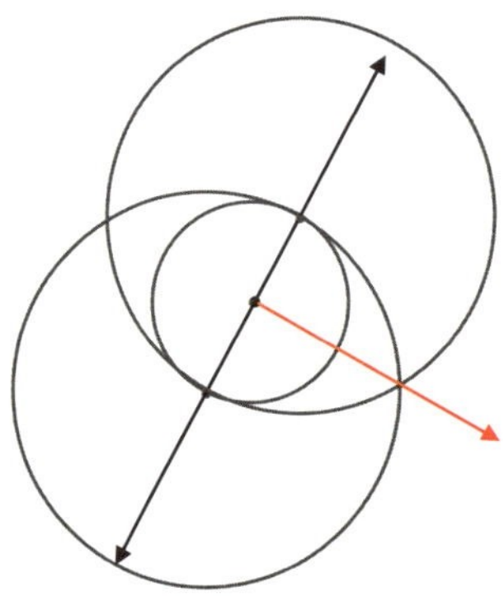

Copy the angle below.

5.

Draw circle B: center B, any radius.
Label the intersections of circle B with the sides of the angle as A and C.
Label the vertex of the original angle as B.
Draw $\overrightarrow{ED}$.
Draw $\overrightarrow{EG}$ as one side of the angle to be drawn.
Draw circle F: center F, radius CA.
Draw circle E: center E, radius BA.
Label the intersection of circle E with $\overrightarrow{EG}$ as F.
Label either intersection of circle E and circle F as D.

Lesson 4: Construct a Perpendicular Bisector

Student Outcome

- Students construct a perpendicular bisector and discover the relationship between symmetry with respect to a line and a perpendicular bisector.

Lesson Notes

In Lesson 4, students learn to construct perpendicular bisectors and apply the construction to problems. Students continue to write precise instructions for constructions. The importance of specific language continues throughout the construction lessons. The steps for constructing an angle bisector from the previous lesson flow nicely into the steps for constructing a perpendicular bisector.

The Opening Exercise is another opportunity for students to critique their work. Students use a rubric to assess the Lesson 3 Problem Set on angle bisectors. Determine where students feel they are making errors (i.e., if they score low on the rubric). In the Discussion, students make a connection between Lesson 3 and Lesson 4 as an angle bisector is linked to a perpendicular bisector. Students should understand that two points are symmetric with respect to a line if and only if the line is the perpendicular bisector of the segment that joins the two points. Furthermore, students should be comfortable with the idea that any point on the perpendicular bisector is equidistant from the endpoints of the segment. Lastly, students extend the idea behind the construction of a perpendicular bisector to construct a perpendicular to a line from a point not on the line.

Classwork

Opening Exercise (5 minutes)

Opening Exercise

Choose one method below to check your Problem Set:

- **Trace your copied angles and bisectors onto patty paper; then, fold the paper along the bisector you constructed. Did one ray exactly overlap the other?**
- **Work with your partner. Hold one partner's work over another's. Did your angles and bisectors coincide perfectly?**

Use the following rubric to evaluate your Problem Set:

Needs Improvement	Satisfactory	Excellent
Few construction arcs visible	Some construction arcs visible	Construction arcs visible and appropriate
Few vertices or relevant intersections labeled	Most vertices and relevant intersections labeled	All vertices and relevant intersections labeled
Lines drawn without straightedge or not drawn correctly	Most lines neatly drawn with straightedge	Lines neatly drawn with straightedge
Fewer than 3 angle bisectors constructed correctly	3 of the 4 angle bisectors constructed correctly	Angle bisector constructed correctly

Mathematical Modeling Exercise (36 minutes)

In addition to the discussion, have students participate in a kinesthetic activity that illustrates the idea of an angle bisector. Ask students to get out of their seats and position themselves at equal distances from two adjacent classroom walls. The students form the bisector of the (likely right) angle formed at the meeting of the adjacent walls.

Discussion

In Lesson 3, we studied how to construct an angle bisector. We know we can verify the construction by folding an angle along the bisector. A correct construction means that one half of the original angle coincides exactly with the other half so that each point of one ray of the angle maps onto a corresponding point on the other ray of the angle.

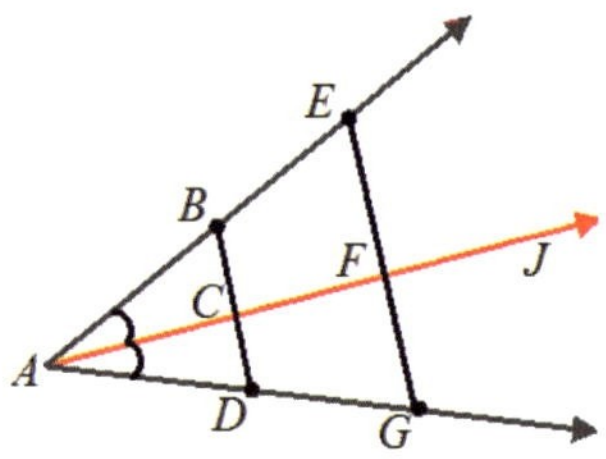

We now extend this observation. Imagine a segment that joins any pair of points that map onto each other when the original angle is folded along the bisector. The figure to the right illustrates two such segments.

Let us examine one of the two segments, $\overline{EG}$. When the angle is folded along $\overrightarrow{AJ}$, E coincides with G. In fact, folding the angle demonstrates that E is the same distance from F as G is from F; $EF = FG$. The point that separates these equal halves of $\overline{EG}$ is F, which is, in fact, the midpoint of the segment and lies on the bisector $\overrightarrow{AJ}$. We can make this case for any segment that falls under the conditions above.

By using geometry facts we acquired in earlier school years, we can also show that the angles formed by the segment and the angle bisector are right angles. Again, by folding, we can show that $\angle EFJ$ and $\angle GFJ$ coincide and must have the same measure. The two angles also lie on a straight line, which means they sum to $180°$. Since they are equal in measure and sum to $180°$, they each have a measure of $90°$.

These arguments lead to a remark about symmetry with respect to a line and the definition of a perpendicular bisector. Two points are symmetric with respect to a line l if and only if l is the perpendicular bisector of the segment that joins the two points. A perpendicular bisector of a segment passes through the ___*midpoint*___ of the segment and forms ___*right angles*___ with the segment.

We now investigate how to construct a perpendicular bisector of a line segment using a compass and a straightedge. Using what you know about the construction of an angle bisector, experiment with your construction tools and the following line segment to establish the steps that determine this construction.

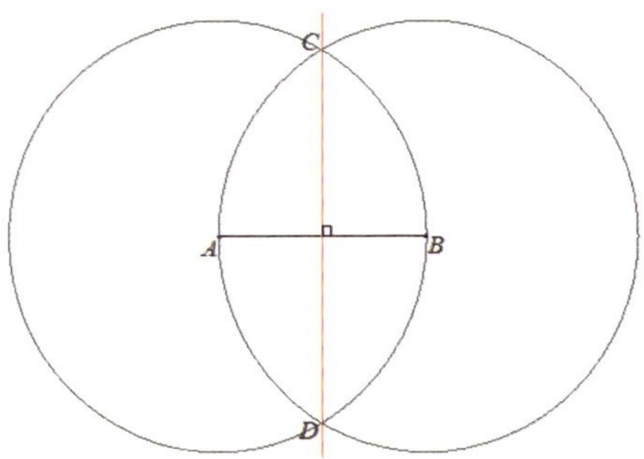

Precisely describe the steps you took to bisect the segment.

1. *Label the endpoints of the segment A and B.*
2. *Draw circle A: center A, radius AB, and circle B: center B, radius BA.*
3. *Label the points of intersections as C and D.*
4. *Draw $\overleftrightarrow{CD}$.*

Now that you are familiar with the construction of a perpendicular bisector, we must make one last observation. Using your compass, string, or patty paper, examine the following pairs of segments:

i. $\overline{AC}, \overline{BC}$

ii. $\overline{AD}, \overline{BD}$

iii. $\overline{AE}, \overline{BE}$

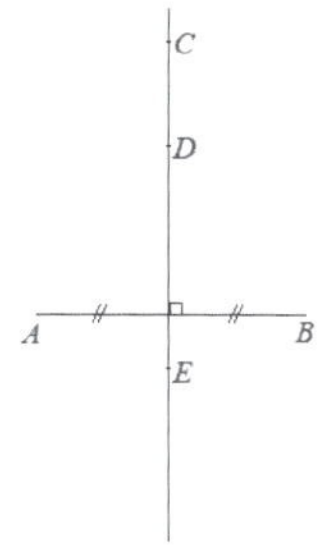

Based on your findings, fill in the observation below.

Observation:

Any point on the perpendicular bisector of a line segment is ___*equidistant*___ from the endpoints of the line segment.

Mathematical Modeling Exercise

MP.5 & MP.6

You know how to construct the perpendicular bisector of a segment. Now, investigate how to construct a perpendicular to a line ℓ from a point A not on ℓ. Think about how you have used circles in constructions so far and *why* the perpendicular bisector construction works the way it does. The first step of the instructions has been provided for you. Discover the construction, and write the remaining steps.

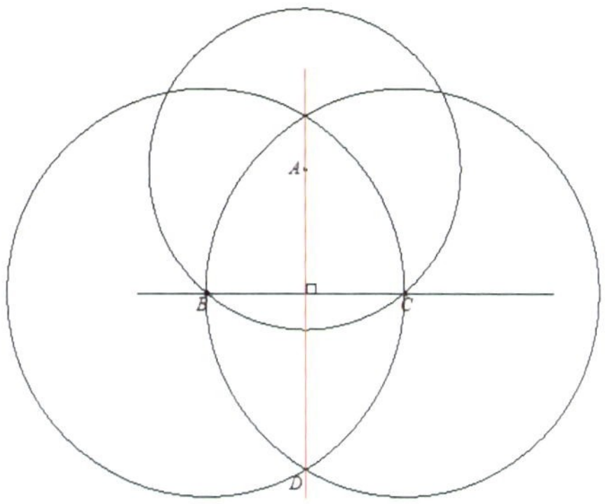

Step 1. Draw circle A: center A with radius so that circle A intersects line ℓ in two points.

Step 2: Label the two points of intersection as B and C.

Step 3: Draw circle B: center B, radius BC.

Step 4: Draw circle C: center C, radius CB.

Step 5: Label one of the unlabeled intersections of circle B and circle C as D.

Step 6: Draw $\overleftrightarrow{AD}$.

Relevant Vocabulary

RIGHT ANGLE: An angle is called a *right angle* if its measure is $90°$.

PERPENDICULAR: Two lines are *perpendicular* if they intersect in one point and if any of the angles formed by the intersection of the lines is a $90°$ (right) angle. Two segments or rays are perpendicular if the lines containing them are perpendicular lines.

EQUIDISTANT: A point A is said to be *equidistant* from two different points B and C if $AB = AC$. A point A is said to be *equidistant* from a point B and a line l if the distance between A and l is equal to AB.

Closing (1 minute)

- How can the perpendicular bisector of a segment be justified as the line of reflection for that segment?
    - *The perpendicular bisector of a segment is symmetric with respect to the endpoints of the segment; the perpendicular bisector passes through the midpoint of the segment and forms right angles with the segment. Any point on the perpendicular bisector is equidistant to the endpoints of the segment.*

Exit Ticket (3 minutes)

Name ______________________________ Date ____________

Lesson 4: Construct a Perpendicular Bisector

Exit Ticket

Divide the following segment AB into four segments of equal length.

Exit Ticket Sample Solution

Divide the following segment AB into four segments of equal length.

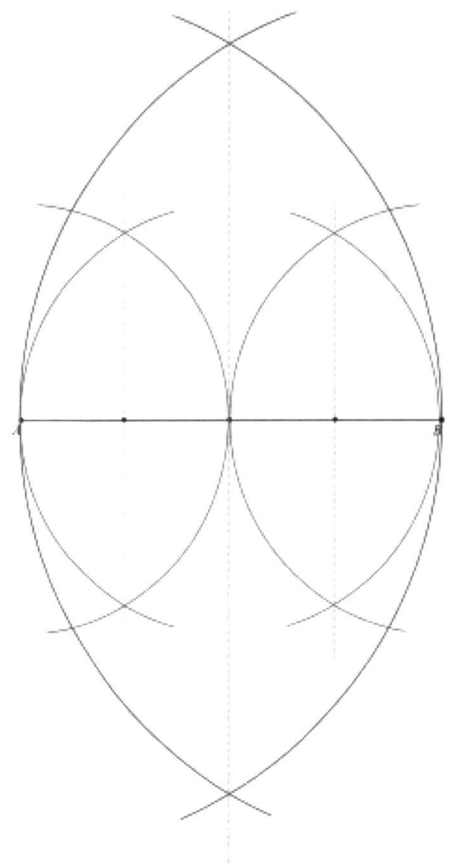

Problem Set Sample Solutions

1. During this lesson, you constructed a perpendicular line to a line ℓ from a point A not on ℓ. We are going to use that construction to construct parallel lines.

   To construct parallel lines ℓ_1 and ℓ_2:

   i. Construct a perpendicular line ℓ_3 to a line ℓ_1 from a point A not on ℓ_1.

   ii. Construct a perpendicular line ℓ_2 to ℓ_3 through point A. *Hint:* Consider using the steps behind Problem 4 in the Lesson 3 Problem Set to accomplish this.

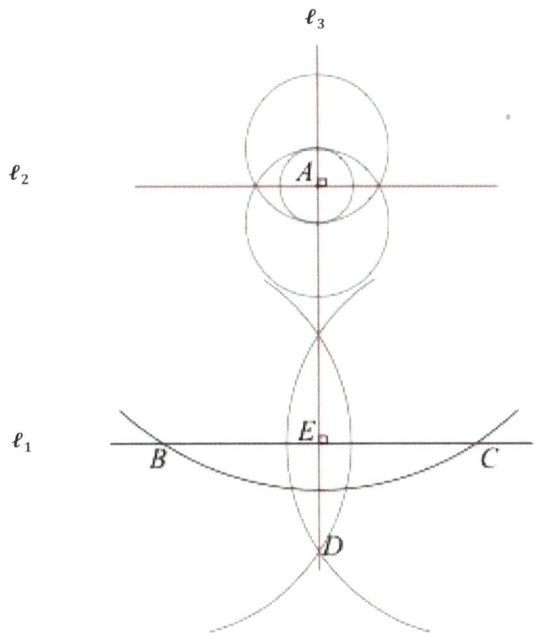

MP.5

2. Construct the perpendicular bisectors of $\overline{AB}$, $\overline{BC}$, and $\overline{CA}$ on the triangle below. What do you notice about the segments you have constructed?

   Students should say that the three perpendicular bisectors pass through a common point. (Students may additionally conjecture that this common point is equidistant from the vertices of the triangle.)

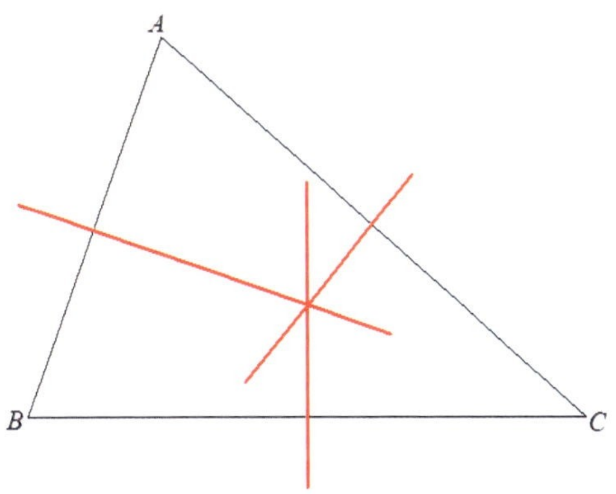

3. Two homes are built on a plot of land. Both homeowners have dogs and are interested in putting up as much fencing as possible between their homes on the land but in a way that keeps the fence equidistant from each home. Use your construction tools to determine where the fence should go on the plot of land. How must the fencing be altered with the addition of a third home?

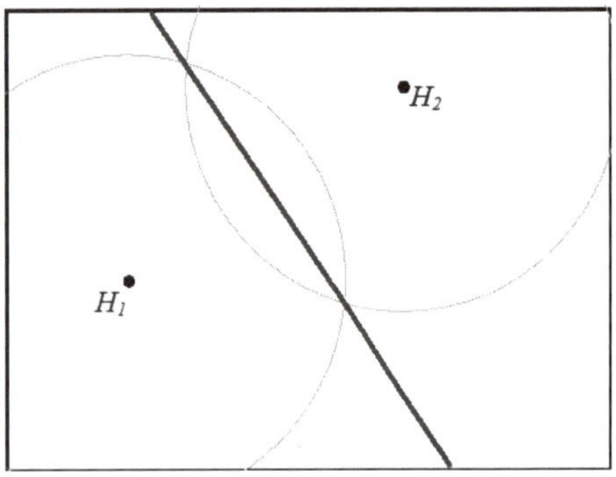

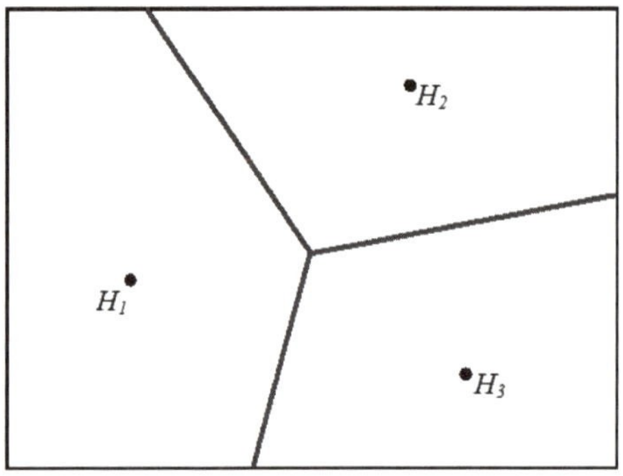

Lesson 5: Points of Concurrencies

Student Outcome

- Students become familiar with vocabulary regarding two points of concurrencies and understand why the points are concurrent.

Lesson Notes

Lesson 5 is an application lesson of the constructions covered so far.

In the Opening Exercise, students construct three perpendicular bisectors of a triangle but this time use a makeshift compass (i.e., a string and pencil). Encourage students to note the differences between the tools and in what manner the tools would change how the steps are written.

The Discussion addresses vocabulary associated with points of concurrencies. The core of the notes presents why the three perpendicular bisectors are concurrent. Students should then make a similar argument explaining why the three angle bisectors of a triangle are also concurrent.

This topic presents an opportunity to incorporate geometry software if available.

Classwork

Opening Exercise (7 minutes)

Students use an alternate method of construction on Lesson 4, Problem Set 2.

MP.5

Opening Exercise

You need a makeshift compass made from string and a pencil.

Use these materials to construct the perpendicular bisectors of the three sides of the triangle below (like you did with Lesson 4, Problem Set 2).

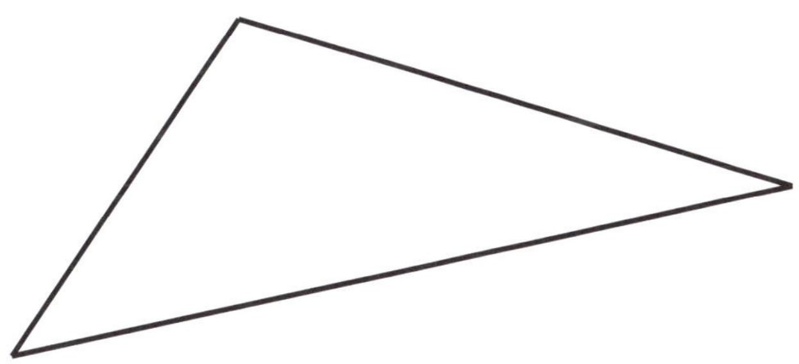

How did using this tool differ from using a compass and straightedge? Compare your construction with that of your partner. Did you obtain the same results?

Exploratory Challenge (36 minutes)

Exploratory Challenge

When three or more lines intersect in a single point, they are _concurrent_, and the point of intersection is the _point of concurrency_.

You saw an example of a point of concurrency in yesterday's Problem Set (and in the Opening Exercise today) when all three perpendicular bisectors passed through a common point.

The point of concurrency of the three perpendicular bisectors is the _circumcenter of the triangle_.

Have students mark the right angles and congruent segments (defined by midpoints) on the triangle.

The circumcenter of $\triangle ABC$ is shown below as point P.

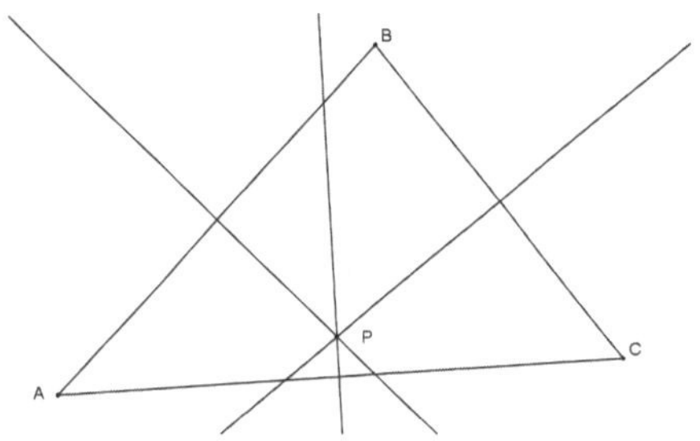

The questions that arise here are WHY are the three perpendicular bisectors concurrent? And WILL these bisectors be concurrent in all triangles? Recall that all points on the perpendicular bisector are equidistant from the endpoints of the segment, which means the following:

a. **P is equidistant from A and B since it lies on the _perpendicular bisector_ of $\overline{AB}$.**

b. **P is also _equidistant_ from B and C since it lies on the perpendicular bisector of $\overline{BC}$.**

c. **Therefore, P must also be equidistant from A and C.**

Hence, $AP = BP = CP$, which suggests that P is the point of _concurrency_ of all three perpendicular bisectors.

You have also worked with angle bisectors. The construction of the three angle bisectors of a triangle also results in a point of concurrency, which we call the _incenter_.

Use the triangle below to construct the angle bisectors of each angle in the triangle to locate the triangle's incenter.

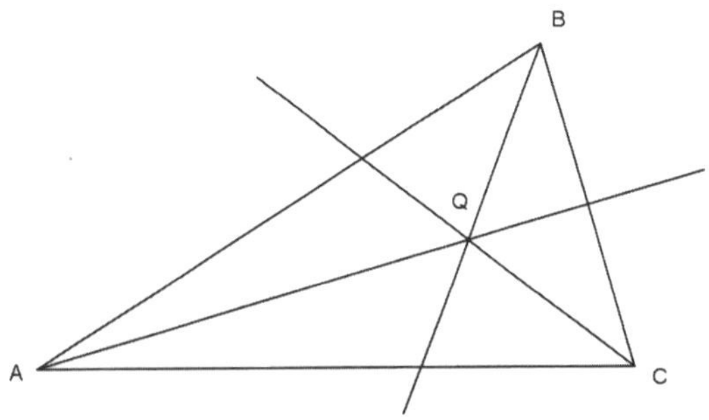

Have students label the congruent angles formed by the angle bisectors.

> *Note to Teacher:*
>
> For the first question, students do not need to re-explain constructions they have established prior to the current construction.

d. **State precisely the steps in your construction above.**

1. *Construct the angle bisectors of $\angle A$, $\angle B$, and $\angle C$.*
2. *Label the point of intersection Q.*

e. **Earlier in this lesson, we explained why the perpendicular bisectors of the sides of a triangle are always concurrent. Using similar reasoning, explain clearly why the angle bisectors are always concurrent at the incenter of a triangle.**

Any point on the angle bisector is equidistant from the rays forming the angle. Therefore, since point Q is on the angle bisector of $\angle ABC$, it is equidistant from $\overrightarrow{BA}$ and $\overrightarrow{BC}$. Similarly, since point Q is on the angle bisector of $\angle BCA$, it is equidistant from $\overrightarrow{CB}$ and $\overrightarrow{CA}$. Therefore, Q must also be equidistant from $\overrightarrow{AB}$ and $\overrightarrow{AC}$, since it lies on the angle bisector of $\angle BAC$. So Q is a point of concurrency of all three angle bisectors.

f. **Observe the constructions below. Point A is the _*circumcenter*_ of $\triangle JKL$. (Notice that it can fall outside of the triangle.) Point B is the _*incenter*_ of $\triangle RST$. The circumcenter of a triangle is the center of the circle that circumscribes that triangle. The incenter of the triangle is the center of the circle that is inscribed in that triangle.**

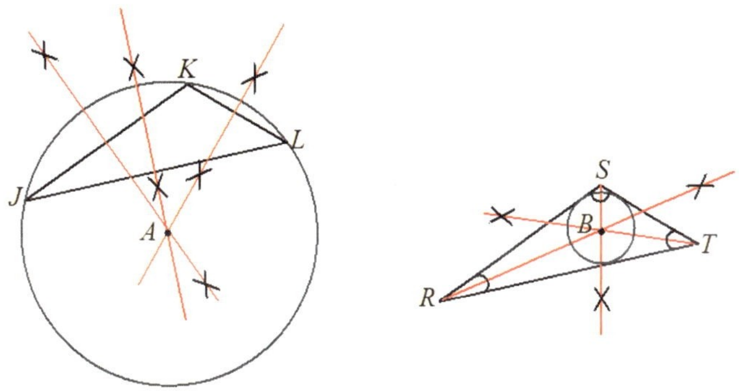

On a separate piece of paper, draw two triangles of your own below and demonstrate how the circumcenter and incenter have these special relationships.

Answers will vary.

g. **How can you use what you have learned in Exercise 3 to find the center of a circle if the center is not shown?**

Inscribe a triangle into the circle and construct the perpendicular bisectors of at least two sides. Where the bisectors intersect is the center of the circle.

Closing (2 minutes)

- What does it mean for lines to be *concurrent*?
  - *Three or more lines are said to be concurrent if all the lines intersect in a single point.*
- For which constructions is the *circumcenter* a point of concurrency? For which constructions is the *incenter* a point of concurrency?
  - *The circumcenter is the point of concurrency of the perpendicular bisectors of a triangle; the incenter is the point of concurrency of the angle bisectors of a triangle.*

Problem Set Sample Solutions

1. Given line segment AB, using a compass and straightedge, construct the set of points that are equidistant from A and B.

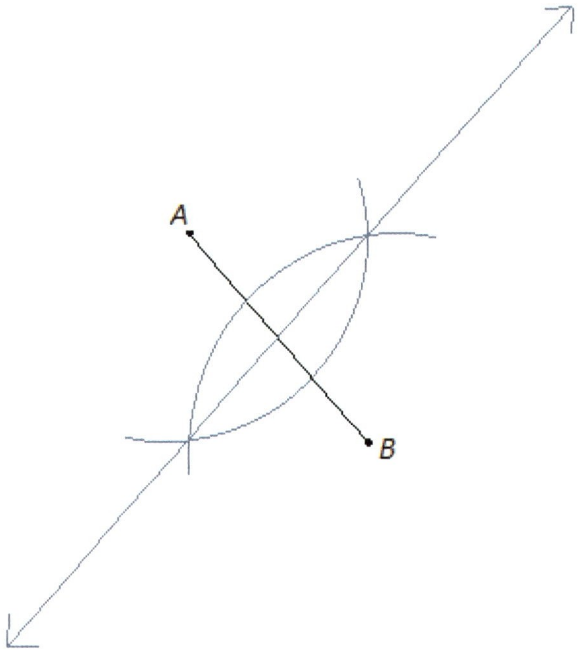

What figure did you end up constructing? Explain.

I ended up drawing the perpendicular bisector of the segment AB. Every point on this line is equidistant from the points A and B.

2. For each of the following, construct a line perpendicular to segment AB that goes through point P.

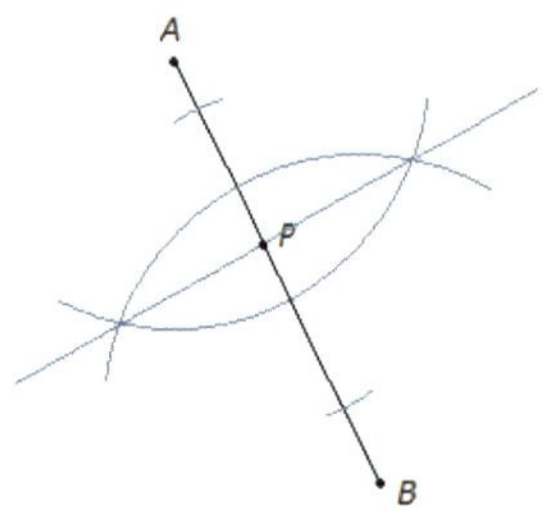

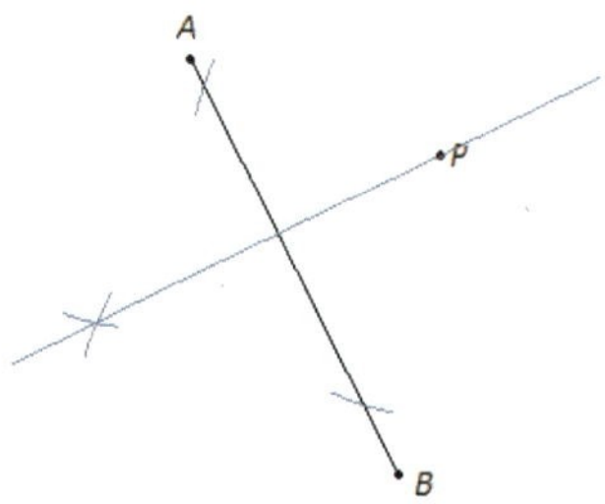

3. Using a compass and straightedge, construct the angle bisector of $\angle ABC$ shown below. What is true about every point that lies on the ray you created?

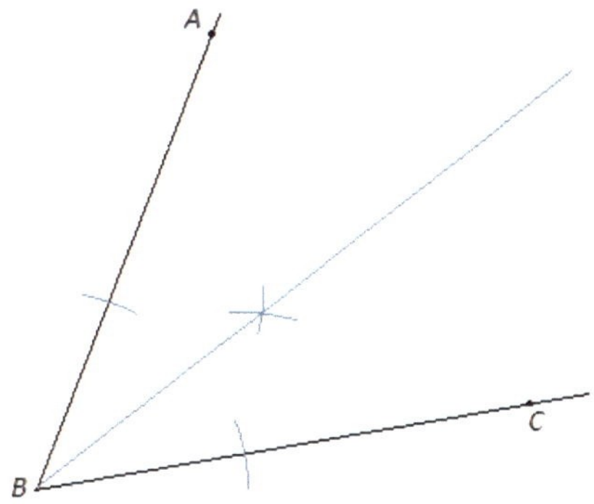

Every point on the ray is equidistant from ray BA and ray BC.

Topic B

Unknown Angles

G-CO.C.9

Focus Standard:	G-CO.C.9	Prove theorems about lines and angles. *Theorems include: vertical angles are congruent; when a transversal crosses parallel lines, alternate interior angles are congruent and corresponding angles are congruent; points on a perpendicular bisector of a line segment are exactly those equidistant from the segment's endpoints.*
Instructional Days:	6	
Lesson 6:	Solve for Unknown Angles—Angles and Lines at a Point (P)[1]	
Lesson 7:	Solve for Unknown Angles—Transversals (P)	
Lesson 8:	Solve for Unknown Angles—Angles in a Triangle (P)	
Lesson 9:	Unknown Angle Proofs—Writing Proofs (P)	
Lesson 10:	Unknown Angle Proofs—Proofs with Constructions (P)	
Lesson 11:	Unknown Angle Proofs—Proofs of Known Facts (P)	

By the time students embark on Topic B, they have seen several of the geometric figures that they studied prior to Grade 8. Topic B incorporates even more of these previously learned figures, such as the special angles created by parallel lines cut by a transversal. As part of the journey to solving proof problems, students begin by solving unknown angle problems in Lessons 6–8. Students develop mastery over problems involving angles at a point, angles in diagrams with parallel lines cut by a transversal, angles within triangles, and all of the above within any given diagram. A base knowledge of how to solve for a given unknown angle lays the groundwork for orchestrating an argument for a proof. In the next phase, Lessons 9–11, students work on unknown angle proofs. Instead of focusing on the computational steps needed to arrive at a particular unknown value, students must articulate the algebraic and geometric concepts needed to arrive at a given relationship. Students continue to use precise language and relevant vocabulary to justify steps in finding unknown angles and to construct viable arguments that defend their method of solution.

[1]Lesson Structure Key: **P**-Problem Set Lesson, **M**-Modeling Cycle Lesson, **E**-Exploration Lesson, **S**-Socratic Lesson

Lesson 6: Solve for Unknown Angles—Angles and Lines at a Point

Student Outcomes

- Students review formerly learned geometry facts and practice citing the geometric justifications in anticipation of unknown angle proofs.

Lesson Notes

Lessons 1–5 serve as a foundation for the main subject of this module, which is congruence. By the end of the unknown angles lessons (Lessons 6–8), students start to develop fluency in two areas: (1) solving for unknown angles in diagrams and (2) justifying each step or decision in the proof-writing process of unknown angle solutions.

The "missing-angle problems" in this topic occur in many American geometry courses and play a central role in some Asian curricula. A missing-angle problem asks students to use several geometric facts together to find angle measures in a diagram. While the simpler problems require good, purposeful recall and application of geometric facts, some problems are complex and may require ingenuity to solve. Historically, many geometry courses have not expected this level of sophistication. Such courses would not have demanded that students use their knowledge constructively but rather to merely regurgitate information. The missing-angle problems are a step up in problem solving. Why do we include them at this juncture in this course? The main focal points of these opening lessons are to recall or refresh and supplement existing conceptual vocabulary, to emphasize that work in geometry involves reasoned explanations, and to provide situations and settings that support the need for reasoned explanations and that illustrate the satisfaction of building such arguments.

Lesson 6 is Problem Set based and focuses on solving for unknown angles in diagrams of angles and lines at a point. By the next lesson, students should be comfortable solving for unknown angles numerically or algebraically in diagrams involving supplementary angles, complementary angles, vertical angles, and adjacent angles at a point. As always, vocabulary is critical, and students should be able to define the relevant terms themselves. It may be useful to draw or discuss counterexamples of a few terms and ask students to explain why they do not fit a particular definition.

As students work on problems, encourage them to show each step of their work and to list the geometric reason for each step (e.g., "Vertical angles have equal measure"). This prepares students to write a reason for each step of their unknown angle proofs in a few days.

A chart of common facts and discoveries from middle school that may be useful for student review or supplementary instruction is included at the end of this lesson. The chart includes abbreviations students may have previously seen in middle school, as well as more widely recognized ways of stating these ideas in proofs and exercises.

Classwork

Opening Exercise (5 minutes)

Ask students to find the missing angles in these diagrams. The exercise reminds students of the basics of determining missing angles that they learned in middle school. Discuss the facts that students recall, and use these as a starting point for the lesson.

Opening Exercise

Determine the measure of the missing angle in each diagram.

A.

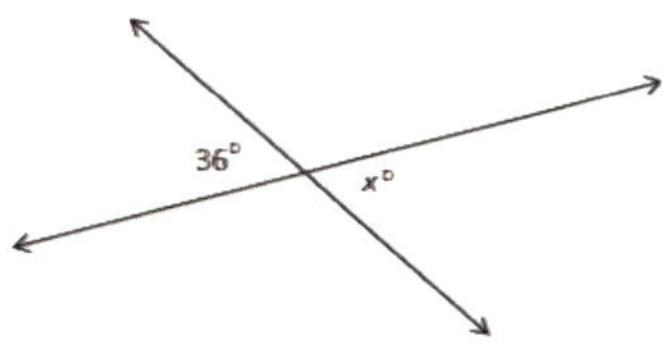

$x = 36$

B.

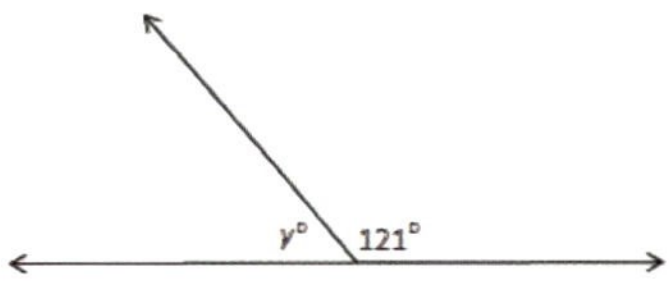

$y = 59$

C.

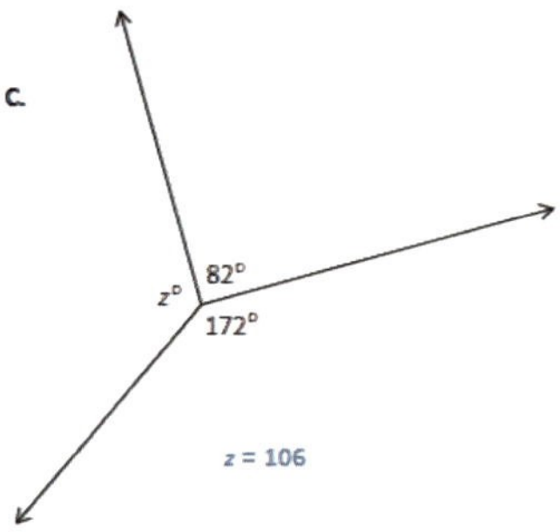

$z = 106$

What facts about angles did you use?

Answers may include the following: vertical angles are equal in measure; linear pairs form supplementary angles; angles at a point sum to $360°$.

Discussion (4 minutes)

Discussion

MP.6

Two angles $\angle AOC$ and $\angle COB$, with a common side $\overrightarrow{OC}$, are *adjacent angles* if C belongs to the interior of $\angle AOB$. The sum of angles on a straight line is $180°$, and two such angles are called a *linear pair*. Two angles are called *supplementary* if the sum of their measures is $\underline{180°}$; two angles are called *complementary* if the sum of their measures is $\underline{90°}$. Describing angles as supplementary or complementary refers only to the measures of their angles. The positions of the angles or whether the pair of angles is adjacent to each other is not part of the definition.

In the figure, line segment AD is drawn.

Find $m\angle DCE$.

$m\angle DCE = 18°$

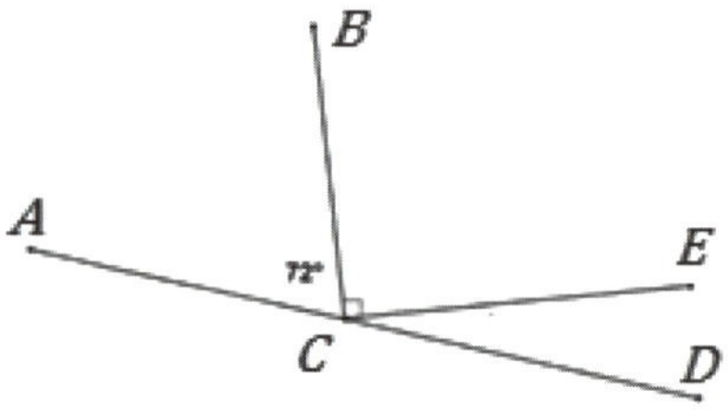

The total measure of adjacent angles around a point is $\underline{360°}$.

Find the measure of $\angle HKI$.

$m\angle HKI = 80°$

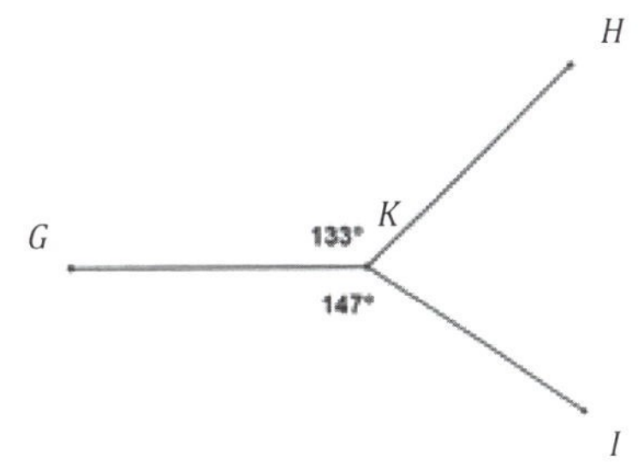

Vertical angles have ___*equal*___ measure. Two angles are vertical if their sides form opposite rays.

Find $m\angle TRV$.

$m\angle TRV = 52°$

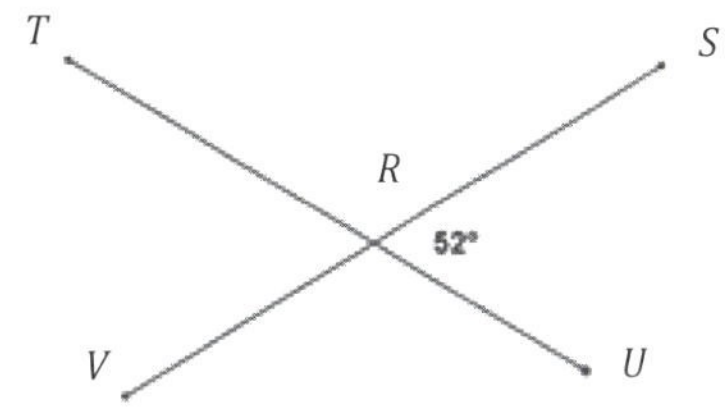

Example (5 minutes)

Example

Find the measure of each labeled angle. Give a reason for your solution.

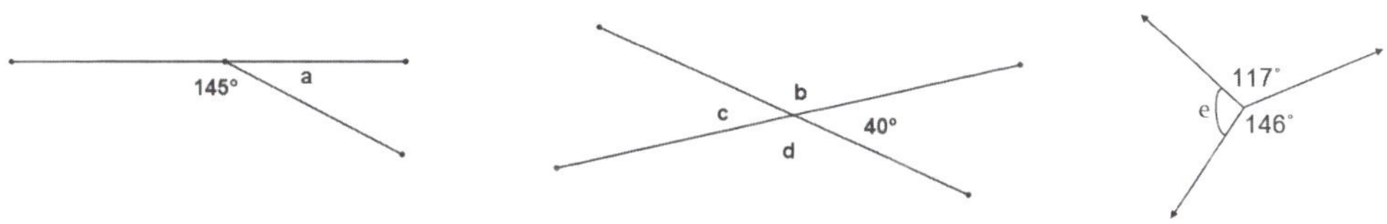

Angle	Angle Measure	Reason
$\angle a$	$35°$	*Linear pairs form supplementary angles.*
$\angle b$	$140°$	*Linear pairs form supplementary angles.*
$\angle c$	$40°$	*Vertical angles are equal in measure.*
$\angle d$	$140°$	*Linear pairs form supplementary angles, or vertical angles are equal in measure.*
$\angle e$	$97°$	*Angles at a point sum to* $360°$.

Exercises (25 minutes)

Exercises

MP.7

In the figures below, $\overline{AB}$, $\overline{CD}$, and $\overline{EF}$ are straight line segments. Find the measure of each marked angle, or find the unknown numbers labeled by the variables in the diagrams. Give reasons for your calculations. Show all the steps to your solutions.

1.

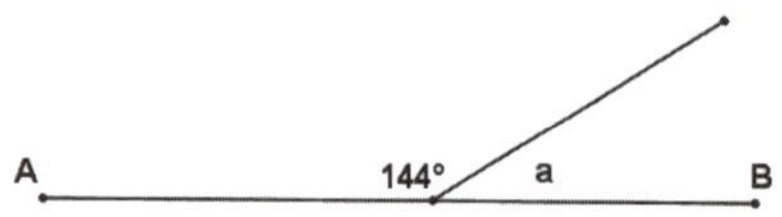

$m\angle a = 36°$
Linear pairs form supplementary angles.

2.

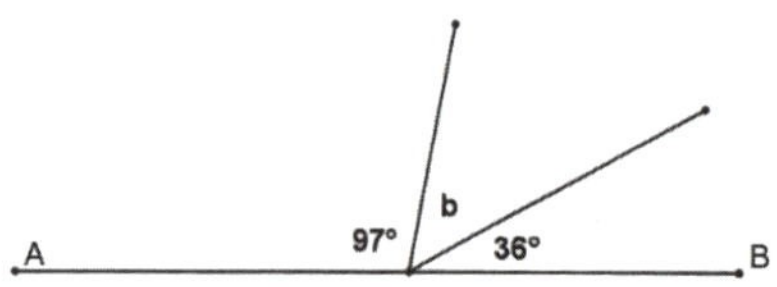

$m\angle b = 47°$
Consecutive adjacent angles on a line sum to $180°$.

3.

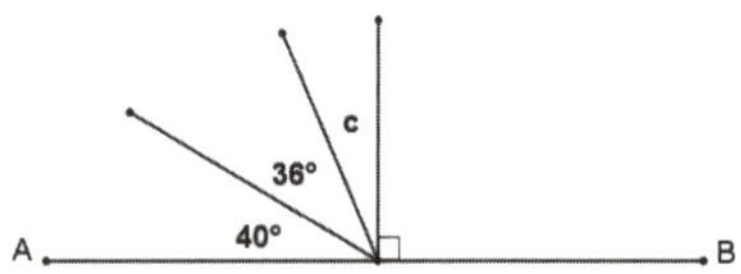

$m\angle c = 14°$
Consecutive adjacent angles on a line sum to $180°$

4.

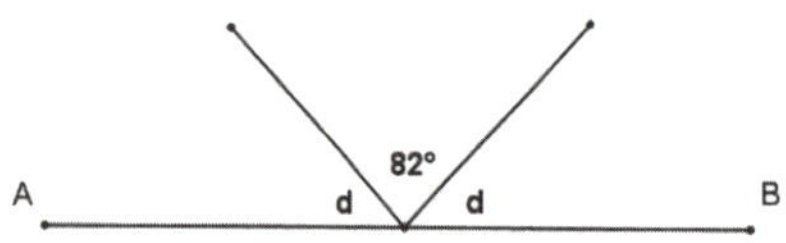

$m\angle d = 49°$
Consecutive adjacent angles on a line sum to $180°$

5.

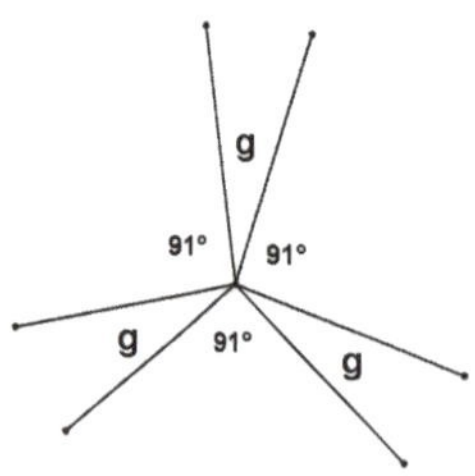

$m\angle g = 29°$
Angles at a point sum to $360°$.

For Exercises 6–12, find the values of x and y. Show all work.

6.

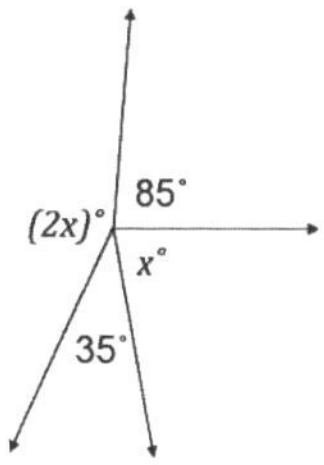

$x = 80$
Angles at a point sum to $360°$.

7.

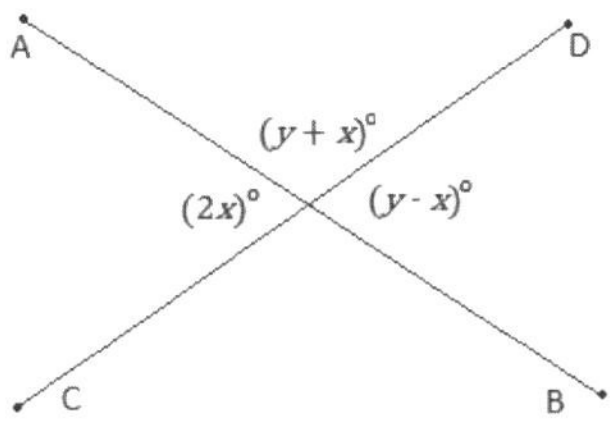

$x = 30$; $y = 90$
Vertical angles are equal in measure.
Angles at a point sum to $360°$.

MP.7

8.

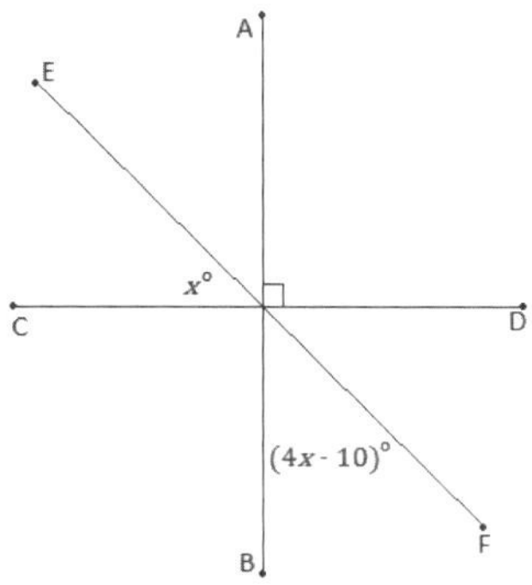

$x = 20$
Vertical angles are equal in measure.
Consecutive adjacent angles on a line sum to $180°$.

9.

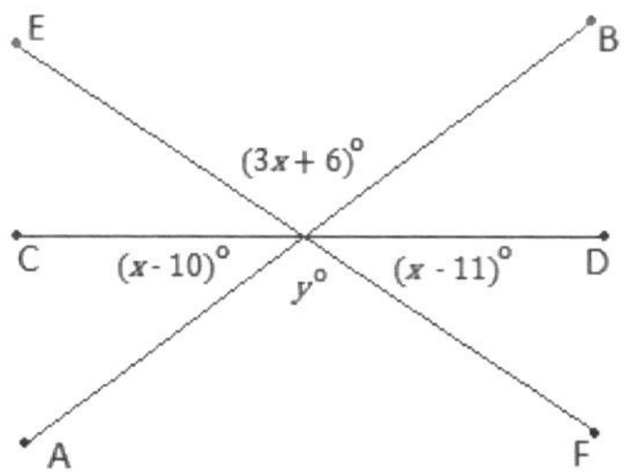

$x = 39$; $y = 123$
Vertical angles are equal in measure. Angles at a point sum to $360°$.

10.

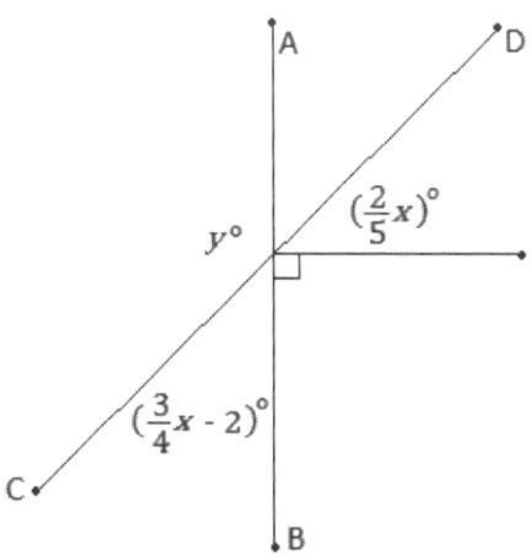

$x = 80$; $y = 122$
Consecutive adjacent angles on a line sum to $180°$.

MP.7

11.

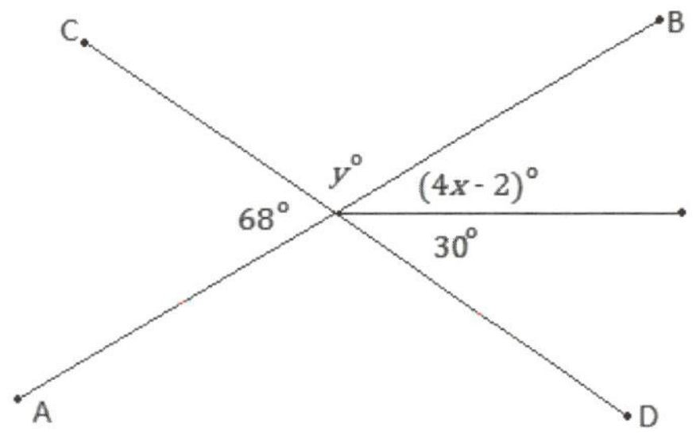

$x = 10$; $y = 112$
Consecutive adjacent angles on a line sum to $180°$. Vertical angles are equal in measure.

12.

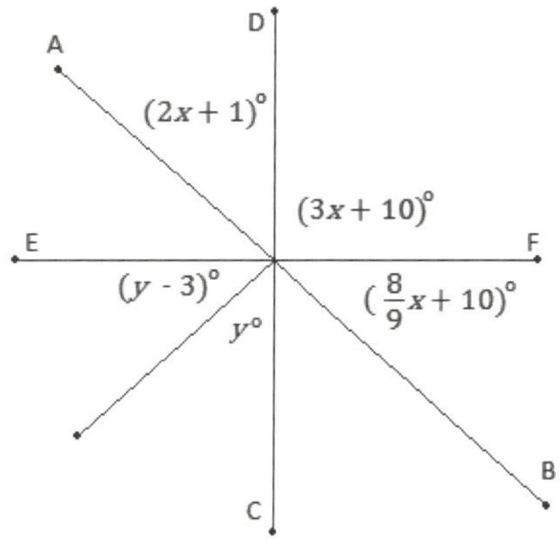

$x = 27$; $y = 47$
Consecutive adjacent angles on a line sum to $180°$. Vertical angles are equal in measure.

Relevant Vocabulary

Relevant Vocabulary

STRAIGHT ANGLE: If two rays with the same vertex are distinct and collinear, then the rays form a line called a *straight angle*.

VERTICAL ANGLES: Two angles are *vertical angles* (or vertically opposite angles) if their sides form two pairs of opposite rays.

Closing (1 minute)

- What is the distinction between complementary and supplementary angles?
  - *Two angles are called complementary if their angle sum is $90°$, and two angles are called supplementary if their angle sum is $180°$. The angles do not have to be adjacent to be complementary or supplementary.*
- If two angles are adjacent and on a straight line, what are they called?
  - *A linear pair*
- What is the angle sum of adjacent angles around a point?
  - *The angle sum of adjacent angles around a point is $360°$.*

Exit Ticket (5 minutes)

EUREKA MATH

Name ______________________________ Date ______________

Lesson 6: Solve for Unknown Angles—Angles and Lines at a Point

Exit Ticket

Use the following diagram to answer the questions below:

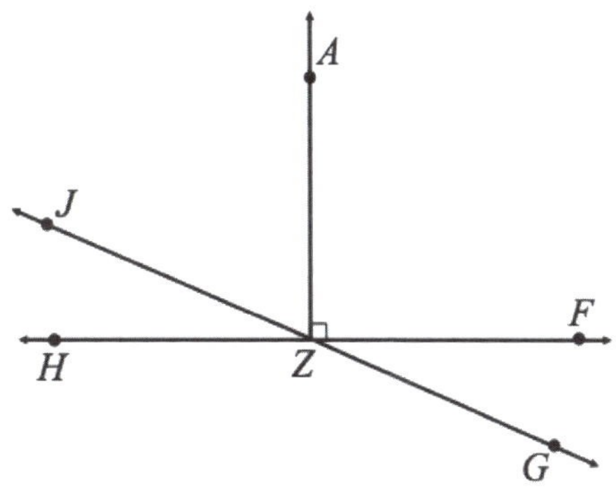

1.
   a. Name an angle supplementary to $\angle HZJ$, and provide the reason for your calculation.

   b. Name an angle complementary to $\angle HZJ$, and provide the reason for your calculation.

2. If $m\angle HZJ = 38°$, what is the measure of each of the following angles? Provide reasons for your calculations.

   a. $m\angle FZG$

   b. $m\angle HZG$

   c. $m\angle AZJ$

Exit Ticket Sample Solutions

Use the following diagram to answer the questions below:

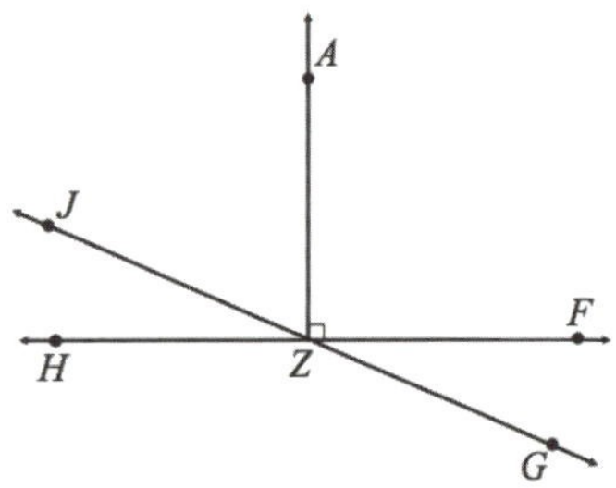

1.

   a. Name an angle supplementary to $\angle HZJ$, and provide the reason for your calculation.

      $\angle JZF$ or $\angle HZG$

      Linear pairs form supplementary angles.

   b. Name an angle complementary to $\angle HZJ$, and provide the reason for your calculation.

      $\angle JZA$

      The angles sum to $90°$.

2. If $m\angle HZJ = 38°$, what is the measure of each of the following angles? Provide reasons for your calculations.

   a. $m\angle FZG$

      $38°$

   b. $m\angle HZG$

      $142°$

   c. $m\angle AZJ$

      $52°$

EUREKA MATH™

Problem Set Sample Solutions

In the figures below, $\overline{AB}$ and $\overline{CD}$ are straight line segments. Find the value of x and/or y in each diagram below. Show all the steps to your solutions, and give reasons for your calculations.

1.

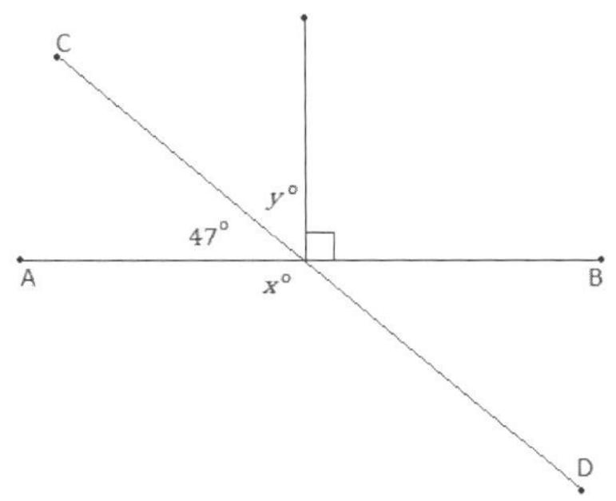

$x = 133$

$y = 43$

Angle addition postulate Linear pairs form supplementary angles.

2.

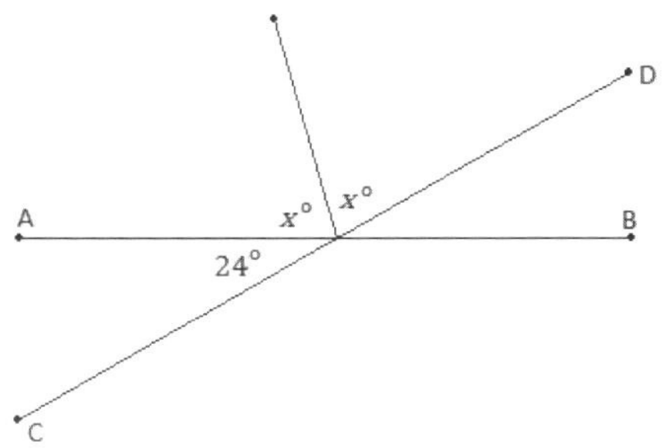

$x = 78$

Consecutive adjacent angles on a line sum to $180°$.

3.

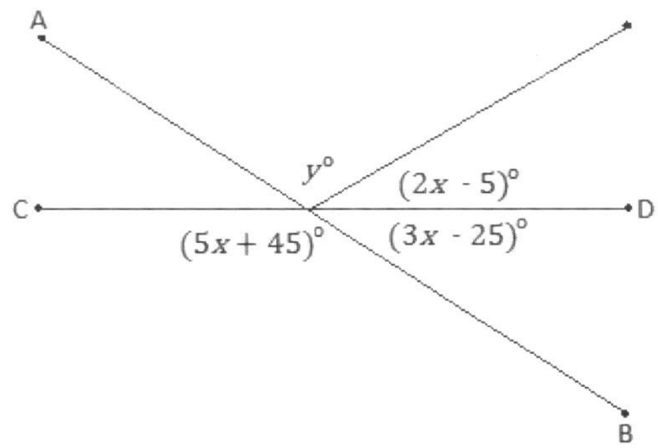

$x = 20$

$y = 110$

Consecutive adjacent angles on a line sum to $180°$.
Vertical angles are equal in measure.

Key Facts and Discoveries from Earlier Grades

Facts (With Abbreviations Used in Grades 4–9)	Diagram/Example	How to State as a Reason in an Exercise or a Proof
Vertical angles are equal in measure. (vert. $\angle$s)	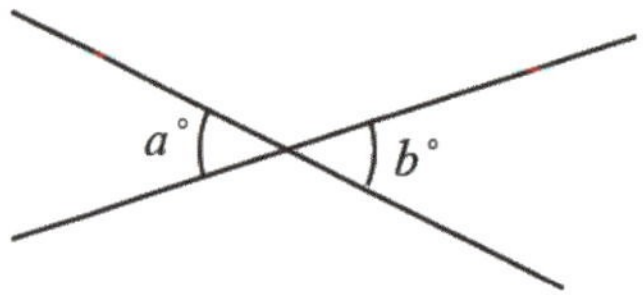 $a^\circ = b^\circ$	"Vertical angles are equal in measure."
If C is a point in the interior of $\angle AOB$, then $m\angle AOC + m\angle COB = m\angle AOB$. ($\angle$s add)	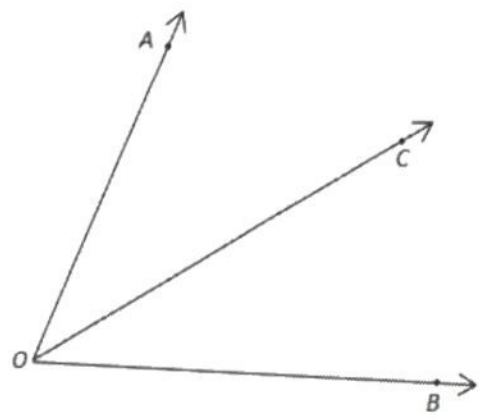 $m\angle AOB = m\angle AOC + m\angle COB$	"Angle addition postulate"
Two angles that form a linear pair are supplementary. ($\angle$s on a line)	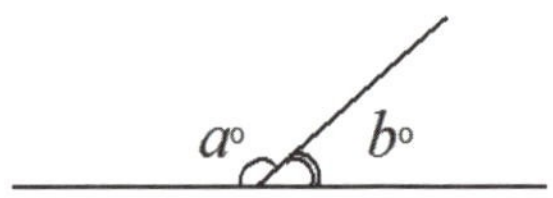 $a^\circ + b^\circ = 180$	"Linear pairs form supplementary angles."
Given a sequence of n consecutive adjacent angles whose interiors are all disjoint such that the angle formed by the first $n-1$ angles and the last angle are a linear pair, then the sum of all of the angle measures is 180°. ($\angle s$ on a line)	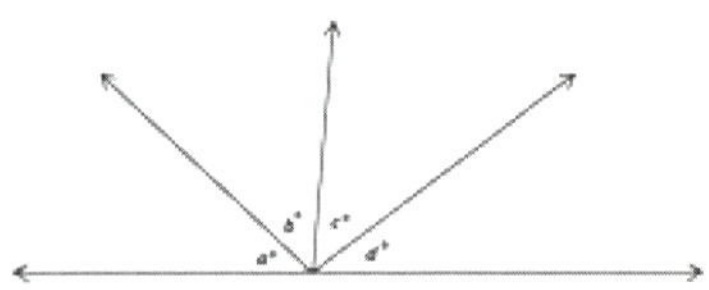 $a^\circ + b^\circ + c^\circ + d^\circ = 180$	"Consecutive adjacent angles on a line sum to 180°."
The sum of the measures of all angles formed by three or more rays with the same vertex and whose interiors do not overlap is 360°. ($\angle$s at a point)	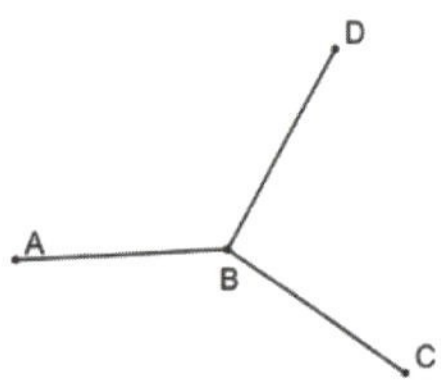 $m\angle ABC + m\angle CBD + m\angle DBA = 360^\circ$	"Angles at a point sum to 360°."

Facts (With Abbreviations Used in Grades 4–9)	Diagram/Example	How to State as a Reason in an Exercise or a Proof
The sum of the 3 angle measures of any triangle is $180°$. ($\angle$ sum of $\triangle$)	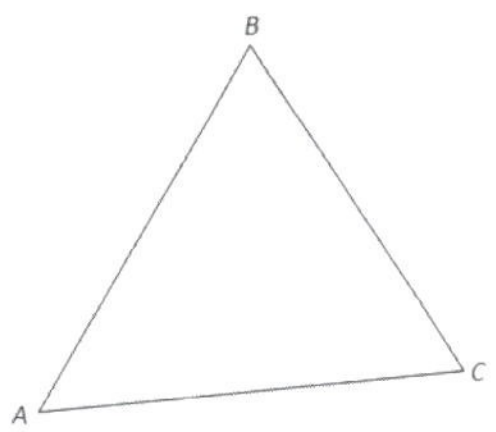 $m\angle A + m\angle B + m\angle C = 180°$	"The sum of the angle measures in a triangle is $180°$."
When one angle of a triangle is a right angle, the sum of the measures of the other two angles is $90°$. ($\angle$ sum of rt. $\triangle$)	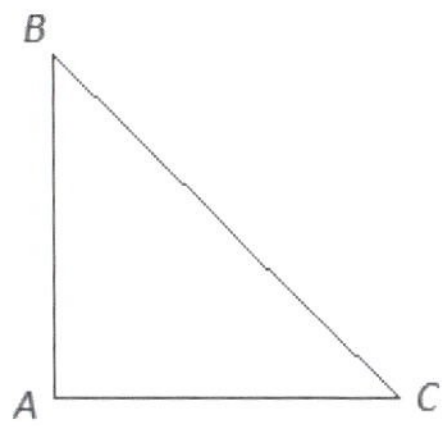 $m\angle A = 90°; m\angle B + m\angle C = 90°$	"Acute angles in a right triangle sum to $90°$."
The sum of each exterior angle of a triangle is the sum of the measures of the opposite interior angles, or the remote interior angles. (ext. $\angle$ of $\triangle$)	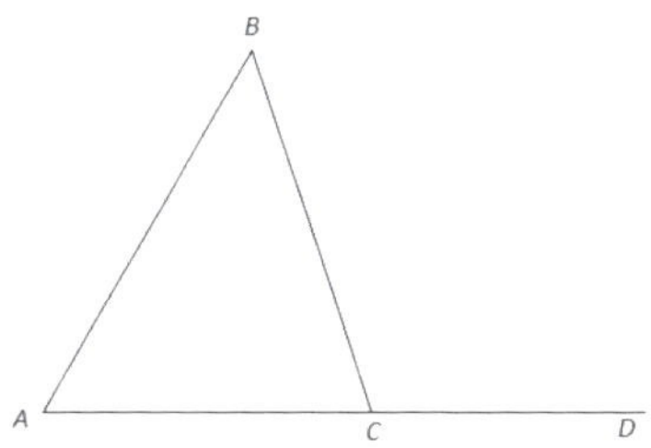 $m\angle BAC + m\angle ABC = m\angle BCD$	"The exterior angle of a triangle equals the sum of the two opposite interior angles."
Base angles of an isosceles triangle are equal in measure. (base $\angle$s of isos. $\triangle$)		"Base angles of an isosceles triangle are equal in measure."
All angles in an equilateral triangle have equal measure. (equilat. $\triangle$)	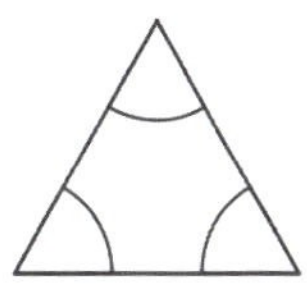	"All angles in an equilateral triangle have equal measure."

Facts (With Abbreviations Used in Grades 4–9)	Diagram/Example	How to State as a Reason in an Exercise or a Proof
If two parallel lines are intersected by a transversal, then corresponding angles are equal in measure. (corr. $\angle$s, $\overline{AB} \parallel \overline{CD}$)	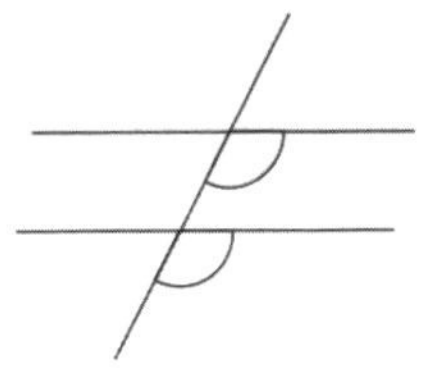	"If parallel lines are cut by a transversal, then corresponding angles are equal in measure."
If two lines are intersected by a transversal such that a pair of corresponding angles are equal in measure, then the lines are parallel. (corr. $\angle$s converse)	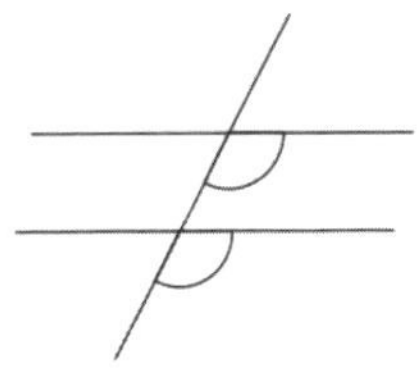	"If two lines are cut by a transversal such that a pair of corresponding angles are equal in measure, then the lines are parallel."
If two parallel lines are intersected by a transversal, then interior angles on the same side of the transversal are supplementary. (int. $\angle$s, $\overline{AB} \parallel \overline{CD}$)	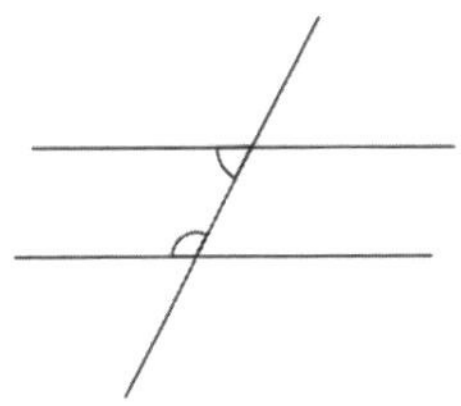	"If parallel lines are cut by a transversal, then interior angles on the same side are supplementary."
If two lines are intersected by a transversal such that a pair of interior angles on the same side of the transversal are supplementary, then the lines are parallel. (int. $\angle$s converse)	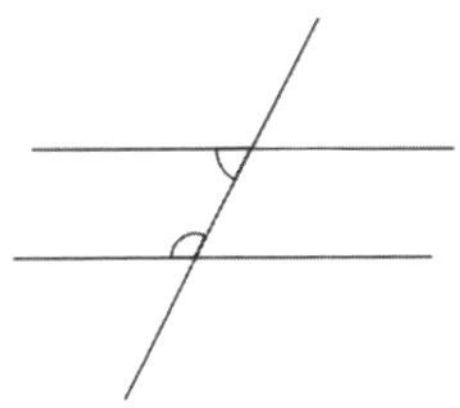	"If two lines are cut by a transversal such that a pair of interior angles on the same side are supplementary, then the lines are parallel."
If two parallel lines are intersected by a transversal, then alternate interior angles are equal in measure. (alt. $\angle$s, $\overline{AB} \parallel \overline{CD}$)	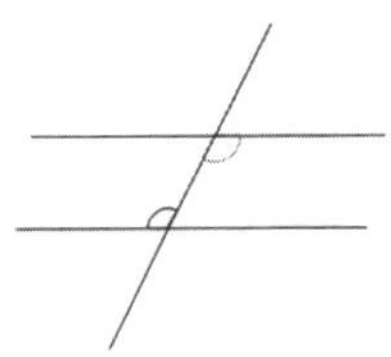	"If parallel lines are cut by a transversal, then alternate interior angles are equal in measure."
If two lines are intersected by a transversal such that a pair of alternate interior angles are equal in measure, then the lines are parallel. (alt. $\angle$s converse)	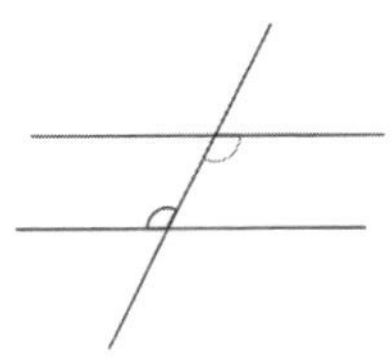	"If two lines are cut by a transversal such that a pair of alternate interior angles are equal in measure, then the lines are parallel."

Lesson 7: Solve for Unknown Angles—Transversals

Student Outcomes

- Students review formerly learned geometry facts and practice citing the geometric justifications in anticipation of unknown angle proofs.

Lesson Notes

The focus of the second day of unknown angle problems is problems with parallel lines crossed by a transversal.

This lesson features one of the main theorems (facts) learned in Grade 8:

1. If two lines are cut by a transversal and corresponding angles are equal, then the lines are parallel.
2. If parallel lines are cut by a transversal, corresponding angles are equal. (This second part is often called the *parallel postulate,* which tells us a property that parallel lines have that cannot be deduced from the definition of parallel lines.)

Of course, students probably remember these two statements as a single fact: For two lines cut by a transversal, the measures of corresponding angles are equal if and only if the lines are parallel. Decoupling these two statements from the unified statement is the work of later lessons.

MP.7 The lesson begins with review material from Lesson 6. In the Discussion and Examples, students review how to identify and apply corresponding angles, alternate interior angles, and same-side interior angles. The key is to make sense of the structure within each diagram.

Before moving on to the Exercises, students learn examples of how and when to use auxiliary lines. Again, the use of auxiliary lines is another opportunity for students to make connections between facts they already know and new information. The majority of the lesson involves solving problems. Gauge how often to prompt and review answers as the class progresses; check to see whether facts from Lesson 6 are fluent. Encourage students to draw in all necessary lines and congruent angle markings to help assess each diagram. The Problem Set should be assigned in the last few minutes of class.

Classwork

Opening Exercise (4 minutes)

Opening Exercise

Use the diagram at the right to determine x and y.
$\overleftrightarrow{AB}$ and $\overleftrightarrow{CD}$ are straight lines.

$x = 30$

$y = 52$

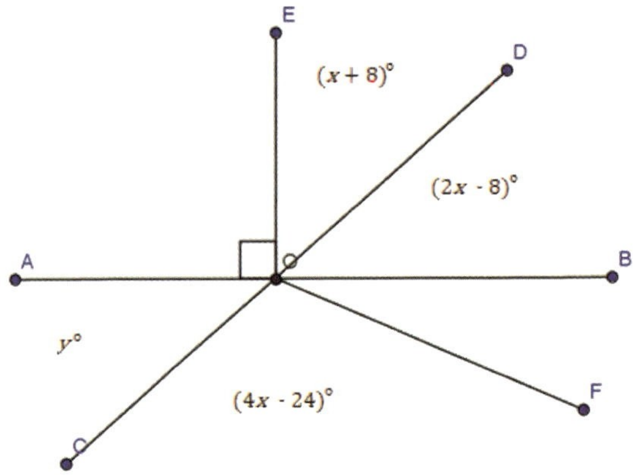

Name a pair of vertical angles:

$\angle AOC$, $\angle DOB$

Find the measure of $\angle BOF$. Justify your calculation.

$m\angle BOF = 32°$ Linear pairs form supplementary angles.

Discussion (4 minutes)

Review the angle facts pertaining to parallel lines crossed by a transversal. Ask students to name examples that illustrate each fact:

Discussion

Given line AB and line CD in a plane (see the diagram below), a third line EF is called a *transversal* if it intersects $\overleftrightarrow{AB}$ at a single point and intersects $\overleftrightarrow{CD}$ at a single but different point. Line AB and line CD are parallel if and only if the following types of angle pairs are congruent or supplementary:

- Corresponding angles are equal in measure.

  $\angle a$ and $\angle e$, $\angle d$ and $\angle h$, etc.

- Alternate interior angles are equal in measure.

  $\angle c$ and $\angle f$, $\angle d$ and $\angle e$

- Same-side interior angles are supplementary.

  $\angle c$ and $\angle e$, $\angle d$ and $\angle f$

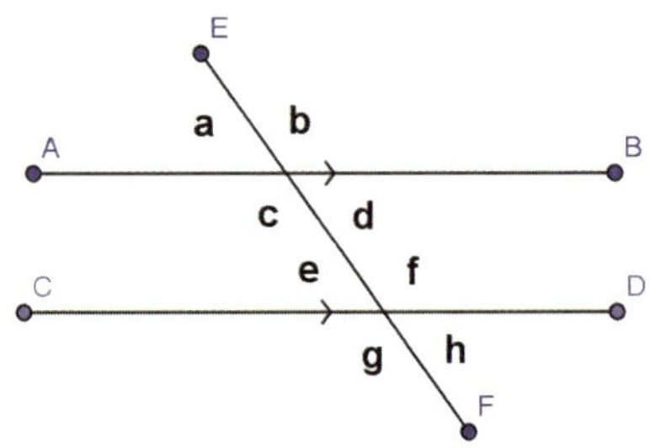

Examples (7 minutes)

Students try examples based on the Discussion; review and then discuss *auxiliary line*.

MP.7

Examples

1.

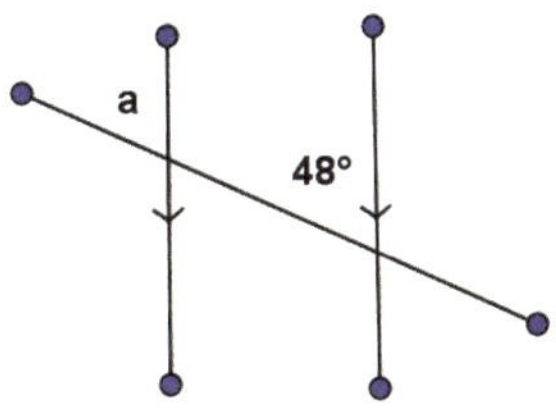

$m\angle a = 48^\circ$

2.

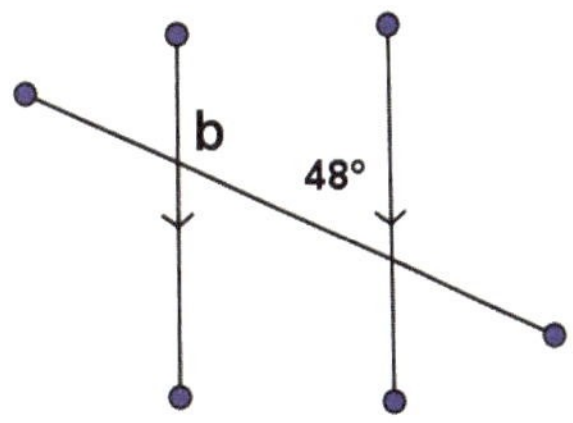

$m\angle b = 132^\circ$

3.

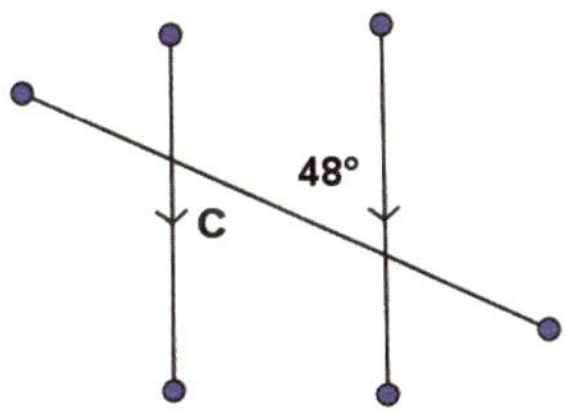

$m\angle c = 48^\circ$

4.

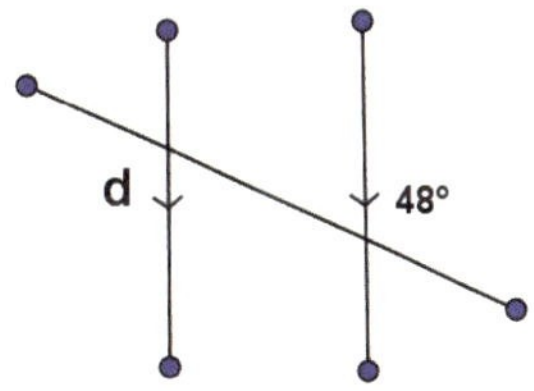

$m\angle d = 48^\circ$

5. An ___*auxiliary line*___ is sometimes useful when solving for unknown angles. In this figure, we can use the auxiliary line to find the measures of $\angle e$ and $\angle f$ (how?) and then add the two measures together to find the measure of $\angle W$. What is the measure of $\angle W$?

$m\angle e = 41^\circ$, $m\angle f = 35^\circ$, $m\angle W = 76^\circ$

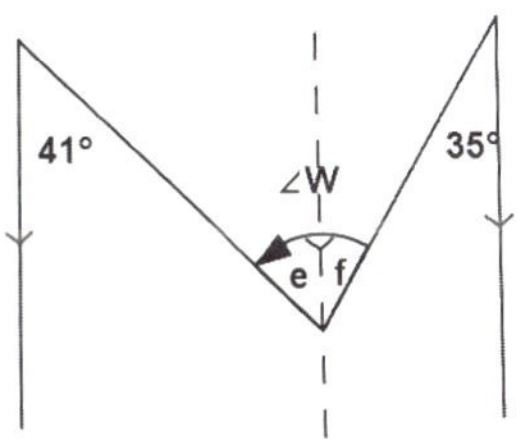

Exercises 1–10 (24 minutes)

Students work on this set of exercises; review periodically.

MP.7

Exercises 1–10

In each exercise below, find the unknown (labeled) angles. Give reasons for your solutions.

1.

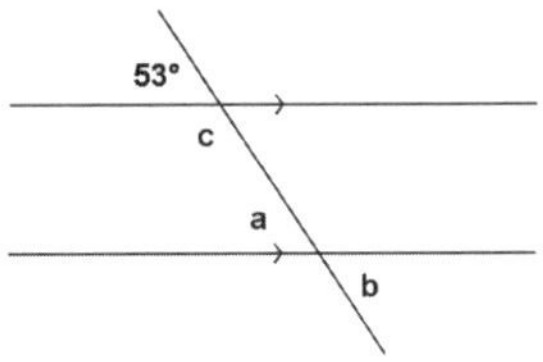

$m\angle a = 53°$
If parallel lines are cut by a transversal, then corresponding angles are equal in measure.

$m\angle b = 53°$
Vertical angles are equal in measure.

$m\angle c = 127°$
If parallel lines are cut by a transversal, then interior angles on the same side are supplementary.

2.

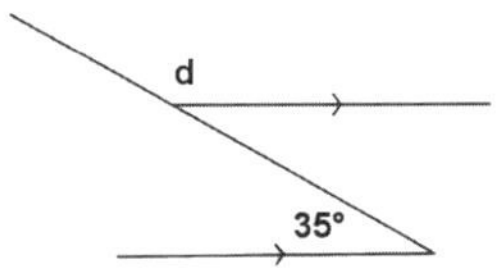

$m\angle d = 145°$
Linear pairs form supplementary angles; if parallel lines are cut by a transversal, then alternate interior angles are equal in measure.

3.

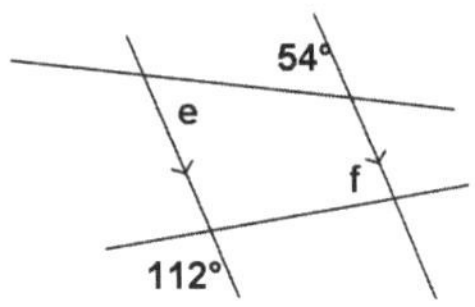

$m\angle e = 54°$
If parallel lines are cut by a transversal, then alternate interior angles are equal in measure.

$m\angle f = 68°$
Vertical angles are equal in measure; if parallel lines are cut by a transversal, then interior angles on the same side are supplementary.

4.

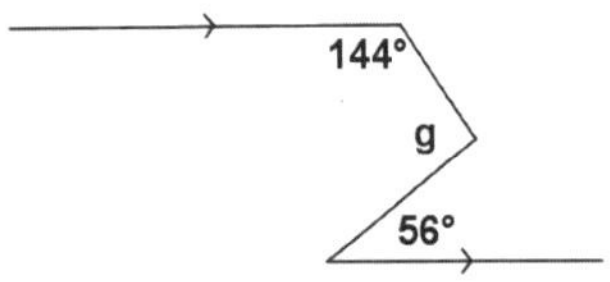

$m\angle g = 92°$
Vertical angles are equal in measure; if parallel lines are cut by a transversal, then interior angles on the same side are supplementary.

5.

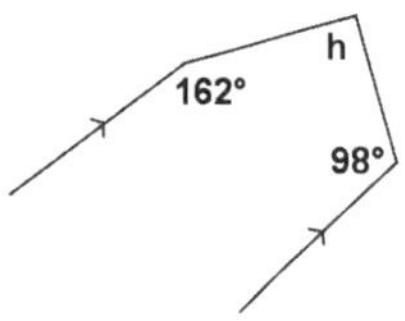

$m\angle h = 100°$
If parallel lines are cut by a transversal, then interior angles on the same side are supplementary.

MP.7

6.

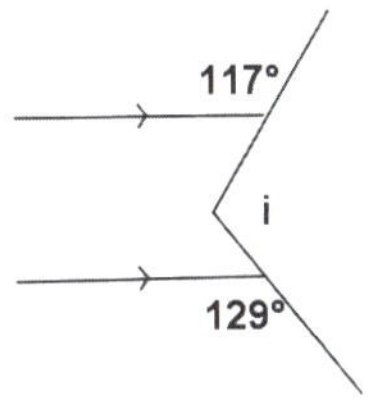

$m\angle i = 114°$
Linear pairs form supplementary angles;
if parallel lines are cut by a transversal, then alternate interior angles are equal in measure.

7.

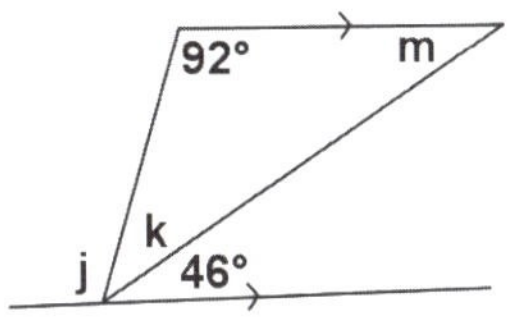

$m\angle j = 92°$
If parallel lines are cut by a transversal, then alternate interior angles are equal in measure.

$m\angle k = 42°$
Consecutive adjacent angles on a line sum to $180°$.

$m\angle m = 46°$
If parallel lines are cut by a transversal, then alternate interior angles are equal in measure.

8.

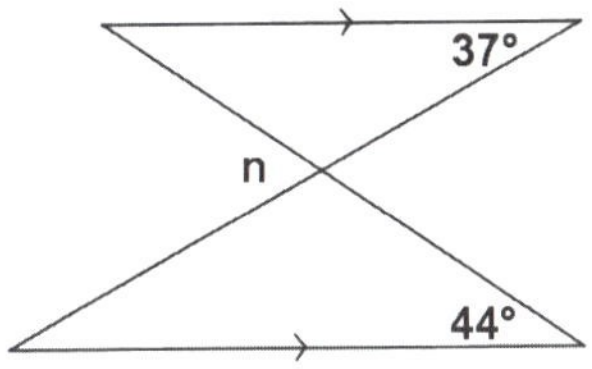

$m\angle n = 81°$
If parallel lines are cut by a transversal, then corresponding angles are equal in measure.

9.

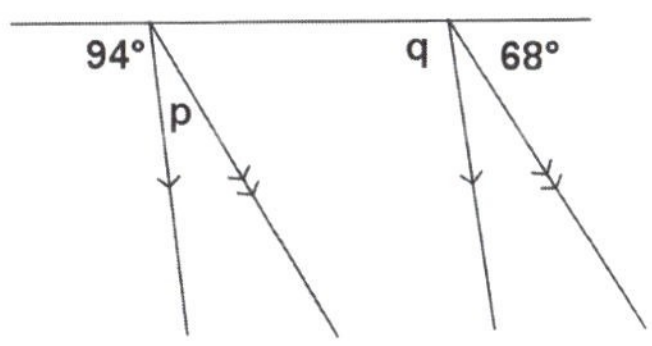

$m\angle p = 18°$
Consecutive adjacent angles on a line sum to $180°$.

$m\angle q = 94°$
If parallel lines are cut by a transversal, then corresponding angles are equal in measure.

10.

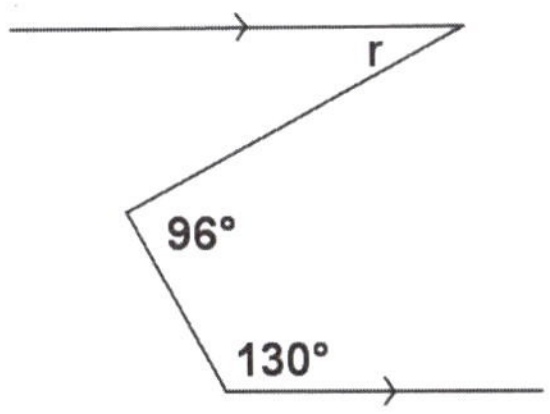

$m\angle r = 46°$
If parallel lines are cut by a transversal, then interior angles on the same side are supplementary; if parallel lines are cut by a transversal, then alternate interior angles are equal in measure.

Relevant Vocabulary

Relevant Vocabulary

ALTERNATE INTERIOR ANGLES: Let line t be a transversal to lines l and m such that t intersects l at point P and intersects m at point Q. Let R be a point on line l and S be a point on line m such that the points R and S lie in opposite half-planes of t. Then $\angle RPQ$ and $\angle PQS$ are called *alternate interior angles* of the transversal t with respect to line m and line l.

CORRESPONDING ANGLES: Let line t be a transversal to lines l and m. If $\angle x$ and $\angle y$ are alternate interior angles and $\angle y$ and $\angle z$ are vertical angles, then $\angle x$ and $\angle z$ are *corresponding angles*.

Closing (1 minute)

- Describe the conditions that tell us when a pair of lines cut by a transversal must be parallel.
  - *Two lines intersected by a transversal must be parallel when corresponding angles are congruent, when alternate interior angles are congruent, and when same-side interior angles are supplementary.*

Exit Ticket (5 minutes)

Name ______________________________ Date ______________

Lesson 7: Solving for Unknown Angles—Transversals

Exit Ticket

Determine the value of each variable.

$x =$ __________

$y =$ __________

$z =$ __________

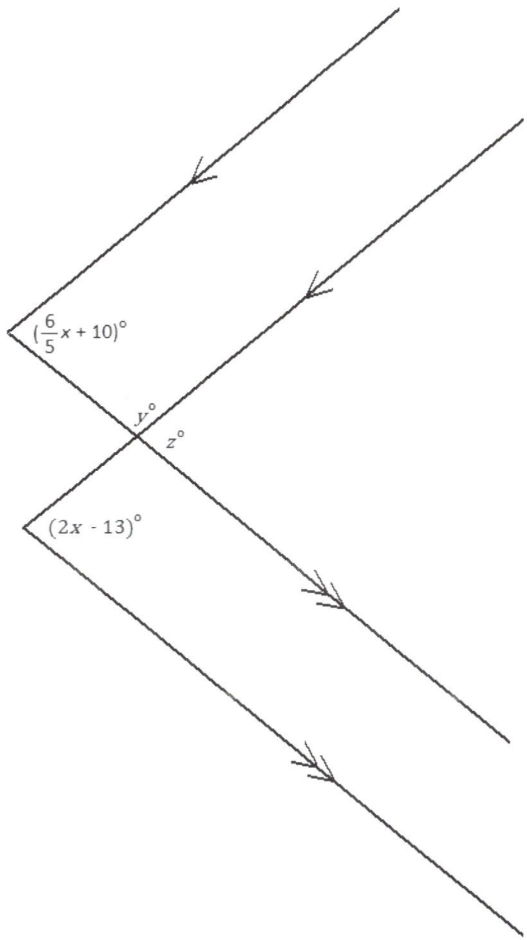

Exit Ticket Sample Solutions

Determine the value of each variable.

$x = \mathbf{28.75}$

$y = \mathbf{135.5}$

$z = \mathbf{44.5}$

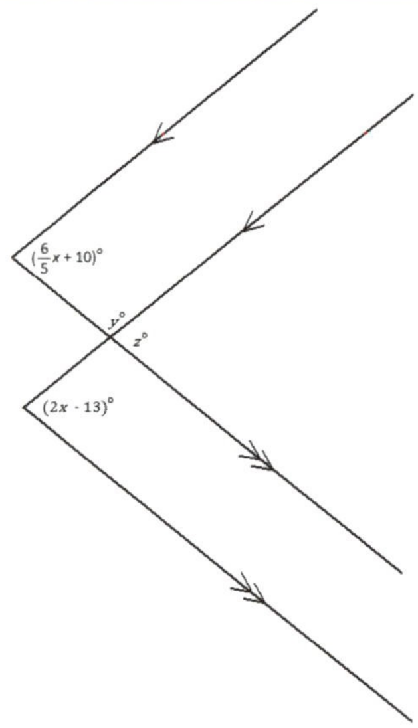

Problem Set Sample Solutions

MP.7

Find the unknown (labeled) angles. Give reasons for your solutions.

1.

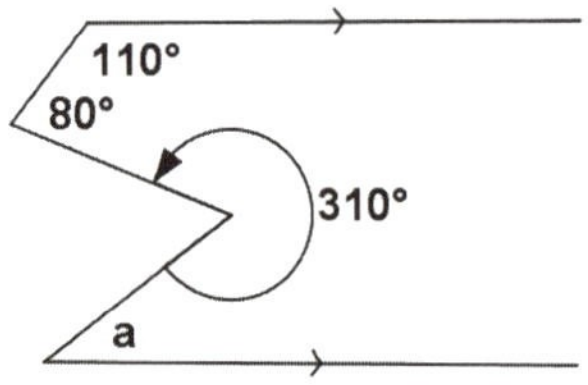

$m\angle a = \mathbf{40^\circ}$

If parallel lines are cut by a transversal, then alternate interior angles are equal in measure.

2.

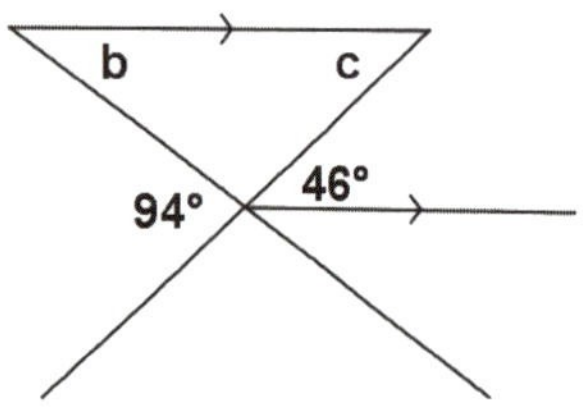

$m\angle b = \mathbf{48^\circ}$

If parallel lines are cut by a transversal, then corresponding angles are equal in measure.

$m\angle c = \mathbf{46^\circ}$

Vertical angles are equal in measure;
if parallel lines are cut by a transversal, then alternate interior angles are equal in measure.

MP.7

3.

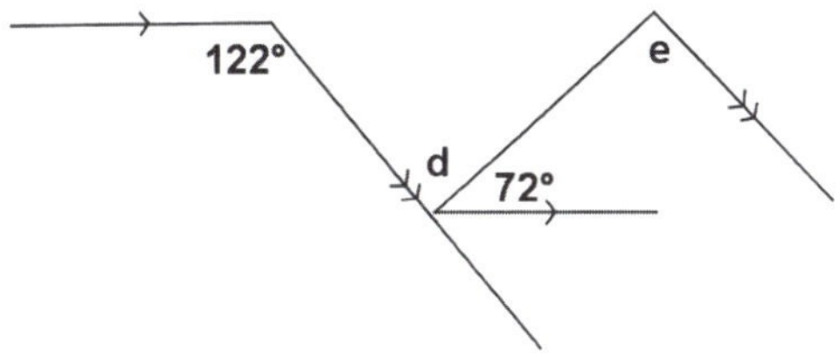

$m\angle d = 50°$

If parallel lines are cut by a transversal, then alternate interior angles are equal in measure.

$m\angle e = 50°$

If parallel lines are cut by a transversal, then alternate interior angles are equal in measure.

4.

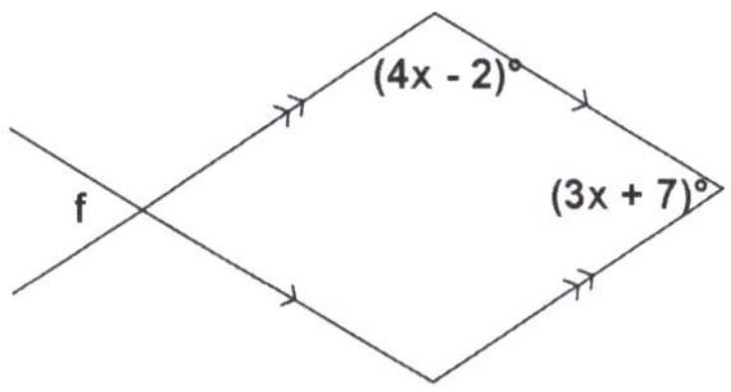

$m\angle f = 82°$

If parallel lines are cut by a transversal, then interior angles on the same side are supplementary; vertical angles are equal in measure.

Lesson 8: Solve for Unknown Angles—Angles in a Triangle

Student Outcome

- Students review formerly learned geometry facts and practice citing the geometric justifications regarding angles in a triangle in anticipation of unknown angle proofs.

Lesson Notes

In Lesson 8, the unknown angle problems expand to include angles in triangles. Knowing how to solve for unknown angles involving lines and angles at a point, angles involving transversals, and angles in triangles, students are prepared to solve unknown angles in a variety of diagrams.

Check the justifications students provide in their answers. The next three lessons on unknown angle proofs depend even more on these justifications.

Classwork

Opening Exercise (5 minutes)

Review the Problem Set from Lesson 7; students also attempt a review question from Lesson 7 below.

MP.7

Opening Exercise

Find the measure of angle x in the figure to the right. Explain your calculations. (Hint: Draw an auxiliary line segment.)

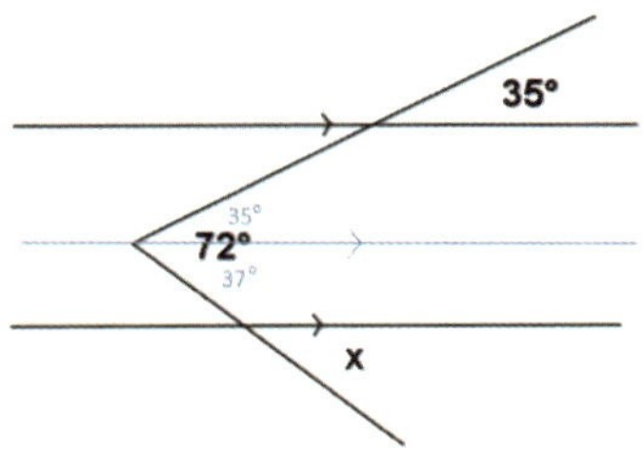

$m\angle x = 37°$

The angle with the measure of $72°$ can be divided two angles. One measures $35°$ (corresponding angles). Then the other angle has a measure of $37°$ (partition property).

Discussion (5 minutes)

Review facts about angles in a triangle.

Discussion

The sum of the 3 angle measures of any triangle is ___$180°$___.

INTERIOR OF A TRIANGLE: A point lies in the *interior of a triangle* if it lies in the interior of each of the angles of the triangle.

In any triangle, the measure of the exterior angle is equal to the sum of the measures of the ___*opposite interior*___ angles. These are sometimes also known as ___*remote interior*___ angles.

Base angles of an ___*isosceles*___ triangle are equal in measure.

Each angle of an ___*equilateral*___ triangle has a measure equal to $60°$.

Relevant Vocabulary (2 minutes)

> **Relevant Vocabulary**
>
> **ISOSCELES TRIANGLE: An *isosceles triangle* is a triangle with at least two sides of equal length.**
>
> **ANGLES OF A TRIANGLE: Every triangle $\triangle ABC$ determines three angles, namely, $\angle BAC$, $\angle ABC$, and $\angle ACB$. These are called the *angles of* $\triangle ABC$.**
>
> **EXTERIOR ANGLE OF A TRIANGLE: Let $\angle ABC$ be an interior angle of a triangle $\triangle ABC$, and let D be a point on $\overleftrightarrow{AB}$ such that B is between A and D. Then $\angle CBD$ is an *exterior angle of the triangle* $\triangle ABC$.**

Use a diagram to remind students that an exterior angle of a triangle forms a linear pair with an adjacent interior angle of the triangle.

Exercises 1–11 (28 minutes)

Students try an example based on the Discussion and review as a whole class.

> **Exercises 1–11**
>
> 1. **Find the measures of angles a and b in the figure to the right. Justify your results.**
>
> $m\angle a = 53°$
>
> $m\angle b = 40°$

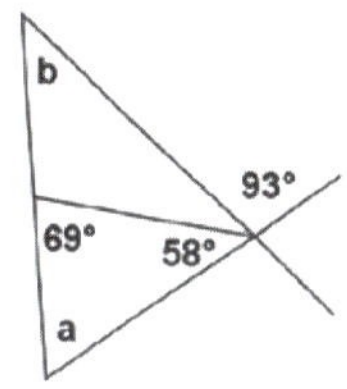

> **In each figure, determine the measures of the unknown (labeled) angles. Give reasons for your calculations.**
>
> 2.

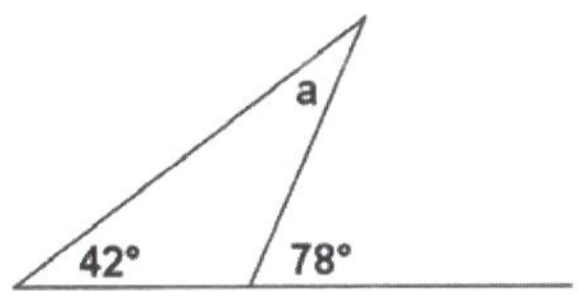

> $m\angle a = 36°$
>
> *The exterior angle of a triangle equals the sum of the two interior opposite angles.*
>
> 3.

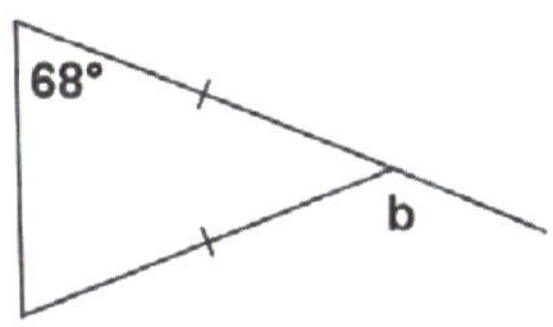

> $m\angle b = 136°$
>
> *The base angles of an isosceles triangle are equal in measure.*
>
> *The sum of the angle measures in a triangle is* $180°$.
>
> *Linear pairs form supplementary angles.*
>
> 4.

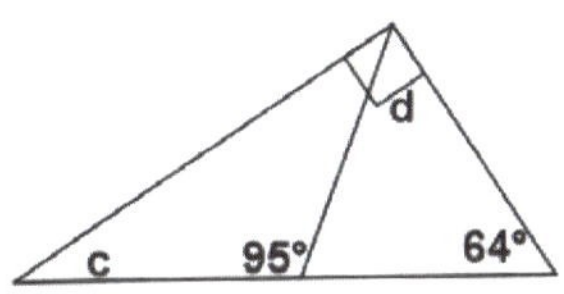

> $m\angle c = 26°$
>
> *The sum of the angle measures in a triangle is* $180°$.
>
> $m\angle d = 31°$
>
> *Linear pairs form supplementary angles.*
>
> *The sum of the angle measures in a triangle is* $180°$.

5.

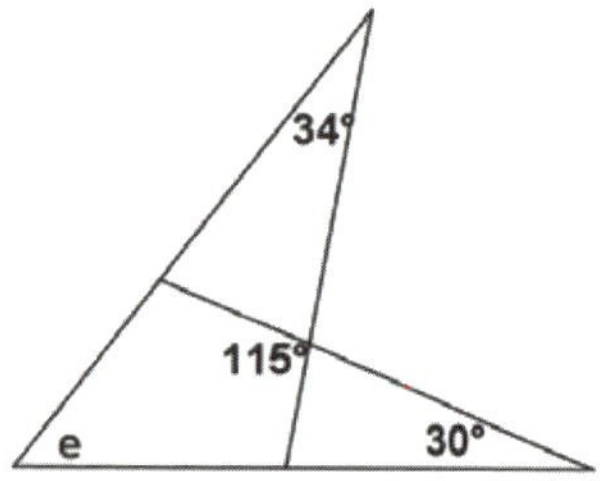

$m\angle e = 51°$

Linear pairs form supplementary angles.

The sum of the angle measures in a triangle is $180°$.

6.

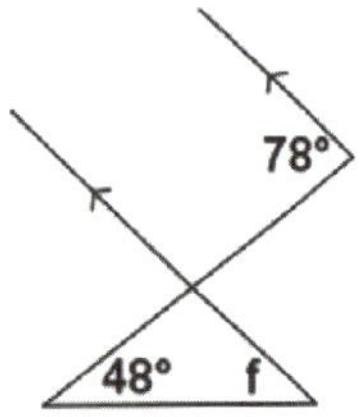

$m\angle f = 30°$

If parallel lines are cut by a transversal, then corresponding angles are equal in measure.

Linear pairs form supplementary angles.

The sum of the angle measures in a triangle is $180°$.

7.

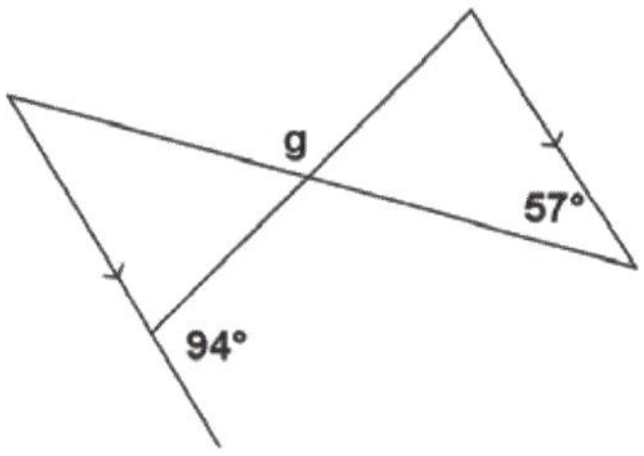

$m\angle g = 143°$

If parallel lines are cut by a transversal, then alternate interior angles are equal in measure.

Linear pairs form supplementary angles.

The sum of the angle measures in a triangle is $180°$.

8.

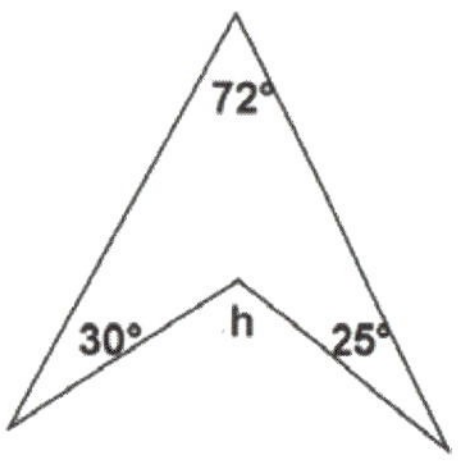

$m\angle h = 127°$

Draw an auxiliary line, and then use the facts that linear pairs form supplementary angles and the sum of the angle measures in a triangle is $180°$.

9.

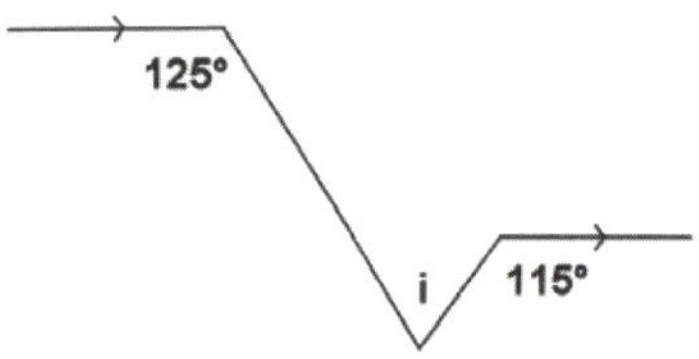

$m\angle i = 60°$

If parallel lines are cut by a transversal, then alternate interior angles are equal in measure.

Linear pairs form supplementary angles (twice).

The sum of the angle measures in a triangle is $180°$.

10.

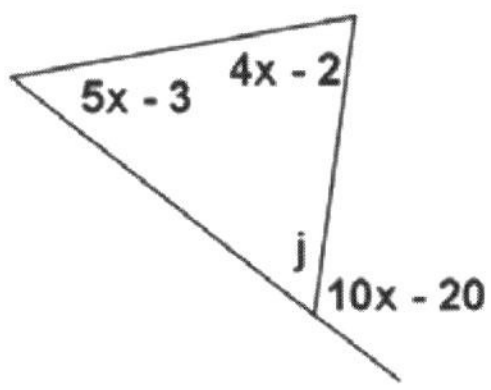

$m\angle j = 50°$

If parallel lines are cut by a transversal, then alternate interior angles are equal in measure.

Linear pairs form supplementary angles.

11.

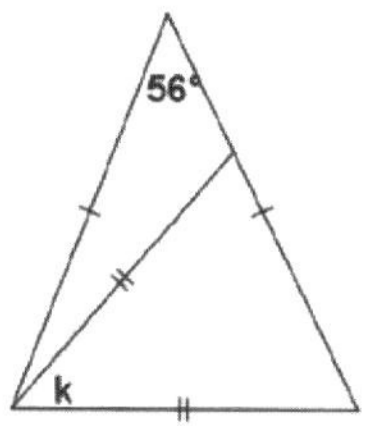

$m\angle k = 56°$

Closing (1 minute)

- What is the sum of angle measures of any triangle?
  - *The sum of angle measures of any triangle is* $180°$.
- Describe the relationship between an exterior angle and the remote interior angles of a triangle.
  - *The measure of the exterior angle of a triangle is equal to the sum of the measures of the* opposite interior *angles.*

Exit Ticket (4 minutes)

Name ______________________________ Date ______________

Lesson 8: Solve for Unknown Angles—Angles in a Triangle

Exit Ticket

Find the value of d and x.

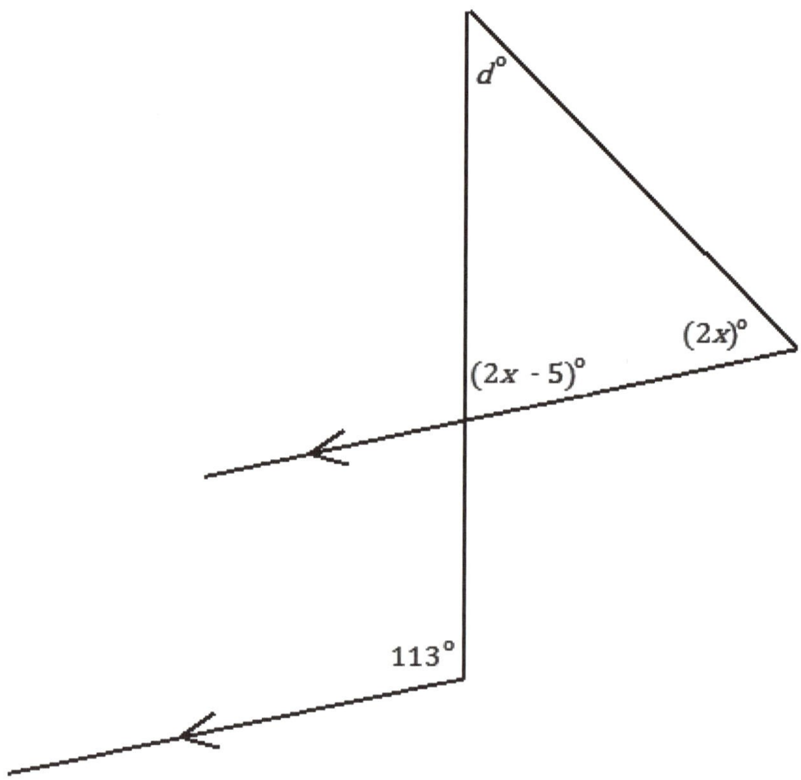

$d =$ ________

$x =$ ________

Exit Ticket Sample Solutions

Find the value of d and x.

$d = 41$

$x = 36$

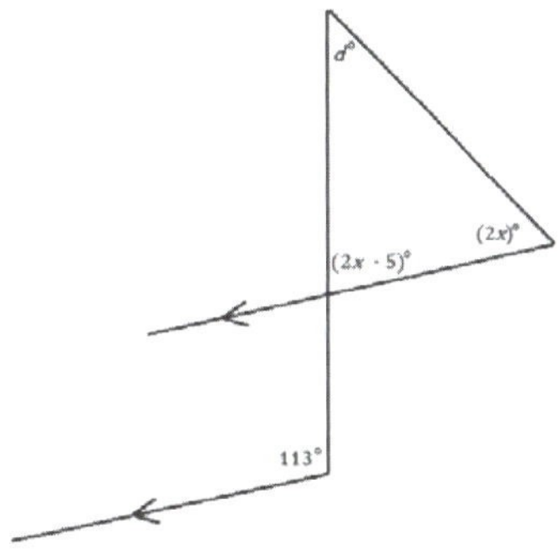

Problem Set Sample Solutions

Find the unknown (labeled) angle in each figure. Justify your calculations.

1.

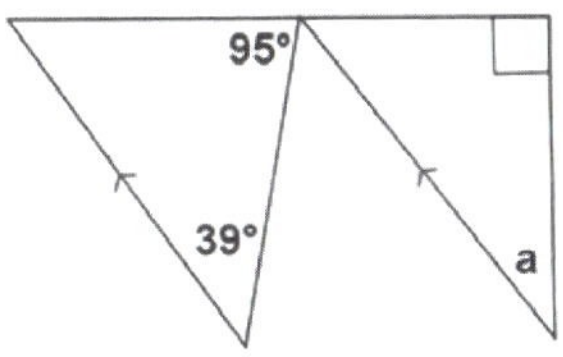

$m\angle a = 44°$

If parallel lines are cut by a transversal, then alternate interior angles are equal in measure.

Linear pairs form supplementary angles.

The sum of the angle measures in a triangle is $180°$.

2.

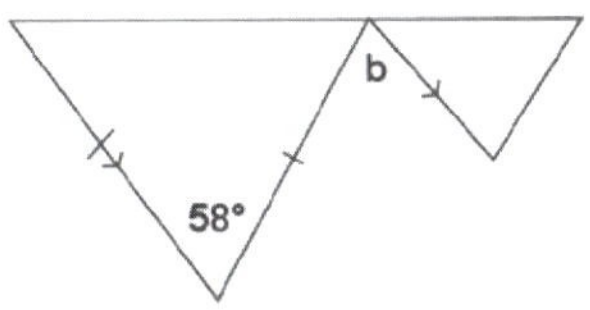

$m\angle b = 58°$

If parallel lines are cut by a transversal, then alternate interior angles are equal in measure.

3.

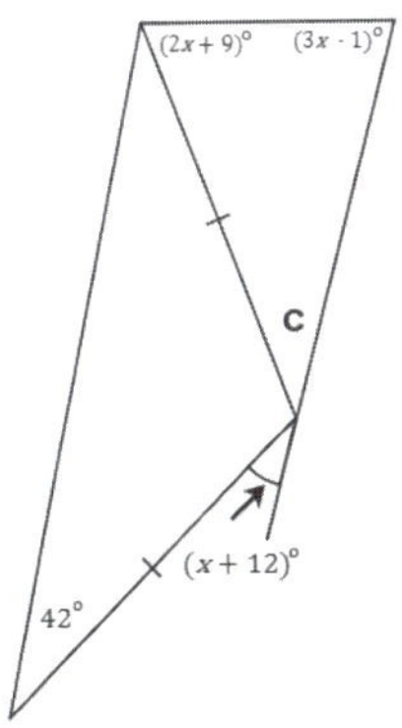

$m\angle c = 47°$

The base angles of an isosceles triangle are equal in measure.

The sum of the angle measures in a triangle is $180°$.

The exterior angle of a triangle equals the sum of the two interior opposite angles.

The sum of the angle measures in a triangle is $180°$.

Lesson 9: Unknown Angle Proofs—Writing Proofs

Student Outcomes

- Students write unknown angle proofs, using already accepted geometry facts.

Lesson Notes

In Lesson 9, students make the transition from unknown angle problems to unknown angle proofs. Instead of solving for a numeric answer, students need to justify a particular relationship. Students are prepared for this as they have been writing a reason for each step of their numeric answers in the last three lessons.

Begin the lesson with a video clip about Sherlock Holmes. Holmes examines a victim and makes deductions about the victim's attacker. He makes each deduction based on several pieces of evidence the victim provides. The video clip sets the stage for the deductive reasoning students must use to write proofs. Each geometric conclusion must be backed up with a concrete reason, a fact that has already been established. Following the video clip, lead the class through the example proof, eliciting the similarities and differences between the sample problem and subsequent proof questions. Emphasize that the questions still draw on the same set of geometric facts used to solve problems, and steps that are purely algebraic (some kind of arithmetic) do not require a justification. Students attempt an example and review together before beginning the exercises.

As students embark on the exercises, teachers can periodically review or ask for student solutions to ensure that they are formulating their steps clearly and providing appropriate reasons.

Note that in writing proofs, students draw upon many of the properties that they learned in middle school; some instruction on these may be necessary. A chart of frequently used properties is provided at the end of this lesson that may be used to supplement instruction or for student reference. Note that although the concept of congruence has not yet been discussed, the first three properties (reflexive, transitive, and symmetric) hold for congruence as well.

Classwork

Opening Exercise (5 minutes)

Students watch video clip:

In this exercise, students watch a video clip (Sherlock Holmes, Master of Deduction Does it Again, https://www.youtube.com/watch?v=o30UY_flFgM&feature=youtu.be) and discuss the connection between Holmes's process of identifying the attacker and the deduction used in geometry.

Emphasize that Holmes makes no guesses and that there is a solid piece of evidence behind each conclusion.

Opening Exercise

One of the main goals in studying geometry is to develop your ability to reason critically, to draw valid conclusions based upon observations and proven facts. Master detectives do this sort of thing all the time. Take a look as Sherlock Holmes uses seemingly insignificant observations to draw amazing conclusions.

Could you follow Sherlock Holmes's reasoning as he described his thought process?

Discussion (10 minutes)

Students examine the similarities and differences between unknown angle problems and proofs.

Remind students that they are drawing on the same set of facts they have been using in the last few days. Tell students that the three dots indicate that the proof has been completed.

Discussion

In geometry, we follow a similar deductive thought process (much like Holmes uses) to prove geometric claims. Let's revisit an old friend—solving for unknown angles. Remember this one?

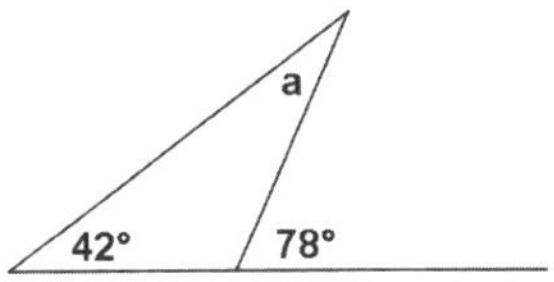

You needed to figure out the measure of a and used the "fact" that an exterior angle of a triangle equals the sum of the measures of the opposite interior angles. The measure of $\angle a$ must, therefore, be 36°.

Suppose that we rearrange the diagram just a little bit.

Instead of using numbers, we use variables to represent angle measures.

Suppose further that we already know that the angles of a triangle sum to 180°. Given the labeled diagram to the right, can we prove that $x + y = z$ (or, in other words, that the exterior angle of a triangle equals the sum of the measures of the opposite interior angles)?

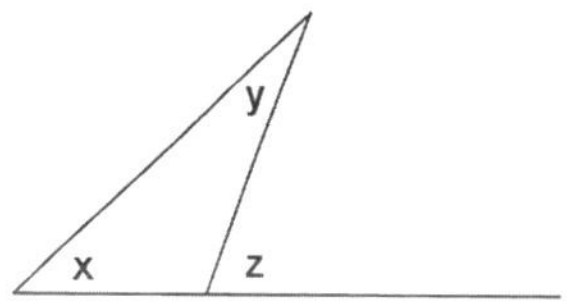

PROOF:

Label $\angle w$, as shown in the diagram.

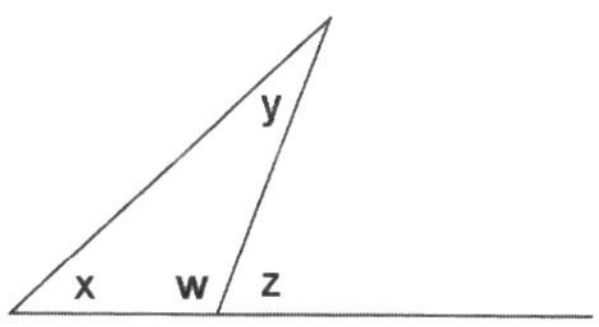

$m\angle x + m\angle y + m\angle w = 180^\circ$	**The sum of the angle measures in a triangle is 180°.**
$m\angle w + m\angle z = 180^\circ$	**Linear pairs form supplementary angles.**
$m\angle x + m\angle y + m\angle w = m\angle w + m\angle z$	**Substitution property of equality**
$\therefore m\angle x + m\angle y = m\angle z$	**Subtraction property of equality**

Notice that each step in the proof was justified by a previously known or demonstrated fact. We end up with a newly proven fact (that an exterior angle of any triangle is the sum of the measures of the opposite interior angles of the triangle). This ability to identify the steps used to reach a conclusion based on known facts is *deductive reasoning* (i.e., the same type of reasoning that Sherlock Holmes used to accurately describe the doctor's attacker in the video clip).

Exercises 1–6 (24 minutes)

Exercises 1–6

1. You know that angles on a line sum to $180°$.

   Prove that vertical angles are equal in measure.

   Make a plan:

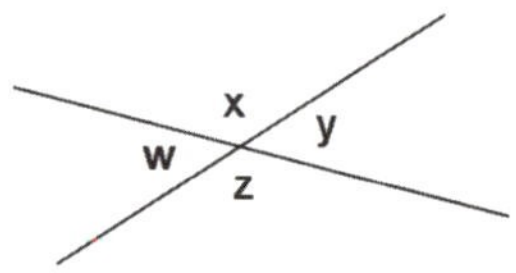

   - What do you know about $\angle w$ and $\angle x$? $\angle y$ and $\angle x$?

     They sum to $180°$.

   - What conclusion can you draw based on both pieces of knowledge?

     $m\angle w = m\angle y$

MP.7 & MP.8

   - Write out your proof:

$m\angle w + m\angle x = 180°$	*Linear pairs form supplementary angles.*
$m\angle y + m\angle x = 180°$	*Linear pairs form supplementary angles.*
$m\angle w + m\angle x = m\angle y + m\angle x$	*Substitution property of equality*
$\therefore\ m\angle w = m\angle y$	*Subtraction property of equality*

Note to Teacher:

There are different ways of notating the *Given* and *Prove*. Alternate pairings include *Hypothesis/Conclusion* and *Suppose/Then*. The point is that we begin with what is observed and end with what is deduced.

2. Given the diagram to the right, prove that $m\angle w + m\angle x + m\angle z = 180°$. (Make a plan first. What do you know about $\angle x$, $\angle y$, and $\angle z$?)

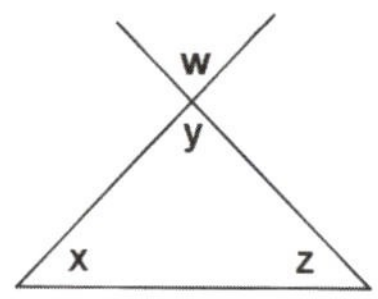

$m\angle y + m\angle x + m\angle z = 180°$	*The sum of the angles of a triangle is* $180°$.
$m\angle y = m\angle w$	*Vertical angles are equal in measure.*
$\therefore\ m\angle w + m\angle x + m\angle z = 180°$	*Substitution property of equality*

Given the diagram to the right, prove that $m\angle w = m\angle y + m\angle z$.

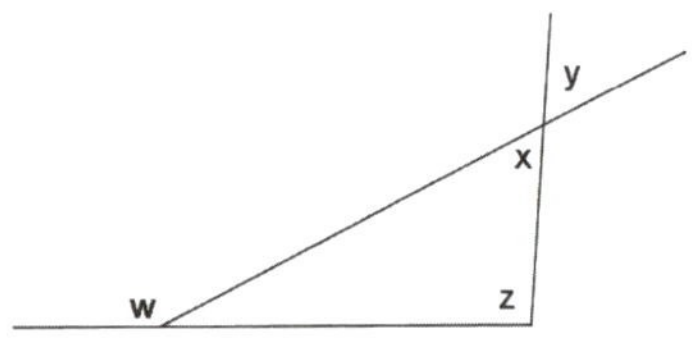

$m\angle w = m\angle x + m\angle z$	*The exterior angle of a triangle equals the sum of the two opposite interior angles.*
$m\angle x = m\angle y$	*Vertical angles are equal in measure.*
$\therefore\ m\angle w = m\angle y + m\angle z$	*Substitution property of equality*

3. In the diagram to the right, prove that $m\angle y + m\angle z = m\angle w + m\angle x$. (You need to write a label in the diagram that is not labeled yet for this proof.)

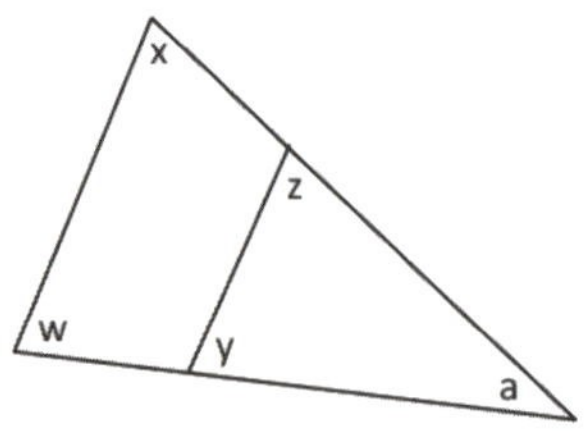

$m\angle a + m\angle x + m\angle w = 180°$	*The sum of the angles of a triangle is* $180°$.
$m\angle a + m\angle z + m\angle y = 180°$	*The sum of the angles of a triangle is* $180°$.
$m\angle a + m\angle x + m\angle w = m\angle a + m\angle z + m\angle y$	*Substitution property of equality*
$\therefore\ m\angle x + m\angle w = m\angle z + m\angle y$	*Subtraction property of equality*

EUREKA MATH™

4. In the figure to the right, $\overline{AB} \parallel \overline{CD}$ and $\overline{BC} \parallel \overline{DE}$.
Prove that $\text{m}\angle ABC = \text{m}\angle CDE$.

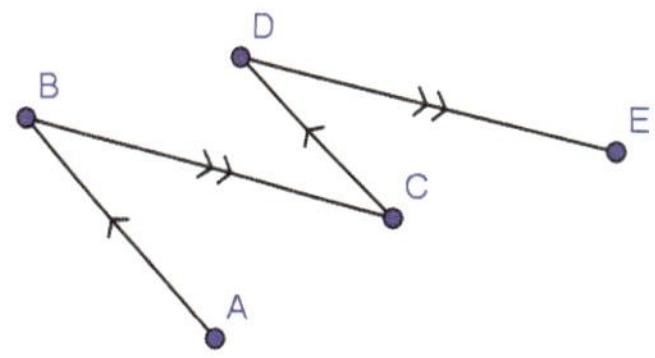

$\overline{AB} \parallel \overline{CD}$, $\overline{BC} \parallel \overline{DE}$	*Given*
$m\angle ABC = m\angle DCB$	*When two parallel lines are cut by a transversal, the alternate interior angles are equal in measure.*
$m\angle DCB = m\angle CDE$	*When two parallel lines are cut by a transversal, the alternate interior angles are equal in measure.*
$\therefore\ m\angle ABC = m\angle CDE$	*Substitution property of equality*

5. In the figure to the right, prove that the sum of the angles marked by arrows is $900°$.
(You need to write several labels in the diagram for this proof.)

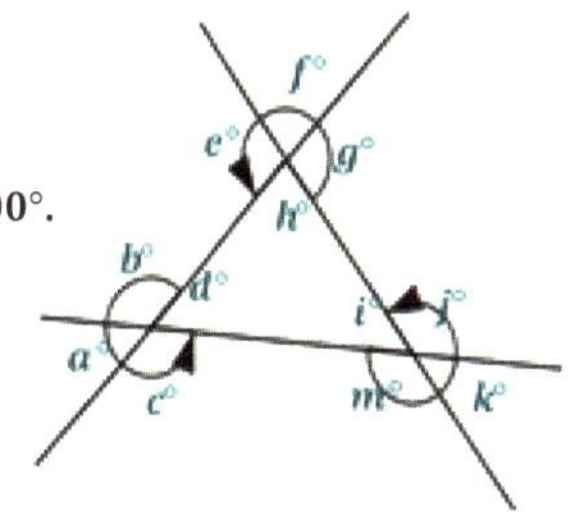

$a + b + c + d = 360$	*Angles at a point sum to* $360°$.
$e + f + g + h = 360$	*Angles at a point sum to* $360°$.
$i + j + k + m = 360$	*Angles at a point sum to* $360°$.
$a + b + c + d + e + f + g + h + i + j + k + m = 1080$	*Addition property of quality*
$d + h + i = 180$	*The sum of the angle measures in a triangle is* $180°$.
$a + b + c + e + f + g + j + k + m + 180 = 1080$	*Substitution property of equality*
$\therefore a + b + c + e + f + g + j + k + m = 900$	*Subtraction property of equality*

6. In the figure to the right, prove that $\overline{DC} \perp \overline{EF}$.
Draw in label Z.

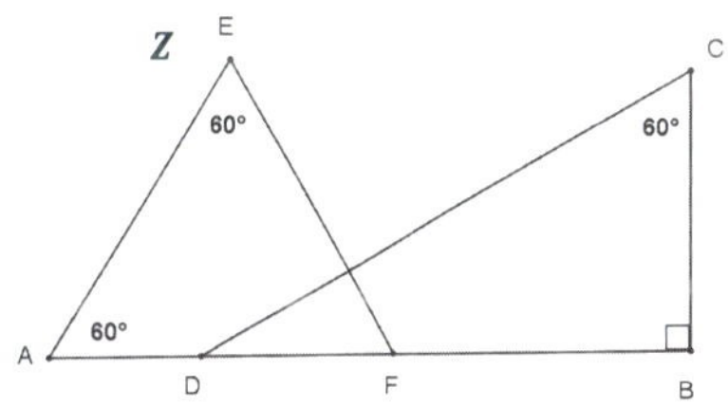

$m\angle E + m\angle A + m\angle EFA = 180°$	*The sum of the angle measures in a triangle is* $180°$.
$m\angle EFA = 60°$	*Subtraction property of equality*
$m\angle B + m\angle C + m\angle CDB = 180°$	*The sum of the angle measures in a triangle is* $180°$.
$m\angle CDB = 30°$	*Subtraction property of equality*
$m\angle CDB + m\angle EFA + m\angle EZC = 180°$	*The sum of the angle measures in a triangle is* $180°$.
$m\angle EZC = 90°$	*Subtraction property of equality*
$\overline{DC} \perp \overline{EF}$	*Perpendicular lines form* $90°$ *angles.*

Closing (1 minute)

- Describe deductive reasoning.
  - *Deductive reasoning is the process of drawing a conclusion by justifying each successive step of an argument with a known fact.*

Exit Ticket (5 minutes)

Name ______________________________ Date ______________

Lesson 9: Unknown Angle Proofs—Writing Proofs

Exit Ticket

In the diagram to the right, prove that the sum of the labeled angles is $180°$.

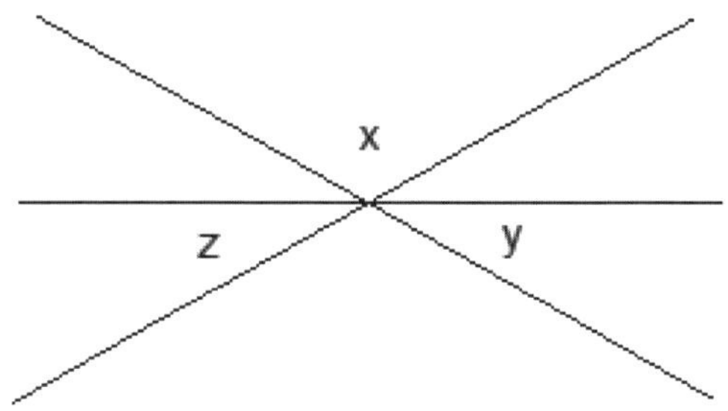

Exit Ticket Sample Solutions

In the diagram to the right, prove that the sum of the labeled angles is $180°$.

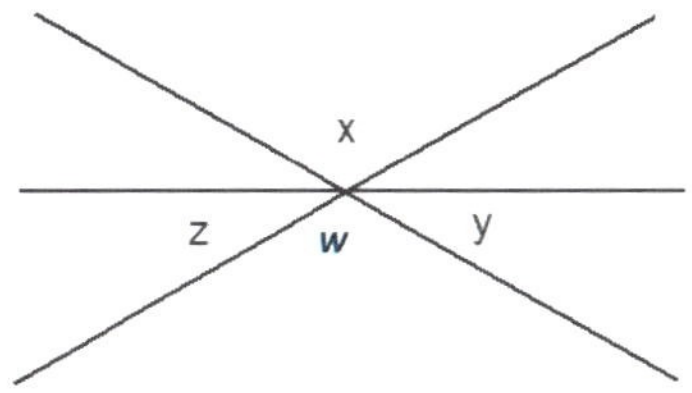

Label $\angle w$, vertical to $\angle x$.

$m\angle w = m\angle x$	*Vertical angles are equal in measure.*
$m\angle z + m\angle w + m\angle y = 180°$	*Angles on a line sum to $180°$.*
$m\angle z + m\angle x + m\angle y = 180°$	*Substitution property of equality*

Problem Set Sample Solutions

1. In the figure to the right, prove that $m \parallel n$.

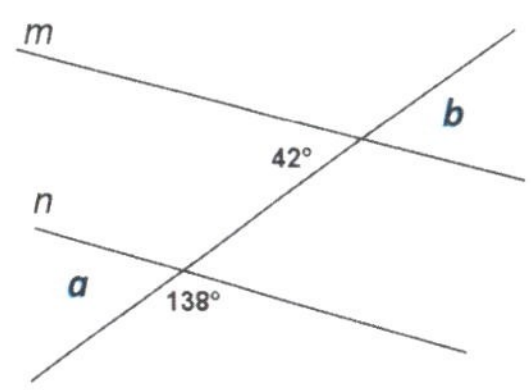

$m\angle a + 138° = 180°$	*Linear pairs form supplementary angles.*
$m\angle b = 42°$	*Vertical angles are equal in measure.*
$m\angle a = 42°$	*Subtraction property of equality*
$m\angle a = m\angle b$	*Substitution property of equality*
$\therefore m \parallel n$	*If two lines are cut by a transversal such that a pair of alternate interior angles are equal in meaure, then the lines are parallel.*

2. In the diagram to the right, prove that the sum of the angles marked by arrows is $360°$.

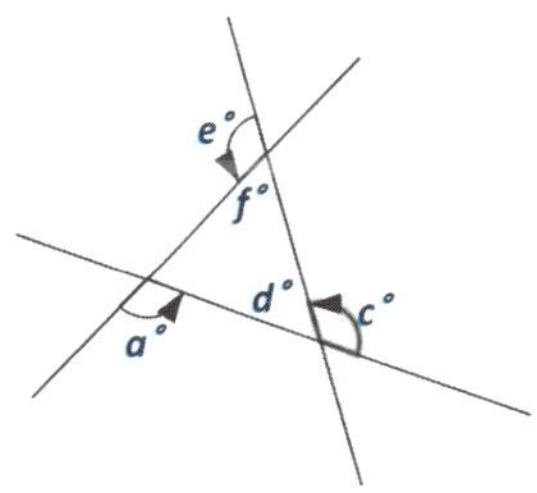

$a + b = 180$	*Linear pairs form supplementary angles.*
$c + d = 180$	*Linear pairs form supplementary angles.*
$e + f = 180$	*Linear pairs form supplementary angles.*
$a + b + c + d + e + f = 540$	*Addition property of equality*
$b + d + f = 180$	*The sum of the angle measures in a triangle is $180°$.*
$\therefore a + c + e = 360$	*Subtraction property of equality*

3. In the diagram to the right, prove that $m\angle a + m\angle d - m\angle b = 180°$.

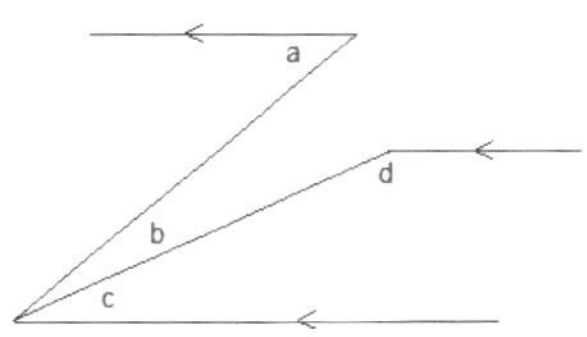

$m\angle a = m\angle b + m\angle c$	*If parallel lines are cut by a transversal, then alternate interior angles are equal in measure.*
$m\angle c + m\angle d = 180°$	*If parallel lines are cut by a transversal, then interior angles on the same side are supplementary.*
$m\angle a - m\angle b + m\angle d = 180°$	*Addition of zero/additive identity property*
$m\angle a + m\angle d - m\angle b = 180°$	*Substitution property of equality*

Basic Properties Reference Chart

Property	Meaning	Geometry Example
Reflexive Property	A quantity is equal to itself.	$AB = AB$
Transitive Property	If two quantities are equal to the same quantity, then they are equal to each other.	If $AB = BC$ and $BC = EF$, then $AB = EF$.
Symmetric Property	If a quantity is equal to a second quantity, then the second quantity is equal to the first.	If $OA = AB$, then $AB = OA$.
Addition Property of Equality	If equal quantities are added to equal quantities, then the sums are equal.	If $AB = DF$ and $BC = CD$, then $AB + BC = DF + CD$.
Subtraction Property of Equality	If equal quantities are subtracted from equal quantities, the differences are equal.	If $AB + BC = CD + DE$ and $BC = DE$, then $AB = CD$.
Multiplication Property of Equality	If equal quantities are multiplied by equal quantities, then the products are equal.	If $m\angle ABC = m\angle XYZ$, then $2(m\angle ABC) = 2(m\angle XYZ)$.
Division Property of Equality	If equal quantities are divided by equal quantities, then the quotients are equal.	If $AB = XY$, then $\frac{AB}{2} = \frac{XY}{2}$.
Substitution Property of Equality	A quantity may be substituted for its equal.	If $DE + CD = CE$ and $CD = AB$, then $DE + AB = CE$.
Partition Property (includes *Angle Addition Postulate*, *Segments Add*, *Betweenness of Points*, etc.)	A whole is equal to the sum of its parts.	If point C is on $\overline{AB}$, then $AC + CB = AB$.

Lesson 10: Unknown Angle Proofs—Proofs with Constructions

Student Outcome

- Students write unknown angle proofs involving auxiliary lines.

Lesson Notes

On the second day of unknown angle proofs, students incorporate the use of constructions, specifically auxiliary lines, to help them solve problems. In this lesson, students refer to the same list of facts they have been working with in the last few lessons. What sets this lesson apart is that necessary information in the diagram may not be apparent without some modification. One of the most common uses for an auxiliary line is in diagrams where multiple sets of parallel lines exist. Encourage students to mark up diagrams until the necessary relationships for the proof become more obvious.

Classwork

Opening Exercise (7 minutes)

Review the Problem Set from Lesson 9. Then, the whole class works through an example of a proof requiring auxiliary lines.

Opening Exercise

In the figure to the right, $\overline{AB} \parallel \overline{DE}$ and $\overline{BC} \parallel \overline{EF}$. Prove that $b = e$.
(Hint: Extend $\overline{BC}$ and $\overline{ED}$.)

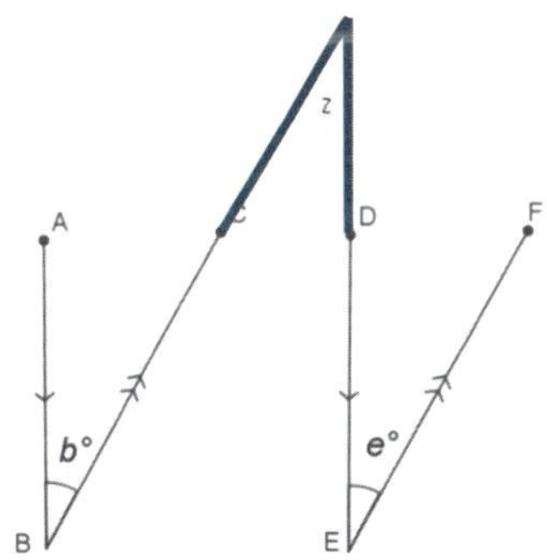

PROOF:

$\overline{AB} \parallel \overline{DE}$, $\overline{BC} \parallel \overline{EF}$ *Given*

$b = z$ *If parallel lines are cut by a transversal, then alternate interior angles are equal in measure.*

$z = e$ *If parallel lines are cut by a transversal, then alternate interior angles are equal in measure.*

$b = e$ *Transitive property*

In the previous lesson, you used deductive reasoning with labeled diagrams to prove specific conjectures. What is different about the proof above?

Drawing or extending segments, lines, or rays (referred to as *auxiliary lines*) is frequently useful in demonstrating steps in the deductive reasoning process. Once $\overline{BC}$ and $\overline{ED}$ were extended, it was relatively simple to prove the two angles congruent based on our knowledge of alternate interior angles. Sometimes there are several possible extensions or additional lines that would work equally well.

For example, in this diagram, there are at least two possibilities for auxiliary lines. Can you spot them both?

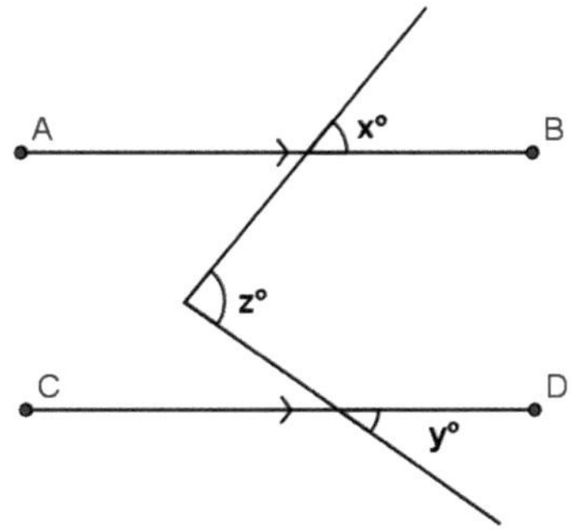

Given: $\overline{AB} \parallel \overline{CD}$.

Prove: $z = x + y$.

Discussion (8 minutes)

Students explore different ways to add auxiliary lines (construction) to the same diagram.

Discussion

Here is one possibility:

Given: $\overline{AB} \parallel \overline{CD}$.

Prove: $z = x + y$.

Extend the transversal as shown by the dotted line in the diagram. Label angle measures v and w, as shown.

What do you know about v and x?

About w and y? How does this help you?

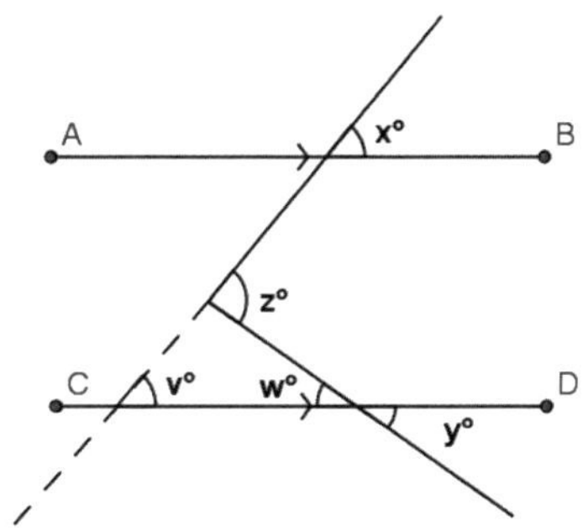

Write a proof using the auxiliary segment drawn in the diagram to the right.

MP.7

$\overline{AB} \parallel \overline{CD}$	*Given.*
$z = v + w$	*The exterior angle of a triangle equals the sum of the two interior opposite angles.*
$x = v$	*If parallel lines are cut by a transversal, then corresponding angles are equal in measure.*
$y = w$	*If parallel lines are cut by a transversal, then corresponding angles are equal in measure.*
$z = v + w$	*Angle addition postulate*
$z = x + y$	*Substitution property of equality*

Another possibility appears here:

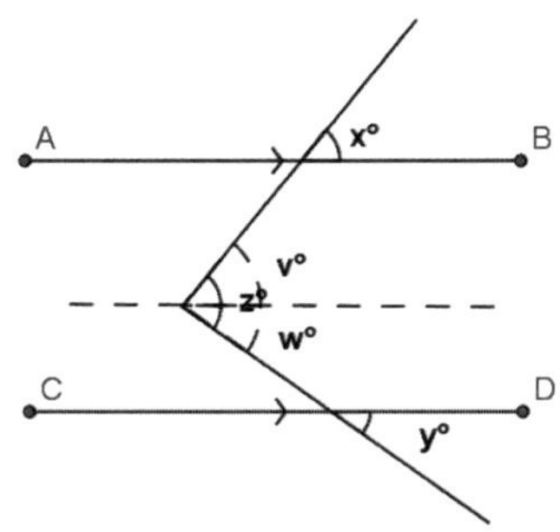

Given: $\overline{AB} \parallel \overline{CD}$.
Prove: $z = x + y$.

Draw a segment parallel to $\overline{AB}$ through the vertex of the angle measuring z degrees. This divides the angle into two parts as shown.

What do you know about v and x?

They are equal since they are corresponding angles of parallel lines crossed by a transversal.

About w and y? How does this help you?

They are also equal in measure since they are corresponding angles of parallel lines crossed by a transversal.

Write a proof using the auxiliary segment drawn in this diagram. Notice how this proof differs from the one above.

$\overline{AB} \parallel \overline{CD}$	*Given*
$x = v$	*If parallel lines are cut by a transversal, the corresponding angles are equal.*
$y = w$	*If parallel lines are cut by a transversal, the corresponding angles are equal.*
$z = v + w$	*Angle addition postulate*
$z = x + y$	*Substitution property of equality*

Examples (24 minutes)

Examples

1. In the figure to the right, $\overline{AB} \parallel \overline{CD}$ and $\overline{BC} \parallel \overline{DE}$.
   Prove that $m\angle ABC = m\angle CDE$.
   (Is an auxiliary segment necessary?)

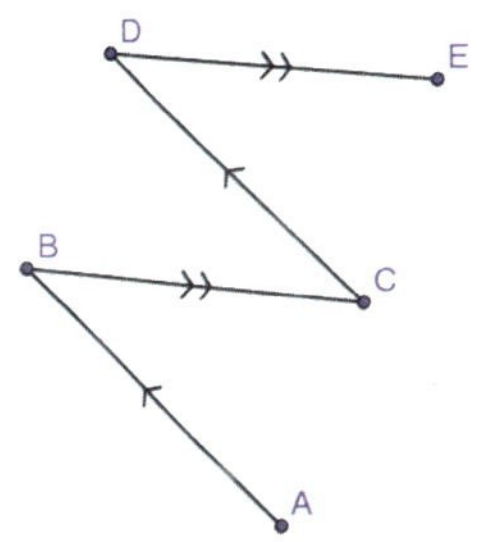

$\overline{AB} \parallel \overline{CD}$, $\overline{BC} \parallel \overline{DE}$	*Given*
$m\angle ABC = m\angle BCD$	*If parallel lines are cut by a transversal, then alternate interior angles are equal in measure.*
$m\angle BCD = m\angle CDE$	*If parallel lines are cut by a transversal, then alternate interior angles are equal in measure.*
$m\angle ABC = m\angle CDE$	*Transitive property*

2. In the figure to the right, $\overline{AB} \parallel \overline{CD}$ and $\overline{BC} \parallel \overline{DE}$.
   Prove that $b + d = 180$.

   Label $c°$.

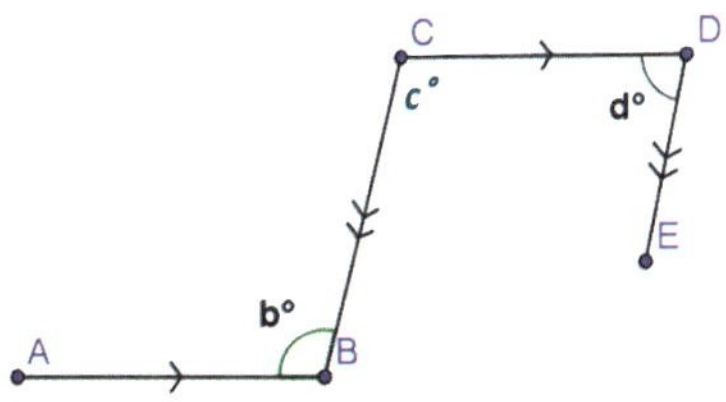

$\overline{AB} \parallel \overline{CD}$, $\overline{BC} \parallel \overline{DE}$	*Given*
$b = c$	*If parallel lines are cut by a transversal, then alternate interior angles are equal in measure.*
$c + d = 180$	*If parallel lines are cut by a transversal, then same-side interior angles are supplementary.*
$b + d = 180$	*Substitution property of equality*

3. In the figure to the right, prove that $d = a + b + c$.

Label Z and z.

$z = b + c$	*The exterior angle of a triangle equals the sum of the two interior opposite angles.*
$d = z + a$	*The exterior angle of a triangle equals the sum of the two interior opposite angles.*
$d = a + b + c$	*Substitution property of equality*

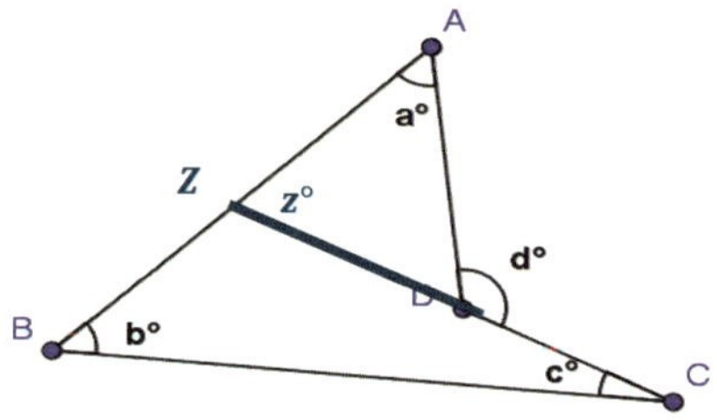

Closing (1 minute)

- How do auxiliary lines help solve for unknown angles?
  - *Sometimes extending a segment, line, or ray, or simply adding one illuminates an intermediate step in a solution to an unknown angle problem or proof.*

Exit Ticket (5 minutes)

Name ____________________ Date ____________

Lesson 10: Unknown Angle Proofs—Proofs with Constructions

Exit Ticket

Write a proof for each question.

1. In the figure to the right, $\overline{AB} \parallel \overline{CD}$. Prove that $a° = b°$.

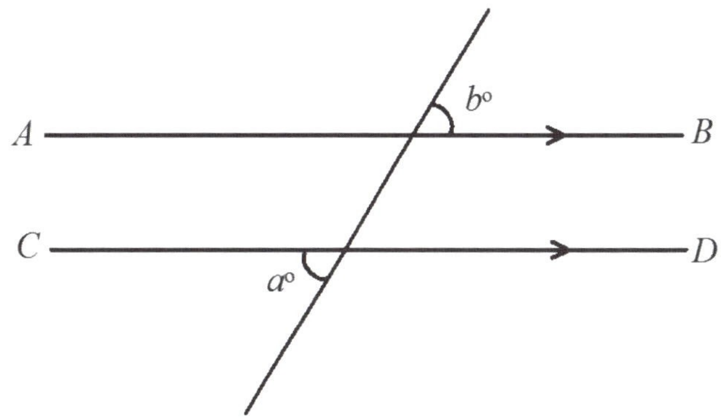

2. Prove $m\angle p = m\angle r$.

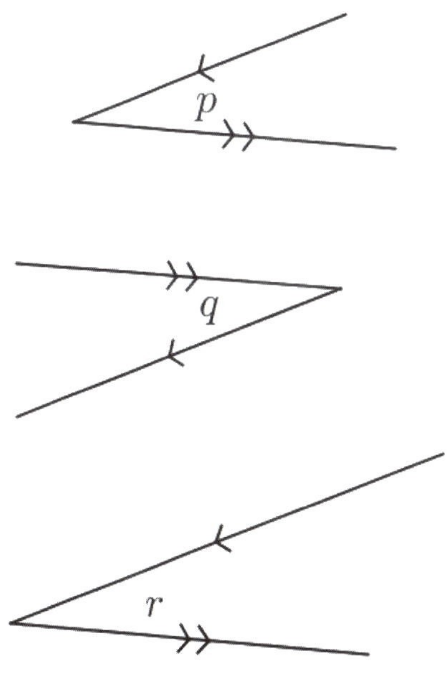

Exit Ticket Sample Solutions

Write a proof for each question.

1. In the figure to the right, $\overline{AB} \parallel \overline{CD}$. Prove that $a° = b°$.

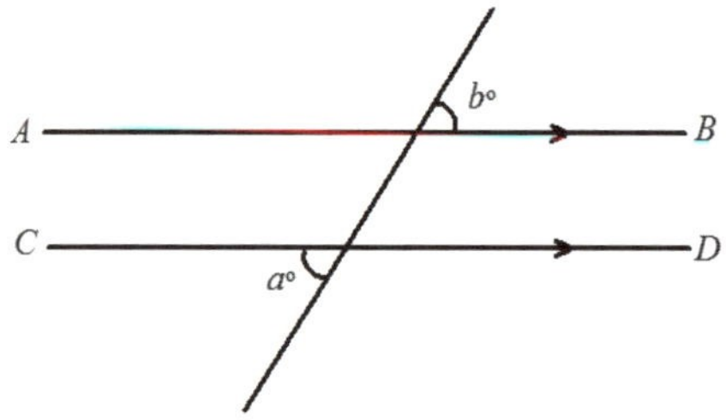

Write in angles c and d.

$\overline{AB} \parallel \overline{CD}$	*Given*
$a° = c°$	*Vertical angles are equal in measure.*
$c° = d°$	*If parallel lines are cut by a transversal, then alternate interior angles are equal in measure.*
$d° = b°$	*Vertical angles are equal in measure.*
$a° = b°$	*Substitution property of equality*

2. Prove $m\angle p = m\angle r$.

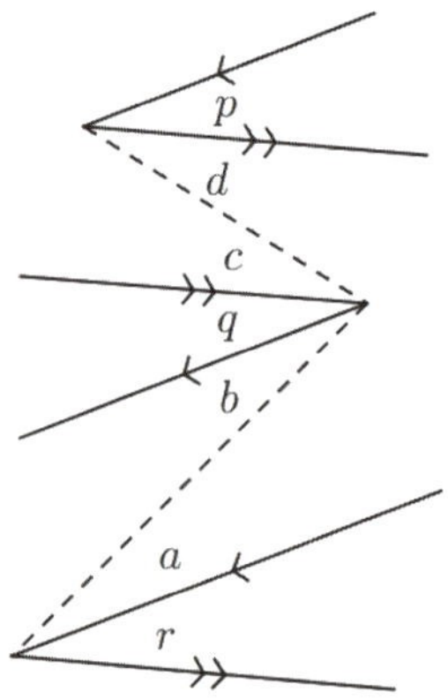

Mark angles a, b, c, and d.

$p \parallel q$	*Given*
$m\angle p + m\angle d = m\angle c + m\angle q$	*If parallel lines are cut by a transversal, then alternate interior angles are equal in measure.*
$m\angle d = m\angle c$	*If parallel lines are cut by a transversal, then alternate interior angles are equal in measure.*
$m\angle p = m\angle q$	*Subtraction property of equality*
$m\angle q + m\angle b = m\angle a + m\angle r$	*If parallel lines are cut by a transversal, then alternate interior angles are equal in measure.*
$m\angle a = m\angle b$	*If parallel lines are cut by a transversal, then alternate interior angles are equal in measure.*
$m\angle q = m\angle r$	*Substitution property of equality and subtraction property of equality*
$m\angle p = m\angle r$	*Substitution property of equality*

Problem Set Sample Solutions

1. In the figure to the right, $\overline{AB} \parallel \overline{DE}$ and $\overline{BC} \parallel \overline{EF}$.
Prove that $m\angle ABC = m\angle DEF$.

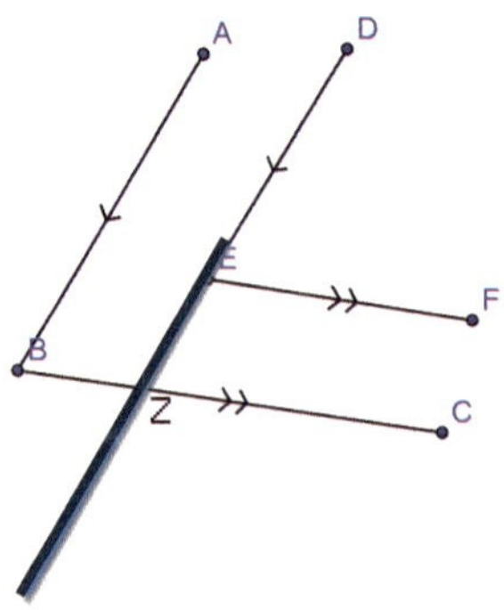

Extend $\overline{DE}$ through $\overline{BC}$, and mark the intersection with $\overline{BC}$ as Z.

$\overline{AB} \parallel \overline{DE}$, $\overline{BC} \parallel \overline{EF}$	*Given*
$m\angle ABC = m\angle EZC$	*If parallel lines are cut by a transversal, then corresponding angles are equal in measure.*
$m\angle EZC = m\angle DEF$	*If parallel lines are cut by a transversal, then corresponding angles are equal in measure.*
$m\angle ABC = m\angle DEF$	*Transitive property*

2. In the figure to the right, $\overline{AB} \parallel \overline{CD}$.
Prove that $m\angle AEC = a^\circ + c^\circ$.

Draw a line through E parallel to $\overline{AB}$ and $\overline{CD}$.

Add point F.

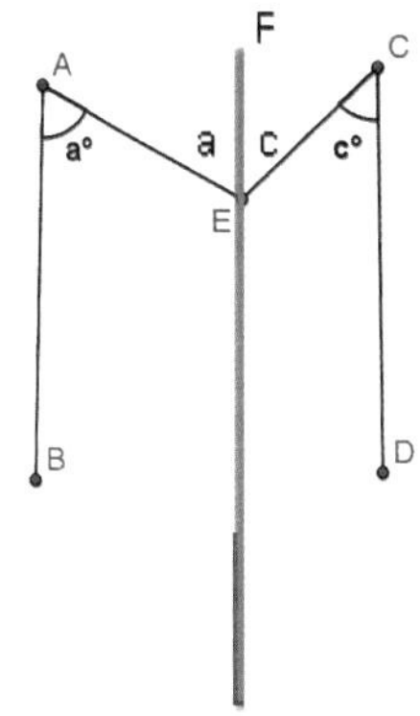

$\overline{AB} \parallel \overline{CD}$	*Given*
$m\angle BAE = m\angle AEF$	*If parallel lines are cut by a transversal, then alternate interior angles are equal in measure.*
$m\angle DCE = m\angle FEC$	*If parallel lines are cut by a transversal, then alternate interior angles are congruent equal in measure.*
$m\angle AEC = a^\circ + c^\circ$	*Angle addition postulate*

Lesson 11: Unknown Angle Proofs—Proofs of Known Facts

Student Outcomes

- Students write unknown angle proofs involving known facts.

Lesson Notes

In this last lesson on unknown angle proofs, students use their proof-writing skills to examine facts already familiar to them (i.e., the sum of angles of a triangle is $180°$, and vertical angles are equal in measure). This offers students a *why* and a *how* to this body of information. Proving known facts also makes students aware of the way this list grows because proving one fact allows the proof of the next fact.

Students begin by reviewing the Problem Set from Lesson 10. Then, they explore a known fact: Opposite angles of parallelograms are equal in measure. After working through the proof as a whole class, the teacher should point out that although we have a body of familiar geometry facts, we have never explored them to make sure they were true. Demonstrate with the examples provided that it is possible to use one basic fact to build to the next. After the notes, share the video clip about Eratosthenes and his use of geometry (especially the use of alternate interior angles in the thought process) to find the circumference of Earth. Then, do Example 1, and also discuss the use of the converse parallel line theorems and their abbreviations. Consider demonstrating why the converse holds true, for example, by asking a helper to hold up a ruler (the transversal) and a second helper to hold two rulers, one in each hand. Ask them to show what the rulers do to test a converse parallel line theorem. End class with the Problem Set assignment.

Classwork

Opening Exercise (8 minutes)

Review the Problem Set from Lesson 10. Students engage in a whole-class discussion about proving known facts, beginning with a specific example.

> *Note to Teacher:*
>
> Remind students that a theorem is often stated as an *if–then*, as
> If *hypothesis*, then *conclusion*.

> **Opening Exercise**
>
> **A *proof* of a mathematical statement is a detailed explanation of how that statement follows logically from other statements already accepted as true.**
>
> **A *theorem* is a mathematical statement with a proof.**

Consider taking a moment to mention that theorems can be stated without reference to any specific, labeled diagram. However, we cannot take steps to prove a statement without a way of referring to parts. Students observe situations where the labels are provided and situations where they must draw diagrams and label parts.

Discussion (15 minutes)

Students use the facts already provided in a list to prove each successive fact.

Discussion

Once a theorem has been proved, it can be added to our list of known facts and used in proofs of other theorems. For example, in Lesson 9, we proved that *vertical angles are of equal measure*, and we know (from earlier grades and by paper cutting and folding) that *if a transversal intersects two parallel lines, alternate interior angles are of equal measure*. How do these facts help us prove that corresponding angles are equal in measure?

Answers may vary.

In the diagram to the right, if you are given that $\overline{AB} \parallel \overline{CD}$, how can you use your knowledge of how vertical angles and alternate interior angles are equal in measure to prove that $x = w$?

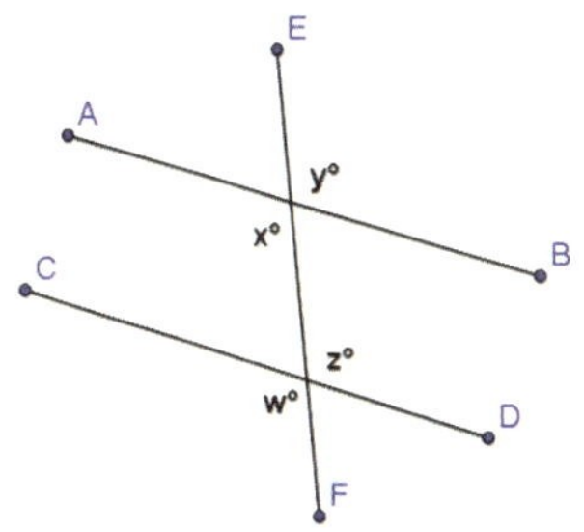

$w = z$ *(Vertical angles are equal in measure.)*
$z = x$ *(Alternate interior angles are equal in measure.)*
$x = w$ *(Substitution)*

You now have available the following facts:

- **Vertical angles are equal in measure.**
- **Alternate interior angles are equal in measure.**
- **Corresponding angles are equal in measure.**

Use any or all of these facts to prove that *interior angles on the same side of the transversal are supplementary*. Add any necessary labels to the diagram below, and then write out a proof including given facts and a statement of what needs to be proved.

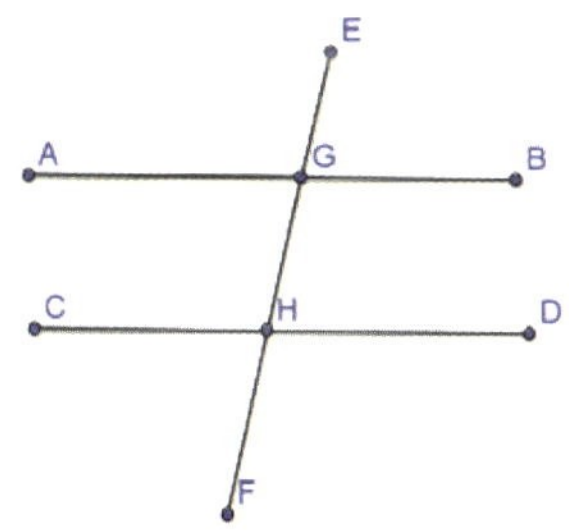

Given: $\overline{AB} \parallel \overline{CD}$, transversal $\overline{EF}$.
Prove: $m\angle BGH + m\angle DHG = 180°$.

$\overline{AB} \parallel \overline{CD}$	*Given*
$m\angle BGH + m\angle AGH = 180°$	*Linear pairs form supplementary angles.*
$m\angle AGH = m\angle DHG$	*If parallel lines are cut by a transversal, then alternate interior angles are equal in measure.*
$m\angle BGH + m\angle DHG = 180°$	*Substitution property of equality*

Now that you have proven this, you may add this theorem to your available facts.

- **Interior angles on the same side of the transversal that intersects parallel lines sum to $180°$.**

MP.7

Use any of these four facts to prove that the three angles of a triangle sum to $180°$. For this proof, you need to draw an auxiliary line parallel to one of the triangle's sides and passing through the vertex opposite that side. Add any necessary labels, and write out your proof.

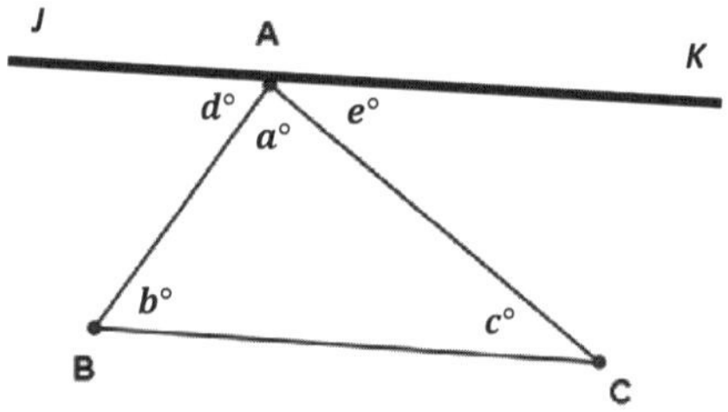

Draw an auxiliary line JK so that $\overleftrightarrow{JK} \parallel \overleftrightarrow{BC}$.

$\overleftrightarrow{JK} \parallel \overleftrightarrow{BC}$	*Construction*
$d + a + e = 180$	*Angles on a line sum to $180°$.*
$d = b$	*If parallel lines are cut by a transversal, then alternate interior angles are equal in measure.*
$e = c$	*If parallel lines are cut by a transversal, then alternate interior angles are equal in measure.*
$a + b + c = 180$	*Substitution property of equality*

Let's review the theorems we have now proven:

- **Vertical angles are equal in measure.**
- **A transversal intersects a pair of lines. The pair of lines is parallel if and only if:**
  - **Alternate interior angles are equal in measure.**
  - **Corresponding angles are equal in measure.**
- **Interior angles on the same side of the transversal add to $180°$. The sum of the degree measures of the angles of a triangle is $180°$.**

- What is the absolute shortest list of facts from which all other facts can be derived?

Side Trip: **Take a moment to take a look at one of those really famous Greek guys we hear so much about in geometry, Eratosthenes. Over $2,000$ years ago, Eratosthenes used the geometry we have just been working with to find the circumference of Earth. He did not have cell towers, satellites, or any other advanced instruments available to scientists today. The only things Eratosthenes used were his eyes, his feet, and perhaps the ancient Greek equivalent to a protractor.**

Watch this video to see how he did it, and try to spot the geometry we have been using throughout this lesson.

https://youtu.be/wnElDaV4esg

Examples 1–2 (15 minutes)

Students try one example and discuss the converse of parallel line theorems.

Example 1

Construct a proof designed to demonstrate the following:

If two lines are perpendicular to the same line, they are parallel to each other.

(a) Draw and label a diagram, (b) state the given facts and the conjecture to be proved, and (c) write out a clear statement of your reasoning to justify each step.

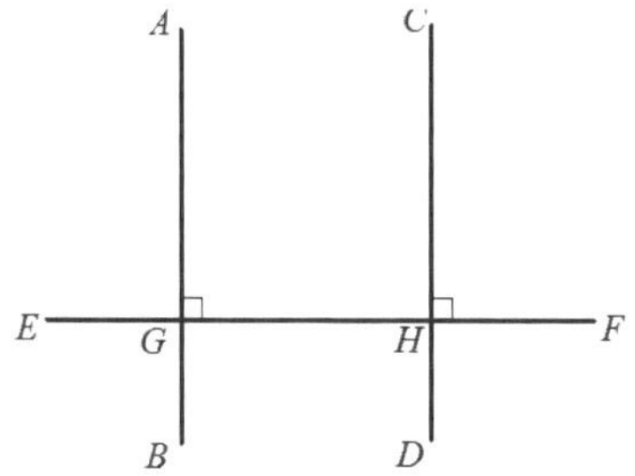

Given: $\overline{AB} \perp \overline{EF}$, $\overline{CD} \perp \overline{EF}$

Prove: $\overline{AB} \parallel \overline{CD}$

$m\angle AGH = 90°$; $m\angle CHF = 90°$	*Perpendicular lines form* $90°$ *angles.*
$m\angle AGH = m\angle CHF$	*Transitive property (since* $m\angle AGH = 90°$ *and* $m\angle CHF = 90°$*)*
$\overline{AB} \parallel \overline{CD}$	*If two lines are cut by a transversal such that a pair of corresponding angles are equal in meaure, then the lines are parallel.*

Discussion

Show a brief example of one parallel line theorem in both directions, the original theorem and its converse, to ensure students understand how to use the converse (one way is using student helpers mentioned in the overview).

Each of the three parallel line theorems has a converse (or reversing) theorem as follows:

Original	Converse
If two parallel lines are cut by a transversal, then alternate interior angles are equal in measure.	If two lines are cut by a transversal such that alternate interior angles are equal in measure, then the lines are parallel.
If two parallel lines are cut by a transversal, then corresponding angles are equal in measure.	If two lines are cut by a transversal such that corresponding angles are equal in measure, then the lines are parallel.
If two parallel lines are cut by a transversal, then interior angles on the same side of the transversal add to $180°$.	If two lines are cut by a transversal such that interior angles on the same side of the transversal add to $180°$, then the lines are parallel.

Notice the similarities between the statements in the first column and those in the second. Think about when you would need to use the statements in the second column, that is, the times when you are trying to prove two lines are parallel.

Example 2

In the figure to the right, $x = y$.
Prove that $\overline{AB} \parallel \overline{EF}$.

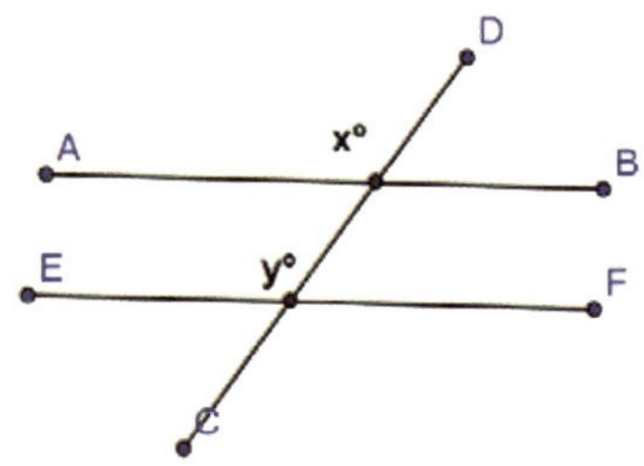

$x = y$	*Given*
$\overline{AB} \parallel \overline{EF}$	*If two lines are cut by a transversal such that a pair of corresponding angles are equal in meaure, then the lines are parallel.*

Closing (2 minutes)

- Review the parallel line theorems by stating the original and converse of each theorem.

Exit Ticket (5 minutes)

Name ____ Date ____________________

Lesson 11: Unknown Angle Proofs—Proofs of Known Facts

Exit Ticket

In the diagram to the right, prove that $m\angle d + m\angle e - m\angle a = 180°$.

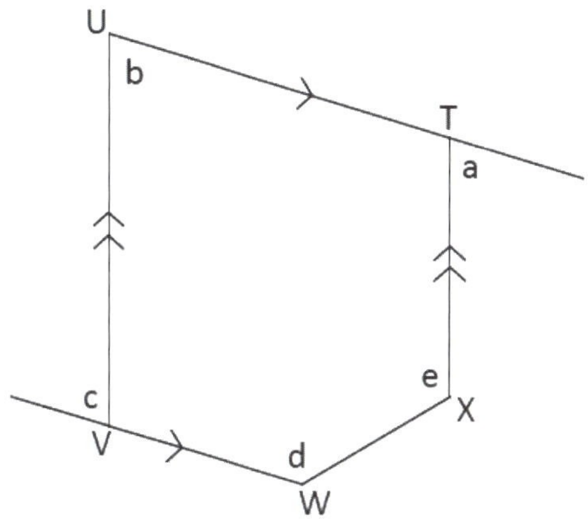

Exit Ticket Sample Solution

In the diagram to the right, prove that $m\angle d + m\angle e - m\angle a = 180°$.

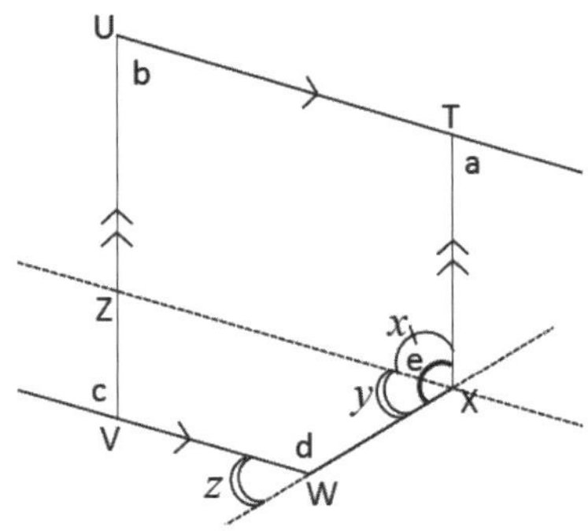

Draw an auxiliary line through X parallel to $\overleftrightarrow{VW}$, and label the intersection of this line with $\overline{UV}$ as Z. Extend $\overline{WX}$, and label angles as shown.

$\overleftrightarrow{TU} \parallel \overleftrightarrow{XZ}$	*Given*
$\overleftrightarrow{XZ} \parallel \overleftrightarrow{VW}$	*Construction*
$m\angle a = m\angle x$	*If parallel lines are cut by a transversal, then alternate interior angles are equal in measure.*
$m\angle y = m\angle e - m\angle x$	*Angle postulate*
$m\angle y = m\angle z$	*If parallel lines are cut by a transversal, then corresponding angles are equal in measure.*
$m\angle d + m\angle z = 180°$	*Linear pairs form supplementary angles.*
$m\angle d + m\angle e - m\angle a = 180°$	*Substitution property of equality*

Problem Set Sample Solutions

1. Given: $\angle C$ and $\angle D$ are supplementary and $m\angle B = m\angle D$
   Prove: $\overline{AB} \parallel \overline{CD}$

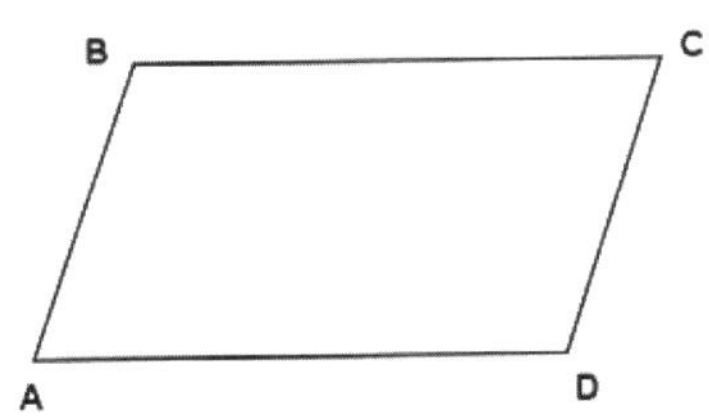

$\angle C$ and $\angle D$ are supplementary.	*Given*
$m\angle C + m\angle D = 180°$	*Definition of supplementary angles*
$m\angle B = m\angle D$	*Given*
$m\angle C + m\angle B = 180°$	*Substitution*
$\overline{AB} \parallel \overline{CD}$	*If two lines are cut by a transversal such that a pair of interior angles on the same side are supplementary, then the lines are parallel.*

2. A theorem states that in a plane, if a line is perpendicular to one of two parallel lines and intersects the other, then it is perpendicular to the other of the two parallel lines.

Prove this theorem. (a) Construct and label an appropriate figure, (b) state the given information and the theorem to be proven, and (c) list the necessary steps to demonstrate the proof.

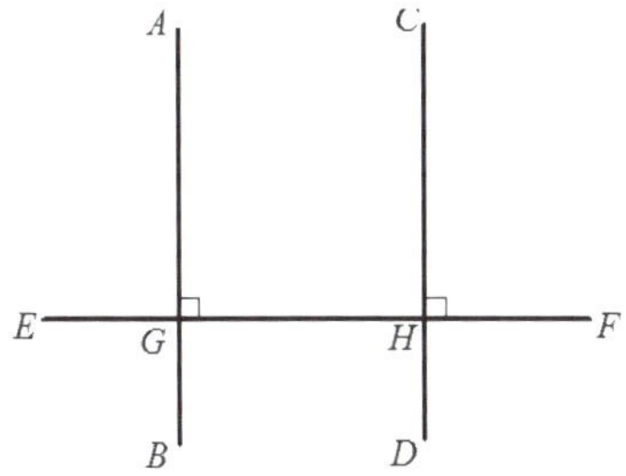

Given: $\overline{AB} \parallel \overline{CD}$, $\overline{EF} \perp \overline{AB}$, $\overline{EF}$ *intersects* $\overline{CD}$

Prove: $\overline{EF} \perp \overline{CD}$

$\overline{AB} \parallel \overline{CD}$, $\overline{EF} \perp \overline{AB}$	*Given*
$m\angle BGH = 90^\circ$	*Definition of perpendicular lines*
$m\angle BGH = m\angle DHF$	*If parallel lines are cut by a transversal, then corresponding angles are equal in measure.*
$\overline{EF} \perp \overline{CD}$	*If two lines intersect to form a right angle, then the two lines are perpendicular.*

Topic C

Transformations/Rigid Motions

G-CO.A.2, G-CO.A.3, G-CO.A.4, G-CO.A.5, G-CO.B.6, G-CO.B.7, G-CO.D.12

Focus Standards:	G-CO.A.2	Represent transformations in the plane using, e.g., transparencies and geometry software; describe transformations as functions that take points in the plane as inputs and give other points as outputs. Compare transformations that preserve distance and angle to those that do not (e.g., translation versus horizontal stretch).
	G-CO.A.3	Given a rectangle, parallelogram, trapezoid, or regular polygon, describe the rotations and reflections that carry it onto itself.
	G-CO.A.4	Develop definitions of rotations, reflections, and translations in terms of angles, circles, perpendicular lines, parallel lines, and line segments.
	G-CO.A.5	Given a geometric figure and a rotation, reflection, or translation, draw the transformed figure using, e.g., graph paper, tracing paper, or geometry software. Specify a sequence of transformations that will carry a given figure onto another.
	G-CO.B.6	Use geometric descriptions of rigid motions to transform figures and to predict the effect of a given rigid motion on a given figure; given two figures, use the definition of congruence in terms of rigid motions to decide if they are congruent.
	G-CO.B.7	Use the definition of congruence in terms of rigid motions to show that two triangles are congruent if and only if corresponding pairs of sides and corresponding pairs of angles are congruent.
	G-CO.D.12	Make formal geometric constructions with a variety of tools and methods (compass and straightedge, string, reflective devices, paper folding, dynamic geometric software, etc.). *Copying a segment; copying an angle; bisecting a segment; bisecting an angle; constructing perpendicular lines, including the perpendicular bisector of a line segment; and constructing a line parallel to a given line through a point not on the line.*
Instructional Days:	10	
Lesson 12:	Transformations—The Next Level (M)[1]	
Lesson 13:	Rotations (E)	
Lesson 14:	Reflections (E)	
Lesson 15:	Rotations, Reflections, and Symmetry (E)	
Lesson 16:	Translations (E)	
Lesson 17:	Characterize Points on a Perpendicular Bisector (S)	

[1]Lesson Structure Key: **P**-Problem Set Lesson, **M**-Modeling Cycle Lesson, **E**-Exploration Lesson, **S**-Socratic Lesson

Lesson 18:	Looking More Carefully at Parallel Lines (S)
Lesson 19:	Construct and Apply a Sequence of Rigid Motions (S)
Lesson 20:	Applications of Congruence in Terms of Rigid Motions (S)
Lesson 21:	Correspondence and Transformations (P)

In Topic C, students are reintroduced to rigid transformations, specifically rotations, reflections, and translations. Students first saw the topic in Grade 8 (**8.G.A.1–3**) and developed an intuitive understanding of the transformations, observing their properties by experimentation. In Topic C, students develop a more exact understanding of these transformations. Beginning with Lesson 12, they discover what they do *not* know about the three motions. The lesson is designed to elicit the gap in students' knowledge, particularly the fact that they need to learn the language of the parameters of each transformation. During this lesson, they also learn to articulate what differentiates rigid motions from non-rigid motions. Students examine each transformation more closely in Lessons 13 through 16, developing precise definitions of each and investigating how rotations and reflections can be used to verify symmetries within certain polygons. In Lessons 17 and 18, students use their construction skills, in conjunction with their understanding of rotations and reflections, to verify properties of parallel lines and perpendicular lines. With a firm grasp of rigid motions, students then define congruence in Lesson 19 in terms of rigid motions. They are able to specify a sequence of rigid motions that map one figure onto another. Topic C closes with Lessons 20 and 21, in which students examine correspondence and its place within the discussion of congruency.

Lesson 12: Transformations—The Next Level

Student Outcome

- Students discover the gaps in specificity regarding their understanding of transformations.
- Students identify the parameters needed to complete any rigid motion.

Lesson Notes

Transformations follow the unknown angles topic. Students enter high school Geometry with an intuitive understanding of transformations, as well as knowing how to illustrate transformations on the coordinate plane. However, they lack a comprehensive understanding of the language of transformations. This topic provides students with the language necessary to speak with precision about transformations and ultimately leads to defining *congruence*.

The Mathematical Modeling Exercise of this lesson builds fluency by reviewing rotations, reflections, and translations in addition to probing for gaps in students' knowledge. During this partner exercise, one partner is provided the details of a given transformation; the other partner has the pre-image and must try to perform the transformation based on his partner's verbal description. (They can pretend as if they are working over the phone.) The purpose of this exercise is to have students realize that they may have an intuitive sense of the effect of a given transformation but that making a precise verbal description that another student can follow requires much more effort. In the Discussion, students have the opportunity to describe what they need to know to make each type of transformation happen. The lesson ends with a Problem Set on rotations as a precursor to Lesson 13.

Classwork

Opening Exercises (11 minutes)

MP.7 & MP.8

Students take a quiz on content from Lessons 6–11. Below are sample responses to the quiz questions.

Opening Exercises

a. **Find the measure of each lettered angle in the figure below.**

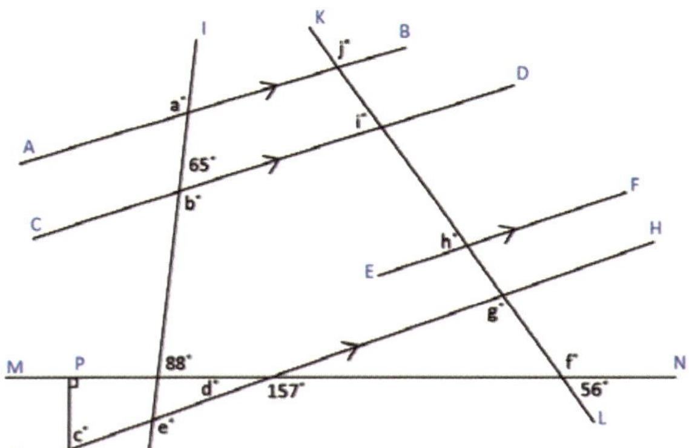

$\overline{AB} \parallel \overline{CD} \parallel \overline{EF} \parallel \overline{GH}$

$m\angle MPO = 90°$

$a = 115$	$b = 115$	$c = 67$	$d = 23$	$e = 115$
$f = 124$	$g = 101$	$h = 79$	$i = 79$	$j = 101$

MP.7 & MP.8

b. Given: $m\angle CDE = m\angle BAC$
Prove: $m\angle DEC = m\angle ABC$

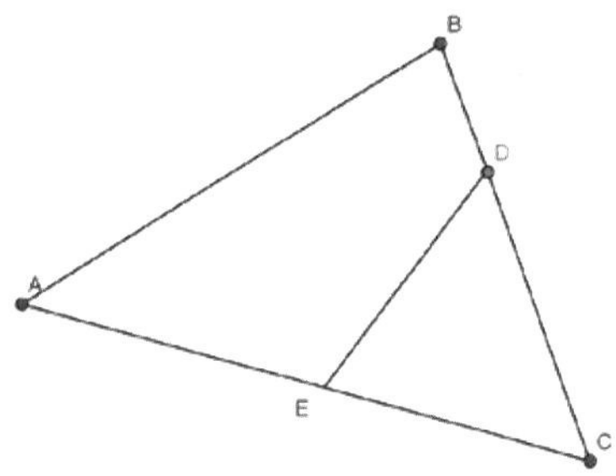

$m\angle CDE = m\angle BAC$	*Given*
$m\angle CDE + m\angle DCE + m\angle DEC = 180°$	*Sum of the angle measures in a triangle is* $180°$
$m\angle BAC + m\angle DCE + m\angle ABC = 180°$	*Sum of the angle measures of a triangle is* $180°$
$m\angle CDE + m\angle DCE + m\angle DEC$	*Substitution property of equality*
$m\angle BAC + m\angle DCE + m\angle ABC$	
$m\angle DEC = m\angle ABC$	*Subtraction property of equality*

Mathematical Modeling Exercise (15 minutes)

Students discover what they do not know about rotations, reflections, and translations through a partner exercise. The cards mentioned are in the Lesson 12 supplement; there are three pairs of cards. Photocopy and split the pair of images from each transformation card. For the first round of the activity, assign Partner A with pre-image/image cards, and assign Partner B with the pre-image cards. Switch card assignments after the first round.

Mathematical Modeling Exercise

You will work with a partner on this exercise and are allowed a protractor, compass, and straightedge.

- **Partner A: Use the card your teacher gives you. Without showing the card to your partner, describe to your partner how to draw the transformation indicated on the card. When you have finished, compare your partner's drawing with the transformed image on your card. Did you describe the motion correctly?**
- **Partner B: Your partner is going to describe a transformation to be performed on the figure on your card. Follow your partner's instructions and then compare the image of your transformation to the image on your partner's card.**

Discussion (15 minutes)

It is critical that students understand transformations as functions that take a set of points as inputs, apply a given rule, and output a new location for the image of the input points. The input points do not move, nor does the plane. (An analogy can be made to documents faxed to another location. The person sending the fax still retains the original. The machine simply sends an image to another location.) In addition, students must understand that two transformations are the same if they produce the same image. Remind students that the function $f(x) = x^2 - 1$ is equivalent to the function $g(x) = (x + 1)(x - 1)$. Although different procedures are followed to obtain the values of $f(x)$ and $g(x)$, given the same input, both output values are identical; therefore, the two functions are equal. This applies to geometric functions as well.

Discussion

Explaining how to transform figures without the benefit of a coordinate plane can be difficult without some important vocabulary. Let's review.

The word *transformation* has a specific meaning in geometry. A transformation F of the plane is a function that assigns to each point P of the plane a unique point $F(P)$ in the plane. Transformations that preserve lengths of segments and measures of angles are called _*basic rigid motions*_. A *dilation* is an example of a transformation that preserves _*angle*_ measures but not the lengths of segments. In this lesson, we work only with rigid transformations.
We call a figure that is about to undergo a transformation the _*pre-image*_, while the figure that has undergone the transformation is called the _*image*_.

Rotation

Reflection

Translation

Using the figures above, identify specific information needed to perform the rigid motion shown.

For a rotation, we need to know:

Center of rotation, direction (Clockwise (CW) or Counterclockwise (CCW)) and number of degrees rotated.

For a reflection, we need to know:

The line of reflection acts as the perpendicular bisector of each segment that joins a given vertex of the pre-image with the respective vertex of the image.

For a translation, we need to know:

The point P to be translated, the length and degree measure of the angle that the vector makes with the line that passes through P, and the endpoint of the vector.

> **Geometry Assumptions**
>
> **We have now done some work with all three basic types of rigid motions (rotations, reflections, and translations). At this point, we need to state our assumptions as to the properties of basic rigid motions:**
>
> a. ***Any basic rigid motion preserves lines, rays, and segments. That is, for a basic rigid motion of the plane, the image of a line is a line, the image of a ray is a ray, and the image of a segment is a segment.***
>
> b. ***Any basic rigid motion preserves lengths of segments and measures of angles.***
>
> **Relevant Vocabulary**
>
> **BASIC RIGID MOTION: A *basic rigid motion* is a rotation, reflection, or translation of the plane.**
>
> **Basic rigid motions are examples of transformations. Given a transformation, the image of a point A is the point the transformation maps A to in the plane.**
>
> **DISTANCE-PRESERVING: A transformation is said to be *distance-preserving* if the distance between the images of two points is always equal to the distance between the pre-images of the two points.**
>
> **ANGLE-PRESERVING: A transformation is said to be *angle-preserving* if (1) the image of any angle is again an angle and (2) for any given angle, the angle measure of the image of that angle is equal to the angle measure of the pre-image of that angle.**

Closing (1 minute)

- Why are reflections, rotations, and translations titled as *basic rigid motions*?
  - *Basic rigid motions preserve lengths and angle measures. In other words, basic rigid motions do not distort figures, lending the term* rigid *to the subcategory of transformations.*

Exit Ticket (3 minutes)

Name ______________________________ Date ________________

Lesson 12: Transformations—The Next Level

Exit Ticket

How are transformations and functions related? Provide a specific example to support your reasoning.

Exit Ticket Sample Solution

How are transformations and functions related? Provide a specific example to support your reasoning.

***Transformations** are **functions**. They take a set of points as inputs, apply a given rule, and output a new location for the image of the input points. Examples are a reflection across a line of reflection and a rotation of 30 degrees around a point of a given figure.*

Problem Set Sample Solutions

An example of a rotation applied to a figure and its image is provided. Use this representation to answer the questions that follow. For each question, a pair of figures (pre-image and image) is given as well as the center of rotation. For each question, identify and draw the following:

i. The circle that determines the rotation, using any point on the pre-image and its image.

ii. An angle, created with three points of your choice, which demonstrates the angle of rotation.

Example of a Rotation:

Pre-image: (solid line)

Image: (dotted line)

Center of rotation: P

Angle of rotation: $\angle APA'$

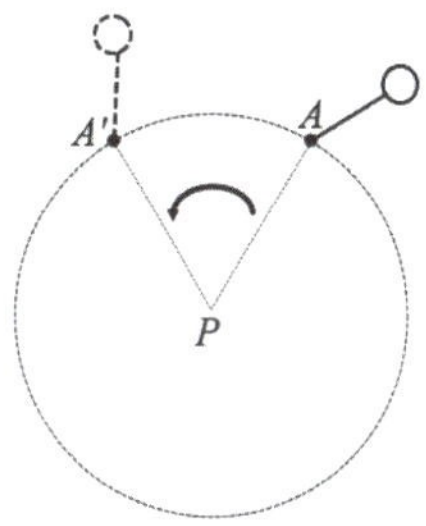

1. Pre-image: (solid line)

   Image: (dotted line)

   Center of rotation: P

   Angle of rotation: <u>$90°$, $270°$ *CW*</u>

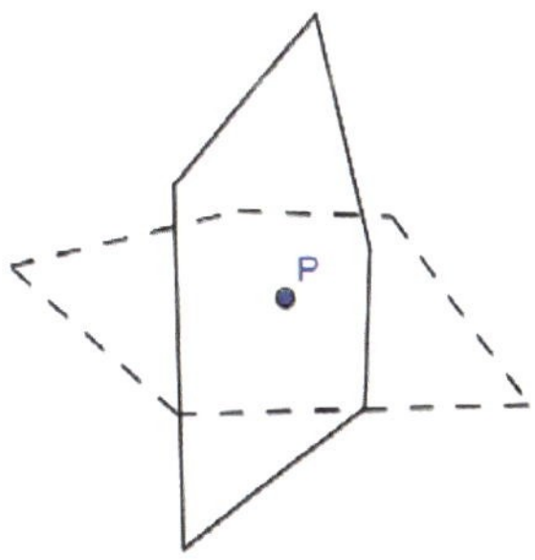

2. Pre-image: $\triangle ABC$

   Image: $\triangle A'B'C'$

   Center: D

   Angle of rotation: <u>$300°$, $60°$ *CW*</u>

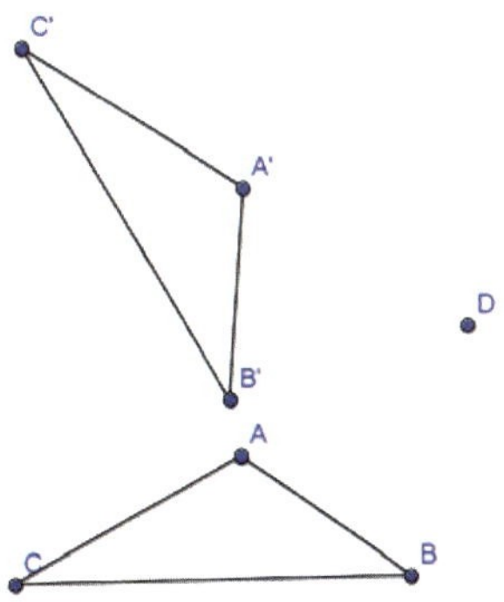

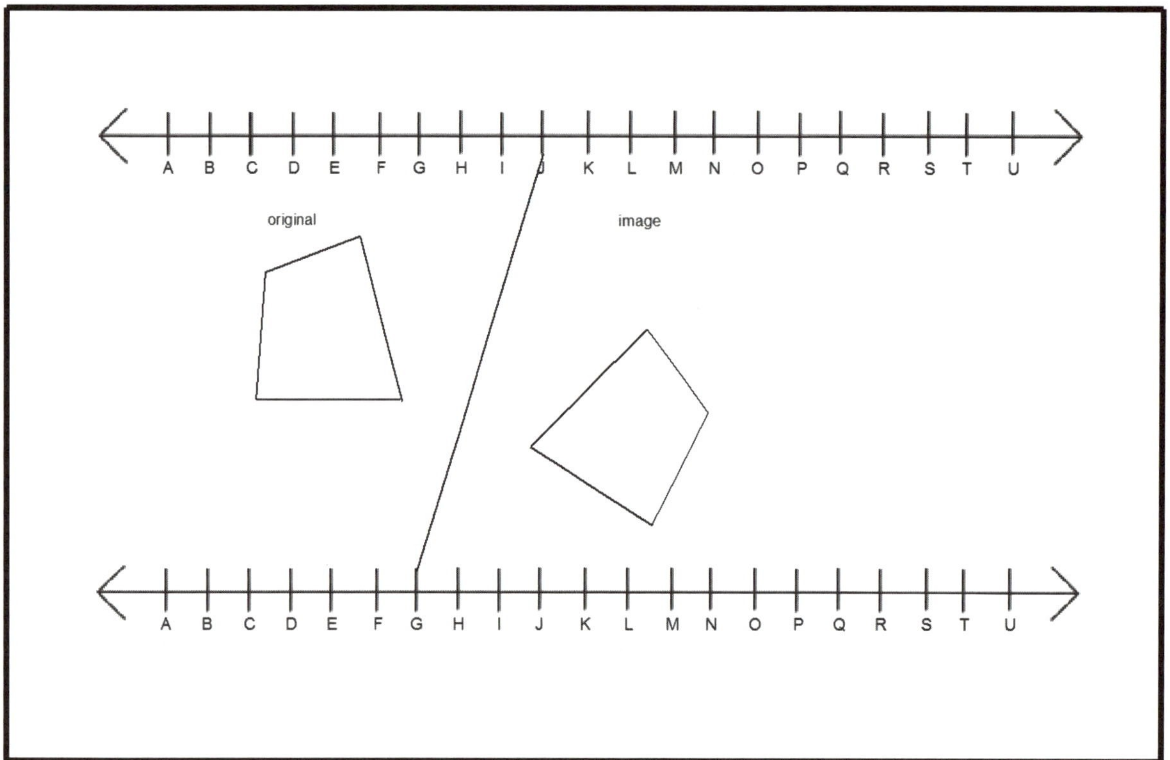
A B C D E F G H I J K L M N O P Q R S T U
original
image
A B C D E F G H I J K L M N O P Q R S T U

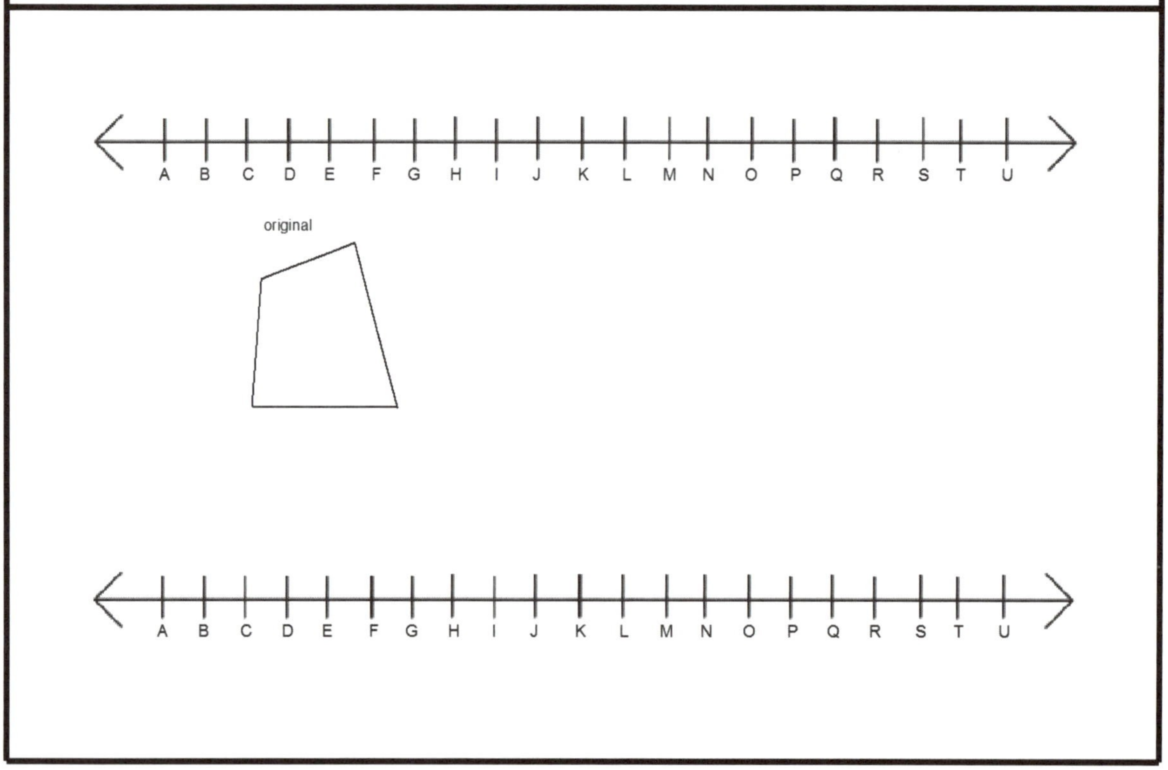
A B C D E F G H I J K L M N O P Q R S T U
original
A B C D E F G H I J K L M N O P Q R S T U

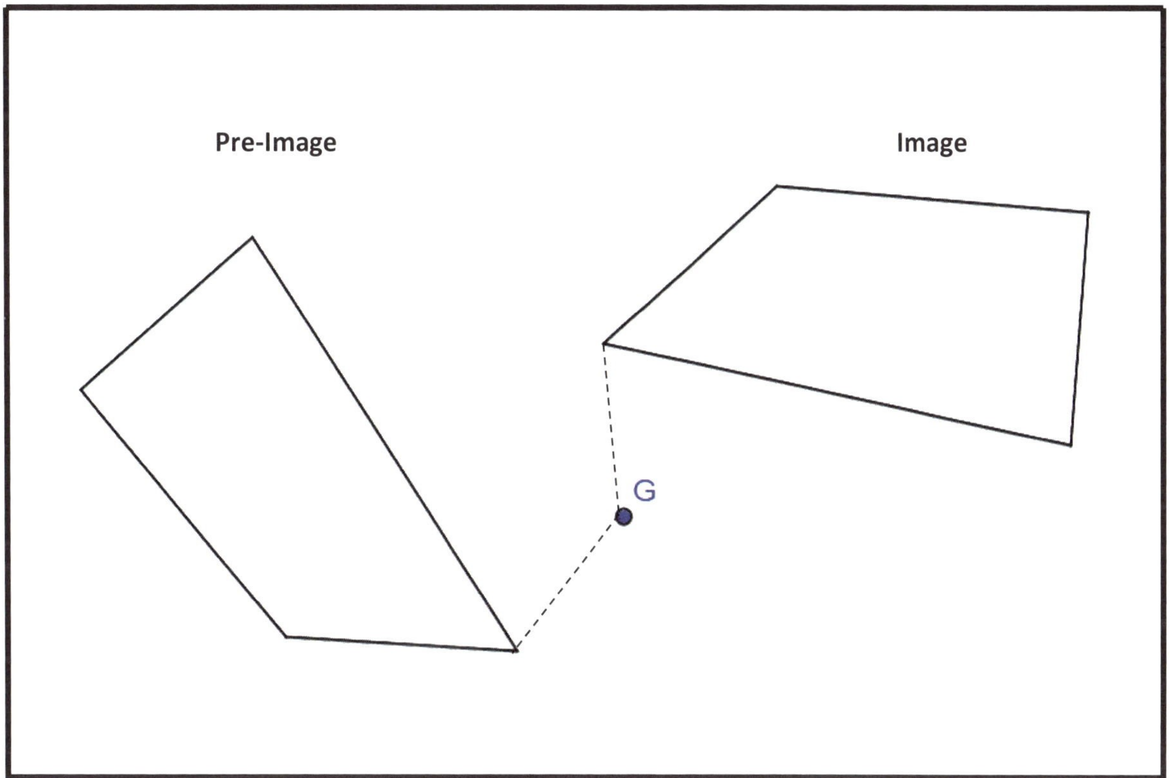
Pre-Image
Image
G

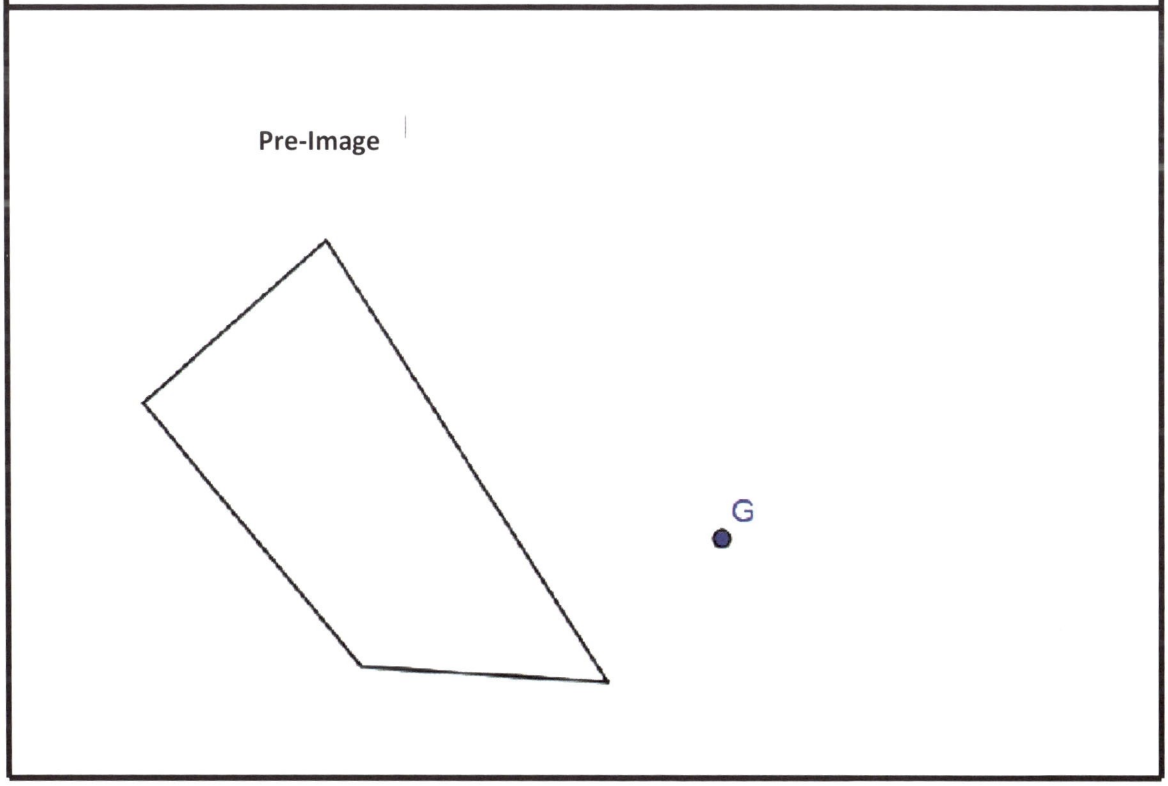
Pre-Image
G

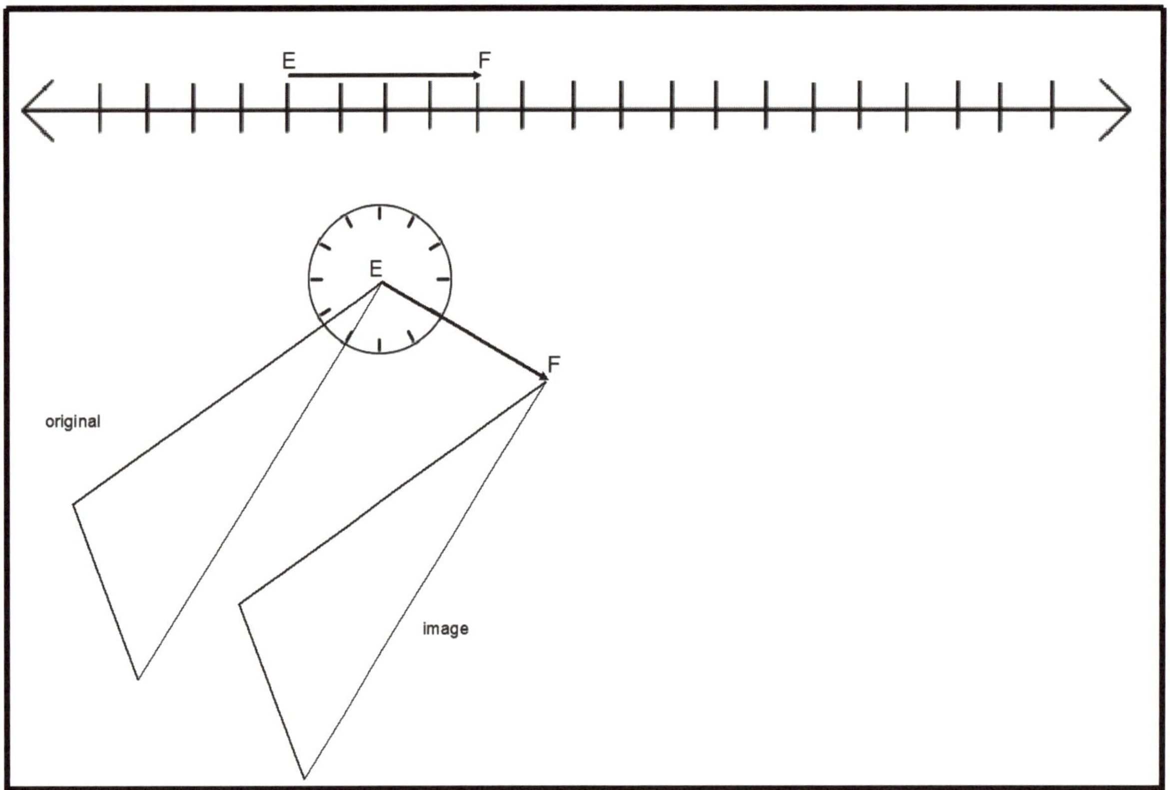
E
F
E
F
original
image

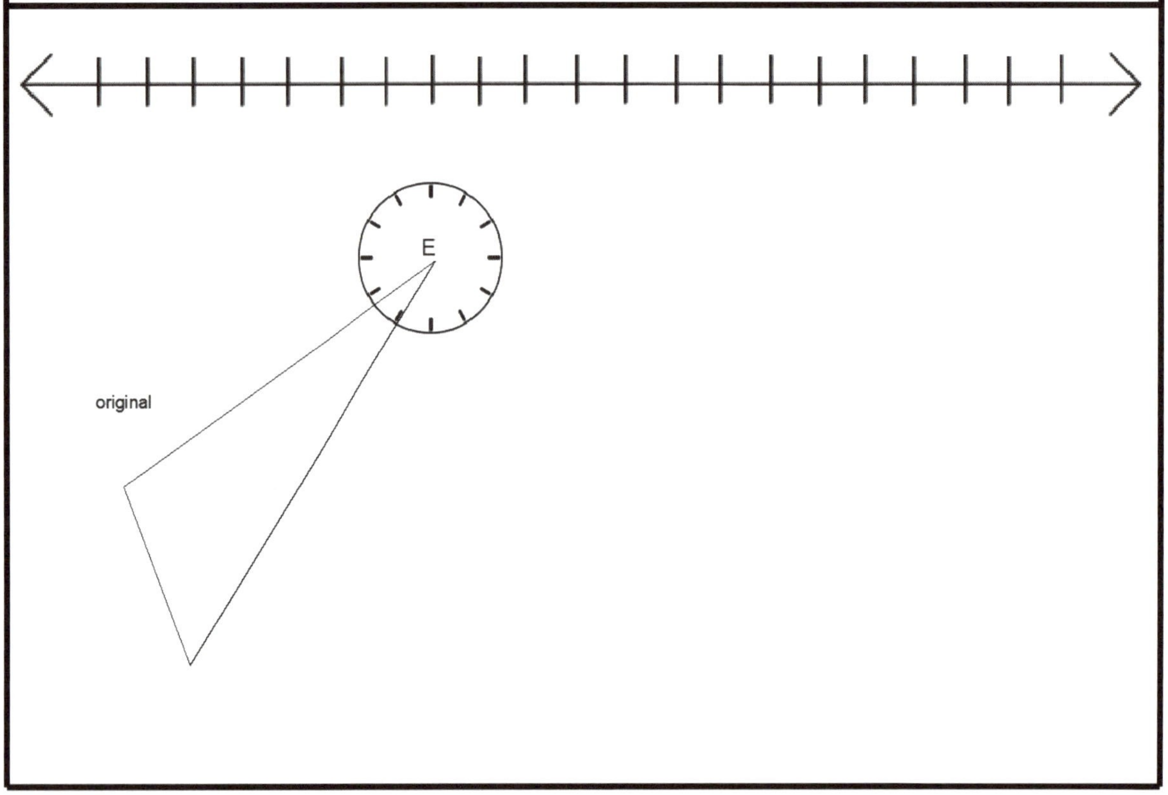
E
original

Lesson 13: Rotations

Student Outcomes

- Students manipulate rotations by each parameter: center of rotation, angle of rotation, and a point under the rotation.

Lesson Notes

Lesson 13 takes a close look at the precise definition of rotation. Students learn how to rotate a figure about a center of rotation for a given number of degrees and in a given direction. Students also learn how to determine the angle of rotation and center of rotation for a given figure and its image.

Teachers should continue to stress the point that rotations preserve the lengths of segments (distance-preserving) and the measures of the angles of the figures being rotated (angle-preserving).

Rotations are one of the three basic rigid motions used to form the definition of one of the main ideas in geometry: congruence. Essential to students' understanding of the definition of congruence is the realization (1) that rotations preserve distances and angle measures and (2) that a rotation of any angle size can be performed at any point in the plane.

The lesson begins with an Exploratory Challenge where students cut out a $75°$ angle to apply to a figure as its degree of rotation. With the help of a ruler, students transform the vertices of the figure to create the image after the rotation. This hands-on exercise demonstrates a basic way of performing a rotation. In the Discussion, students use a point on a given figure, the center of rotation, and the respective point in the image to determine the angle of rotation. They test several sets of points to verify that the angle measure is preserved. Then, students learn how to find the center of rotation using what they know about perpendicular. Students practice these two new skills before learning the precise definition of rotation. This definition incorporates a center of rotation, an angle of rotation, and exactly how the points move across the plane (along a circular path of a given radius). Students practice their updated understanding of rotations in the Problem Set.

Note that the study of transformations over the next several lessons involves significant use of compass and straightedge constructions. This is done to build deep understanding of transformations and also to lend coherence between clusters within the G-CO domain, connecting transformations (**G-CO.A**), congruence (**G-CO.B**), and transformations (**G-CO.D**). Additionally, students develop in their ability to persist through challenging problems (MP.1). However, if students are struggling, it may be necessary to modify the exercises to include the use of graph paper, patty paper, or geometry software (such as the freely available Geogebra).

Classwork

Exploratory Challenge (10 minutes)

MP.5 Students apply a $75°$ rotation to a figure using the cutout of an angle and a ruler.

MP.6 & MP.7

- Consider whether all the tools in the exercise are necessary. How could the exercise be modified?
  - *With the use of a compass or protractor instead of the given angle*
- Try the rotation again using a different center of rotation, either one of the vertices of the pre-image or a point Q on the opposite side of the figure from point P. Discuss the effects of changing the center of rotation.

Exploratory Challenge

You need a pair of scissors and a ruler.

Cut out the $75°$ angle on the right, and use it as a guide to rotate the figure below $75°$ counterclockwise around the given center of rotation (Point P).

- **Place the vertex of the $75°$ angle at point P.**
- **Line up one ray of the $75°$ angle with vertex A on the figure. Carefully measure the length from point P to vertex A.**
- **Measure that same distance along the other ray of the reference angle, and mark the location of your new point, A'.**
- **Repeat these steps for each vertex of the figure, labeling the new vertices as you find them.**
- **Connect the six segments that form the sides of your rotated image.**

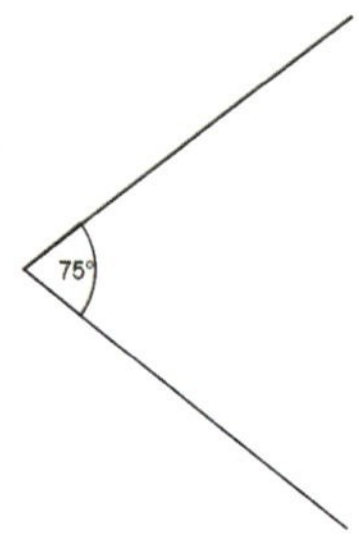

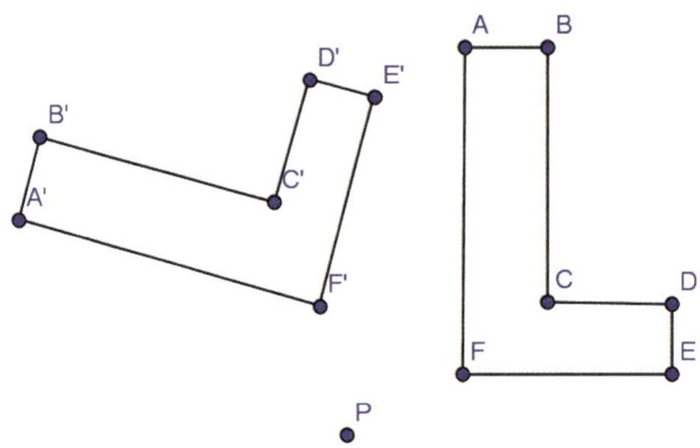

Discussion (12 minutes)

Discussion

In Grade 8, we spent time developing an understanding of what happens in the application of a rotation by participating in hands-on lessons. Now, we can define rotation precisely.

The notion of the entire plane being subject to the transformation is new and should be reinforced. However, note that neither the figure nor the plane is actually moved in the transformation. The image represents the output after applying the transformation *rule* to the input points.

In the definition below, students may benefit from some discussion using a visual of a clock. Discuss the intuitive notion of directions on a clock before the more formal definition.

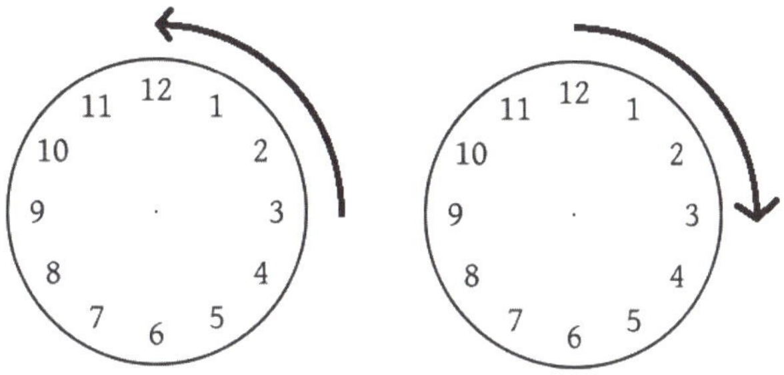

Counterclockwise Clockwise

First, we need to talk about the direction of the rotation. If you stand up and spin in place, you can either spin to your left or spin to your right. This spinning to your left or right can be rephrased using what we know about analog clocks: spinning to your left is spinning in a counterclockwise direction, and spinning to your right is spinning in a clockwise direction. We need to have the same sort of notion for rotating figures in the plane. It turns out that there is a way to always choose a *counterclockwise half-plane* for any ray: The counterclockwise half-plane of $\overrightarrow{CP}$ is the half-plane of $\overleftrightarrow{CP}$ that lies to the left as you move along $\overrightarrow{CP}$ in the direction from C to P. (The *clockwise half-plane* is then the half-plane that lies to the right as you move along $\overrightarrow{CP}$ in the direction from C to P.) We use this idea to state the definition of rotation.

For $0° < \theta° < 180°$, the rotation of θ degrees around the center C is the transformation $R_{C,\theta}$ of the plane defined as follows:

1. For the center point C, $R_{C,\theta}(C) = C$, and
2. For any other point P, $R_{C,\theta}(P)$ is the point Q that lies in the counterclockwise half-plane of $\overrightarrow{CP}$, such that $CQ = CP$ and $m\angle PCQ = \theta°$.

A rotation of 0 degrees around the center C is the identity transformation (i.e., for all points A in the plane, it is the rotation defined by the equation $R_{C,0}(A) = A$).

A rotation of $180°$ around the center C is the composition of two rotations of $90°$ around the center C. It is also the transformation that maps every point P (other than C) to the other endpoint of the diameter of a circle with center C and radius CP.

- A rotation leaves the center point C fixed. $R_{C,\theta}(C) = C$ states exactly that. The rotation function R with center point C that moves everything else in the plane $\theta°$ leaves only the center point itself unmoved.
- Any other point P in the plane moves the exact same degree arc along the circle defined by the center of rotation and the angle $\theta°$.
- Then $R_{C,\theta}(P)$ is the point Q that lies in the counterclockwise half-plane of ray $\overrightarrow{CP}$ such that $CQ = CP$ and such that $m\angle PCQ = \theta°$. Visually, you can imagine rotating the point P in a counterclockwise arc around a circle with center C and radius CP to find the point Q.
- All positive angle measures θ assume a counterclockwise motion; if citing a clockwise rotation, the answer should be labeled with *CW*.

A composition of two rotations applied to a point is the image obtained by applying the second rotation to the image of the first rotation of the point. In mathematical notation, the image of a point A after *a composition of two rotations of* $90°$ *around the center* C can be described by the point $R_{C,90}\left(R_{C,90}(A)\right)$. The notation reads, "Apply $R_{C,90}$ to the point $R_{C,90}(A)$." So, we lose nothing by defining $R_{C,180}(A)$ to be that image. Then, $R_{C,180}(A) = R_{C,90}\left(R_{C,90}(A)\right)$ for all points A in the plane.

In fact, we can generalize this idea to define a rotation by any positive degree: For $\theta° > 180°$, a *rotation of $\theta°$ around the center C* is any composition of three or more rotations, such that each rotation is less than or equal to a $90°$ rotation and whose angle measures sum to $\theta°$. For example, a rotation of $240°$ is equal to the composition of three rotations by $80°$ about the same center, the composition of five rotations by $50°$, $50°$, $50°$, $50°$, and $40°$ about the same center, or the composition of 240 rotations by $1°$ about the same center.

Notice that we have been assuming that all rotations rotate in the counterclockwise direction. However, the inverse rotation (the rotation that *undoes* a given rotation) can be thought of as rotating in the clockwise direction. For example, rotate a point A by $30°$ around another point C to get the image $R_{C,30}(A)$. We can *undo* that rotation by rotating by $30°$ in the clockwise direction around the same center C. Fortunately, we have an easy way to describe a *rotation in the clockwise direction*. If all positive degree rotations are in the counterclockwise direction, then we can define a negative degree rotation as a rotation in the clockwise direction (using the clockwise half-plane instead of the counterclockwise half-plane). Thus, $R_{C,-30}$ is a $30°$ rotation in the clockwise direction around the center C. Since a composition of two rotations around the same center is just the sum of the degrees of each rotation, we see that

$$R_{C,-30}\left(R_{C,30}(A)\right) = R_{C,0}(A) = A,$$

for all points A in the plane. Thus, we have defined how to perform a rotation for any number of degrees—positive or negative.

As this is our first foray into close work with rigid motions, we emphasize an important fact about rotations. Rotations are one kind of rigid motion or transformation of the plane (a function that assigns to each point P of the plane a unique point $F(P)$) that preserves lengths of segments and measures of angles. Recall that Grade 8 investigations involved manipulatives that modeled rigid motions (e.g., transparencies) because you could actually *see* that a figure was not altered, as far as length or angle was concerned. It is important to hold onto this idea while studying all of the rigid motions.

Constructing rotations precisely can be challenging. Fortunately, computer software is readily available to help you create transformations easily. Geometry software (such as Geogebra) allows you to create plane figures and rotate them a given number of degrees around a specified center of rotation. The figures in the exercises were rotated using Geogebra. Determine the angle and direction of rotation that carries each pre-image onto its (dashed-line) image. Assume both angles of rotation are positive. The center of rotation for Exercise 1 is point D and for Figure 2 is point E.

(Remind students that identifying either CCW or CW degree measures is acceptable; if performing a $30°$ rotation in the clockwise direction, students can label their answers as $30°$ *CW* or $-30°$.)

EUREKA MATH™

Exercises 1–3 (9 minutes)

Exercises 1–3

1.

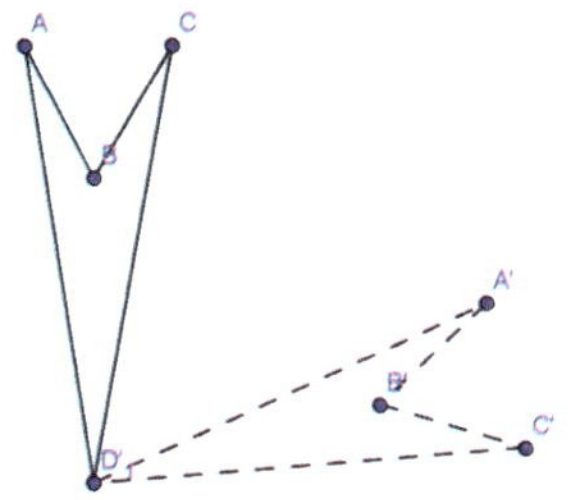

$285°$ *or* $75°$ *CW or* $-75°$

To determine the angle of rotation, you measure the angle formed by connecting corresponding vertices to the center point of rotation. In Exercise 1, measure $\angle AD'A'$. What happened to $\angle D$? Can you see that D is the center of rotation, therefore, mapping D' onto itself? Before leaving Exercise 1, try drawing $\angle BD'B'$. Do you get the same angle measure? What about $\angle CD'C'$?

Try finding the angle and direction of rotation for Exercise 2 on your own.

Remind students that the solid-line figure is the pre-image, and the dotted-line figure is the image.

2.

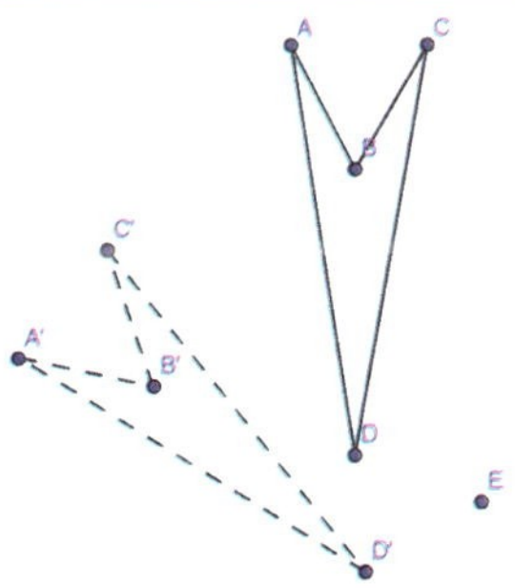

$60°$ *or* $300°$ *CW or* $-300°$

Did you draw $\angle DED'$ or $\angle CEC'$?

Now that you can find the angle of rotation, let's move on to finding the center of rotation. Follow the directions below to locate the center of rotation, taking the figure at the top right to its image at the bottom left.

MP.5

3.

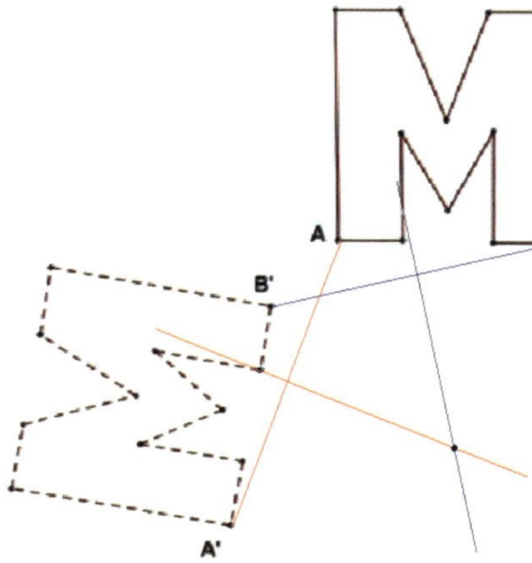

a. Draw a segment connecting points A and A'.

b. Using a compass and straightedge, find the perpendicular bisector of this segment.

c. Draw a segment connecting points B and B'.

d. Find the perpendicular bisector of this segment.

e. The point of intersection of the two perpendicular bisectors is the center of rotation. Label this point P.

Justify your construction by measuring $\angle APA'$ and $\angle BPB'$. Did you obtain the same measure?

Yes

This method works because a *chord* of a circle is a segment joining two points on a circle. The endpoints of the chord are equidistant from the center of the circle. The perpendicular bisector of a chord (being the set of *all* points equidistant from the endpoints) includes the center of the circle. Since students may have had limited experience studying circles, they may have difficulty understanding why this works. Consider pointing out or sketching the circle that has the center of rotation as its center, for each of the examples, to supply some justification for why it works.

Exercises 4–5 (8 minutes)

Exercises 4–5

Find the centers of rotation and angles of rotation for Exercises 4 and 5.

4.

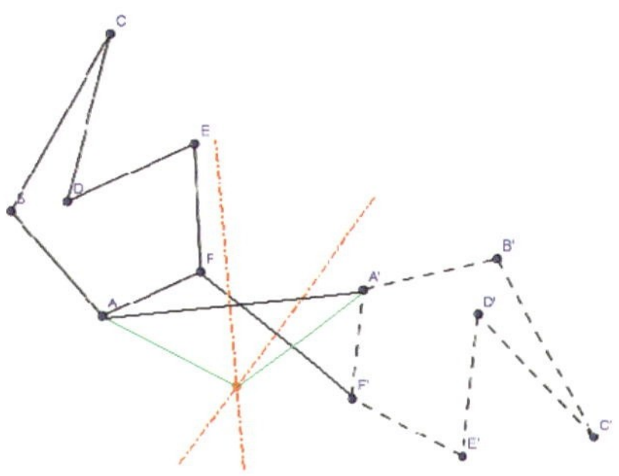

Either 244° or 116° CW or -116°

5.

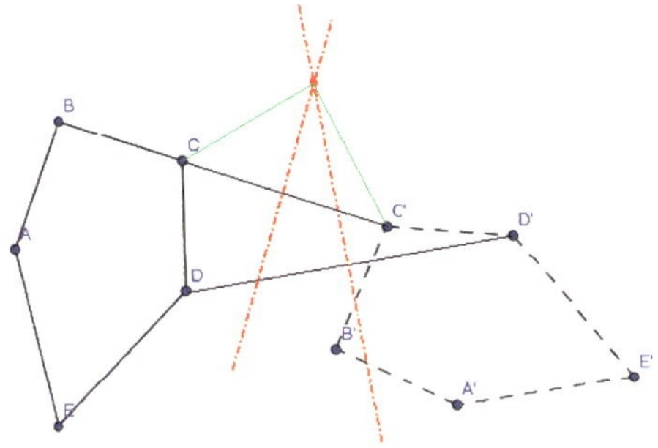

$90°$ *or* $270°$ *CW or* $-270°$

Closing (1 minute)

- How does rotation R of degree θ about a center C affect the points of the plane?
  - *A rotation leaves the center point C unchanged. Any other point P in the plane moves the exact same degree arc along the circle defined by the center of rotation and the angle $\theta°$.*
- How is the direction of a rotation indicated?
  - *Rotations are applied in the counterclockwise direction, unless the degree measure is negative or the rotation is notated as* CW*.*

> **Lesson Summary**
>
> **A rotation carries segments onto segments of equal length.**
>
> **A rotation carries angles onto angles of equal measure.**

Exit Ticket (5 minutes)

Name ______________________________ Date______________

Lesson 13: Rotations

Exit Ticket

Find the center of rotation and the angle of rotation for the transformation below that carries A onto B.

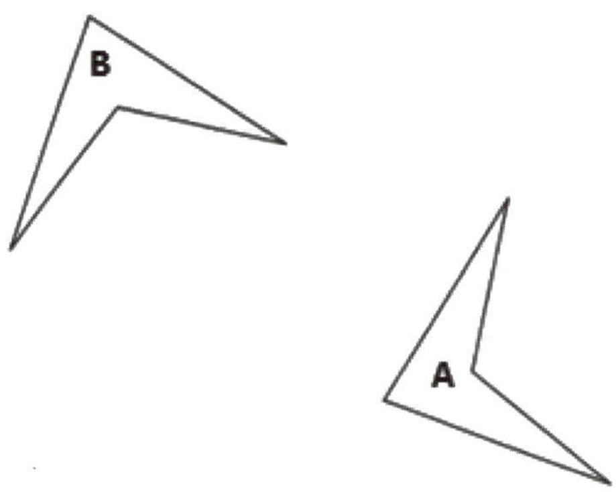

Exit Ticket Sample Solution

Find the center of rotation and the angle of rotation for the transformation below that carries A onto B.

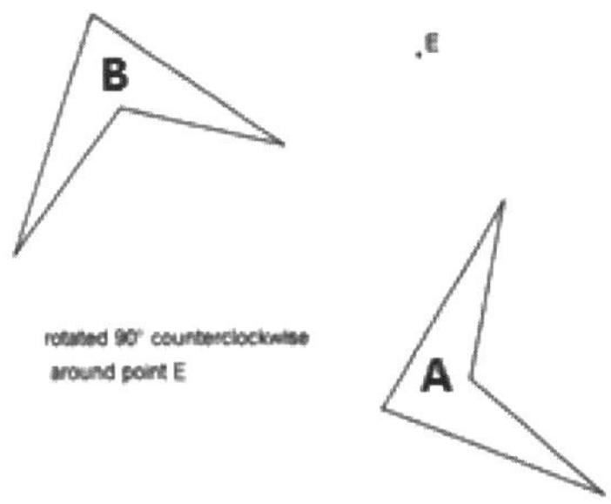

270° *or* 90° *CW or* -90°

Problem Set Sample Solutions

1. Rotate triangle ABC 60° around point F using a compass and straightedge only.

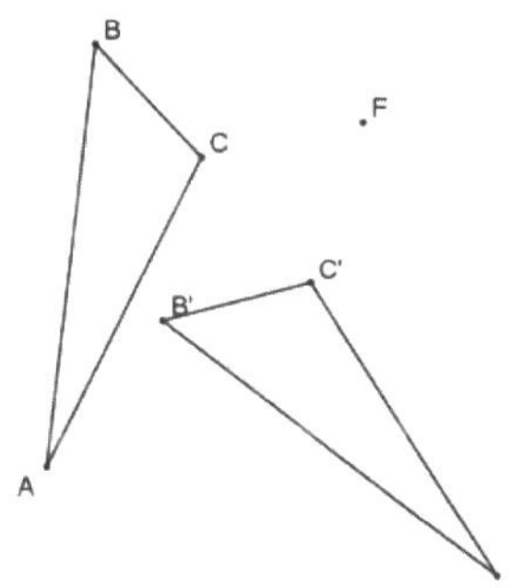

2. Rotate quadrilateral $ABCD$ 120° around point E using a straightedge and protractor.

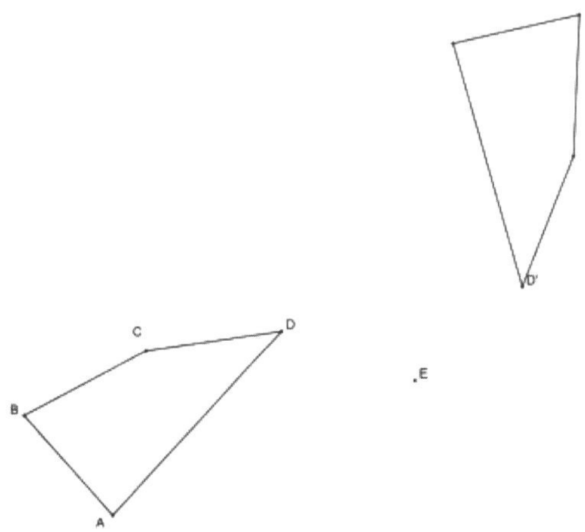

3. On your paper, construct a $45°$ angle using a compass and straightedge. Rotate the angle $180°$ around its vertex, again using only a compass and straightedge. What figure have you formed, and what are its angles called?

   The figure formed is an X, and the angles are called vertical angles.

4. Draw a triangle with angles $90°$, $60°$, and $30°$ using only a compass and straightedge. Locate the midpoint of the longest side using your compass. Rotate the triangle $180°$ around the midpoint of the longest side. What figure have you formed?

   The figure formed is a rectangle.

5. On your paper, construct an equilateral triangle. Locate the midpoint of one side using your compass. Rotate the triangle $180°$ around this midpoint. What figure have you formed?

   The figure formed is a rhombus.

6. Use either your own initials (typed using WordArt in Microsoft Word) or the initials provided below. If you create your own WordArt initials, copy, paste, and rotate to create a design similar to the one below. Find the center of rotation and the angle of rotation for your rotation design.

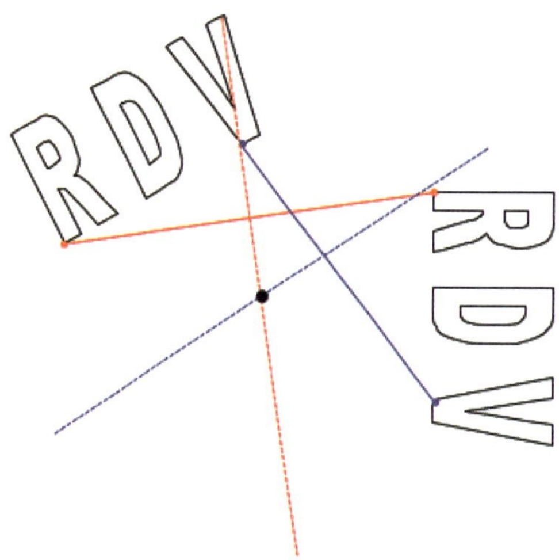

Lesson 14: Reflections

Student Outcomes

- Students learn the precise definition of a reflection.
- Students construct the line of reflection of a figure and its reflected image. Students construct the image of a figure when provided the line of reflection.

Lesson Notes

In Lesson 14, students precisely define a reflection and construct reflections using a perpendicular bisector and circles. Students continue focusing on their use of vocabulary throughout the lesson with their discussion of the constructions. The exploratory nature of this lesson allows for students to discover uses for the skills they have learned in previous construction lessons in addition to the vocabulary they have been working on.

Teachers should continue to stress that reflections preserve the lengths of segments (distance-preserving) and the measures of the angles of the figures being reflected (angle-preserving).

Reflections are one of the three basic rigid motions used to form the definition of one the main ideas in geometry, which is congruence. Essential to students' understanding of the definition of congruence is the realization (1) that reflections preserve distances and angle measures and (2) that a reflection can be performed across *any line* in the plane.

Note that in many cases, it is assumed that the prime notation indicates the image of a figure after a transformation (e.g., $\triangle A'B'C'$ is the image of $\triangle ABC$).

Classwork

Exploratory Challenge (10 minutes)

Students discuss that each of the perpendicular bisectors they drew lined up exactly with the line of reflection. The class can discuss whether they think this will always be the case and why the distance to the perpendicular bisector from each point is equivalent. Help students create a set of guidelines for constructing reflections using precise vocabulary.

Note to Teacher:

Due to space limitations, only the perpendicular bisector of $\overline{CC'}$ has been shown here.

Exploratory Challenge

Think back to Lesson 12 where you were asked to describe to your partner how to reflect a figure across a line. The greatest challenge in providing the description was using the precise vocabulary necessary for accurate results. Let's explore the language that yields the results we are looking for.

$\triangle ABC$ is reflected across $\overline{DE}$ and maps onto $\triangle A'B'C'$.

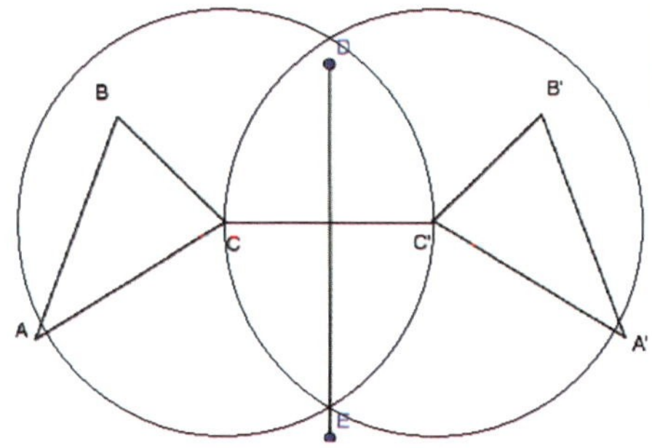

Use your compass and straightedge to construct the perpendicular bisector of each of the segments connecting A to A', B to B', and C to C'. What do you notice about these perpendicular bisectors?

Label the point at which $\overline{AA'}$ intersects $\overline{DE}$ as point O. What is true about AO and $A'O$? How do you know this is true?

$AO = A'O$. I constructed the perpendicular bisector, and O is the point where the perpendicular bisector crosses $\overline{AA'}$, so it is halfway between A and A'.

Examples 1–5 (30 minutes)

Discussion

You just demonstrated that the line of reflection between a figure and its reflected image is also the perpendicular bisector of the segments connecting corresponding points on the figures.

In the Exploratory Challenge, you were given the pre-image, the image, and the line of reflection. For your next challenge, try finding the line of reflection provided a pre-image and image.

Example 1

MP.5

Construct the segment that represents the line of reflection for quadrilateral $ABCD$ and its image $A'B'C'D'$.

What is true about each point on $ABCD$ and its corresponding point on $A'B'C'D'$ with respect to the line of reflection?

Each pair of corresponding points is equidistant from the line of reflection.

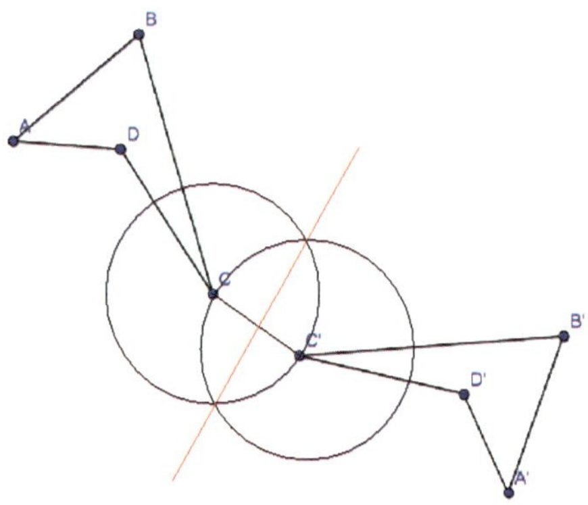

Notice one very important fact about reflections. Every point in the original figure is carried to a corresponding point on the image by the same rule—a reflection across a specific line. This brings us to a critical definition:

MP.6

REFLECTION: For a line l in the plane, a *reflection across* l is the transformation r_l of the plane defined as follows:

1. For any point P on the line l, $r_l(P) = P$, and
2. For any point P not on l, $r_l(P)$ is the point Q so that l is the perpendicular bisector of the segment PQ.

MP.6

If the line is specified using two points, as in $\overleftrightarrow{AB}$, then the reflection is often denoted by $r_{\overline{AB}}$. Just as we did in the last lesson, let's examine this definition more closely:

- ***A transformation of the plane*—the entire plane is transformed; what was once on one side of the line of reflection is now on the opposite side;**
- **$r_l(P) = P$ *means that the points on line l are left fixed*—the only part of the entire plane that is left fixed is the line of reflection itself;**
- **$r_l(P)$ is the point Q—the transformation r_l maps the point P to the point Q;**
- **The line of reflection l is the perpendicular bisector of the segment PQ—to find Q, first construct the perpendicular line m to the line l that passes through the point P. Label the intersection of l and m as N. Then locate the point Q on m on the other side of l such that $PN = NQ$.**

Examples 2–3

Construct the line of reflection across which each image below was reflected.

2.

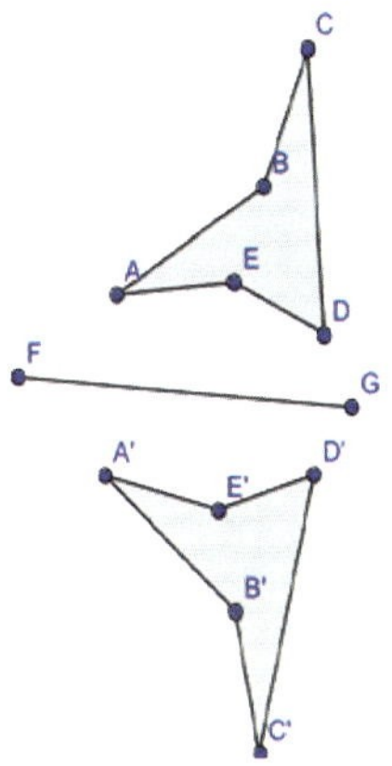

3.

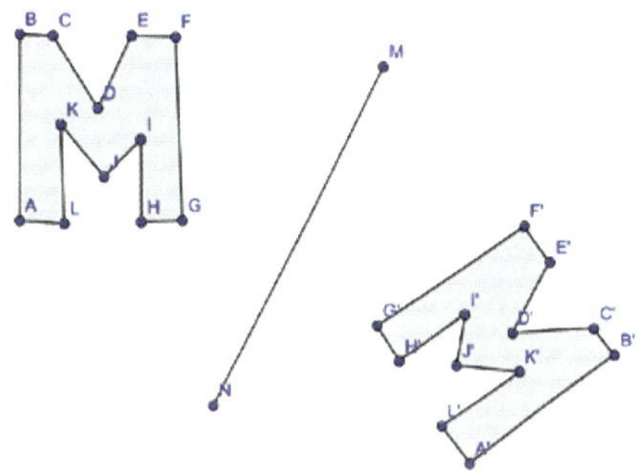

Next, students complete a reflection using circles. The teacher may wish to go through the steps with students or give the steps to students and have them work independently. As students work, encourage them to think and discuss why using circles allows them to construct a reflection. Remind them of what they discovered in the Exploratory Challenge as well as Euclid's use of circles when constructing equilateral triangles. Consider also asking students to confirm the properties of reflections and conclude that they preserve the lengths of segments and the measures of the angles of the figures being reflected.

You have shown that a line of reflection is the perpendicular bisector of segments connecting corresponding points on a figure and its reflected image. You have also constructed a line of reflection between a figure and its reflected image. Now we need to explore methods for constructing the reflected image itself. The first few steps are provided for you in this next stage.

MP.6

Example 4

The task at hand is to construct the reflection of $\triangle ABC$ over $\overline{DE}$. Follow the steps below to get started; then complete the construction on your own.

1. **Construct circle A: center A, with radius such that the circle crosses $\overline{DE}$ at two points (labeled F and G).**
2. **Construct circle F: center F, radius FA, and circle G: center G, radius GA. Label the (unlabeled) point of intersection between circles F and G as point A'. This is the reflection of vertex A across $\overline{DE}$.**
3. **Repeat steps 1 and 2 for vertices B and C to locate B' and C'.**
4. **Connect A', B', and C' to construct the reflected triangle.**

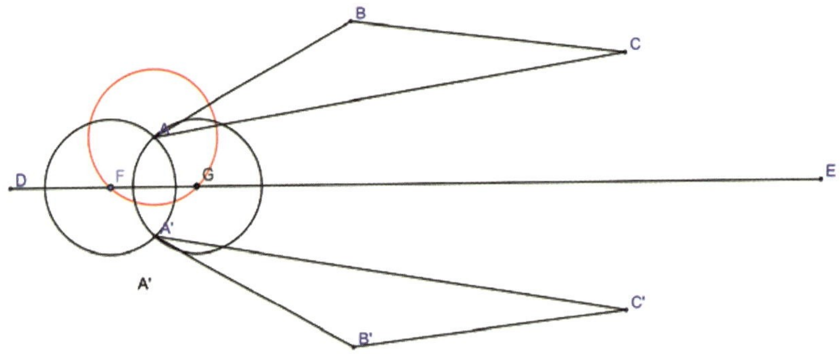

Things to consider:

When you found the line of reflection earlier, you did this by constructing perpendicular bisectors of segments joining two corresponding vertices. How does the reflection you constructed above relate to your earlier efforts at finding the line of reflection itself? Why did the construction above work?

Example 5

Now try a slightly more complex figure. Reflect $ABCD$ across $\overline{EF}$.

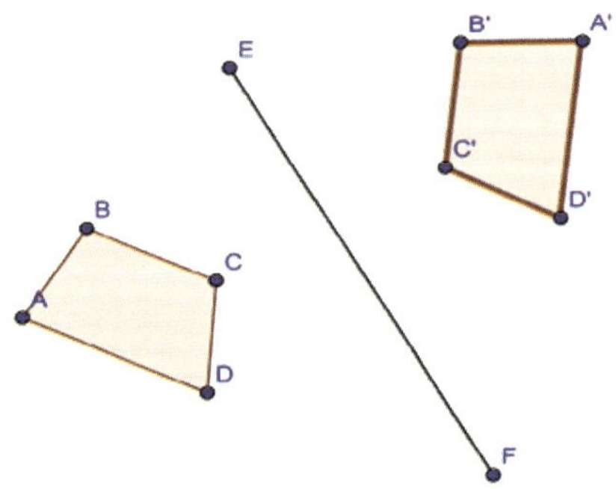

EUREKA MATH™

Closing (1 minute)

- How does a reflection r across line l affect the points of the plane?
    - *The points on the line of reflection l are unchanged. For all other points P, $r_l(P)$ is the point Q so that l is the perpendicular bisector of the segment PQ.*

> **Lesson Summary**
>
> **A reflection carries segments onto segments of equal length.**
>
> **A reflection carries angles onto angles of equal measure.**

Exit Ticket (4 minutes)

Name ______________________________ Date______________

Lesson 14: Reflections

Exit Ticket

1. Construct the line of reflection for the figures.

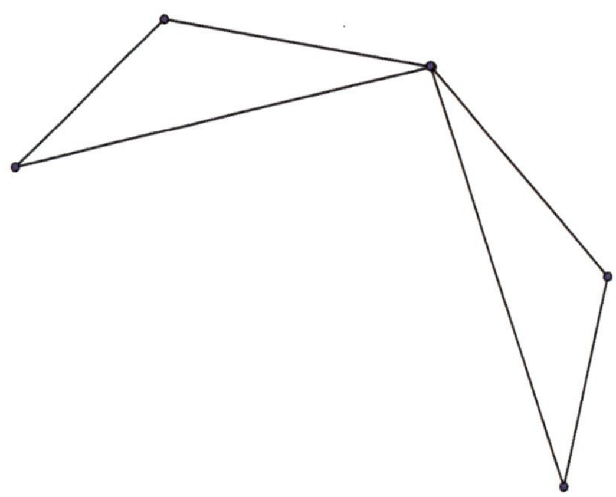

2. Reflect the given pre-image across the line of reflection provided.

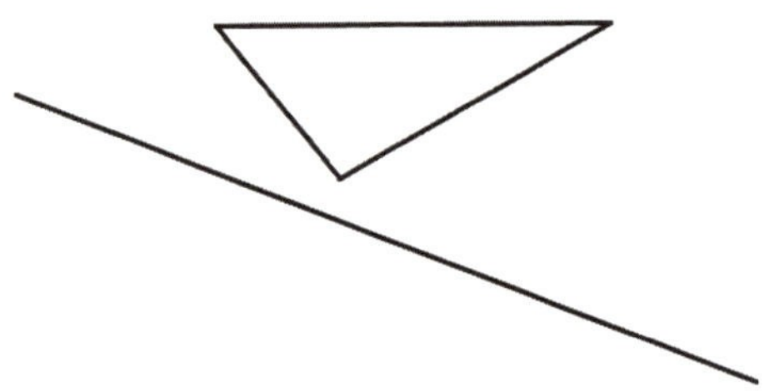

Exit Ticket Sample Solutions

1. **Construct the line of reflection for the figures.**

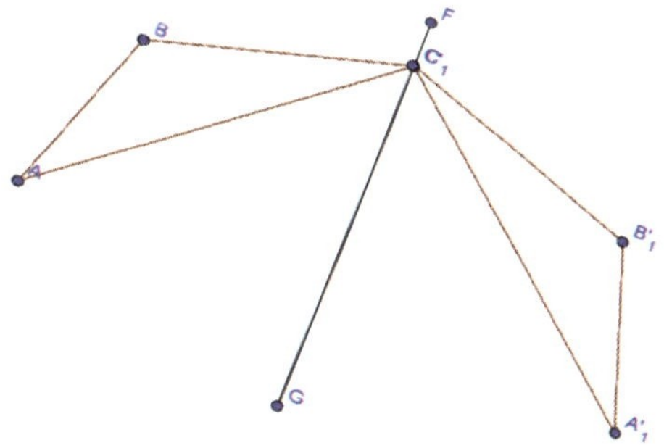

2. **Reflect the given pre-image across the line of reflection provided.**

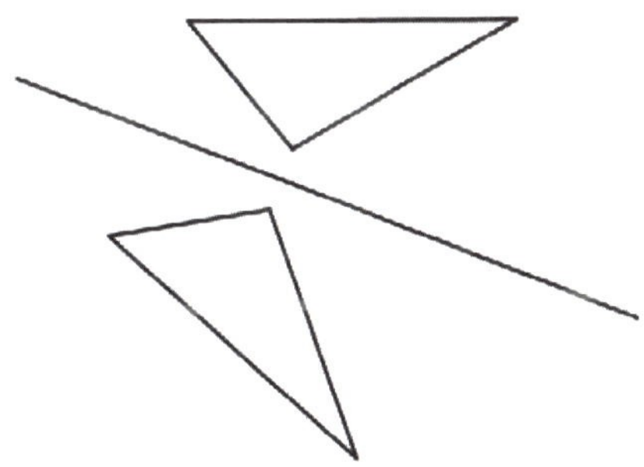

Problem Set Sample Solutions

Construct the line of reflection for each pair of figures below.

1.

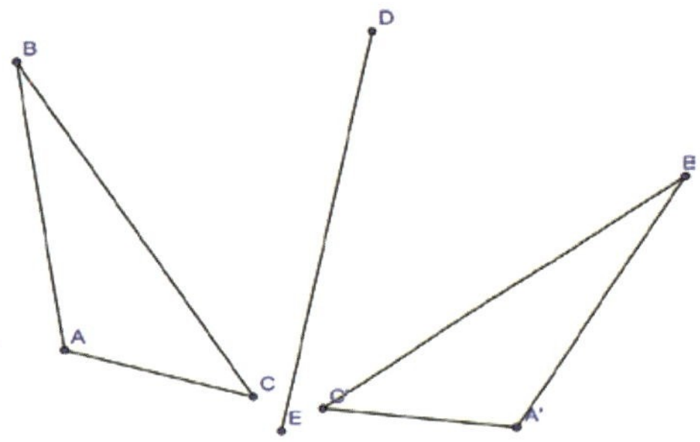

2.

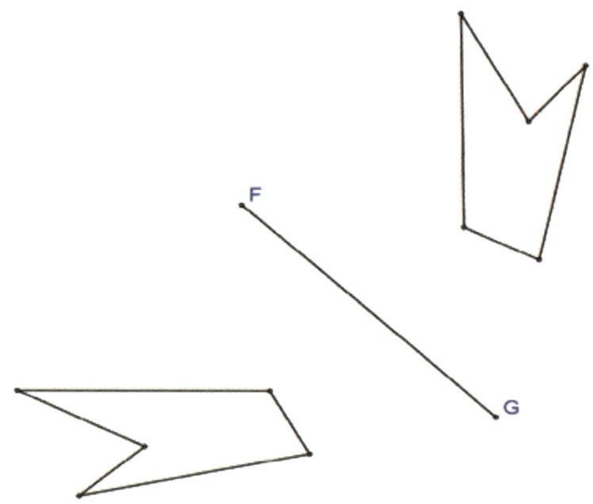

3.

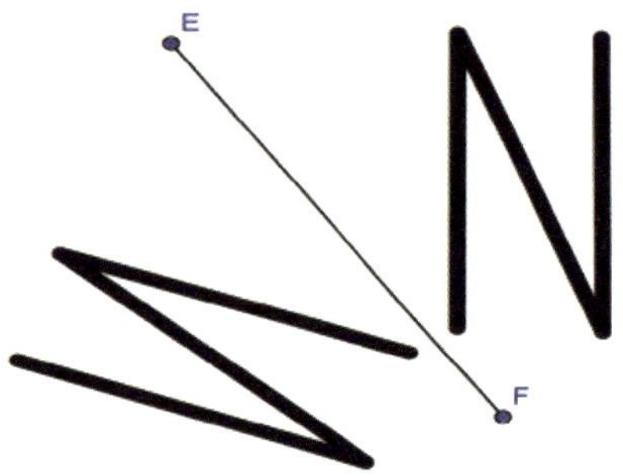

4. **Reflect the given pre-image across the line of reflection provided.**

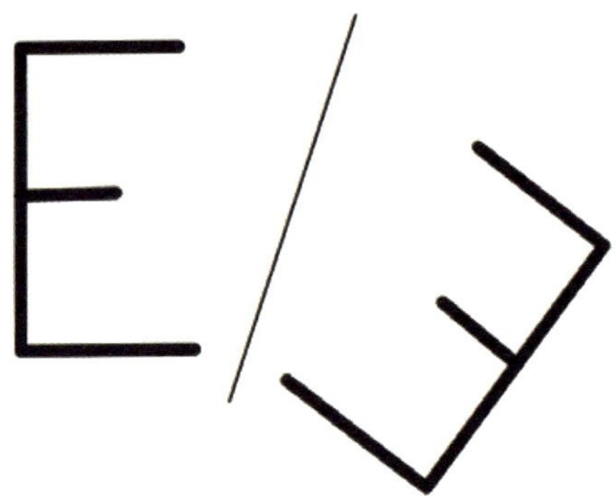

5. **Draw a triangle ABC. Draw a line l through vertex C so that it intersects the triangle at more than just the vertex. Construct the reflection across l.**

Answers will vary.

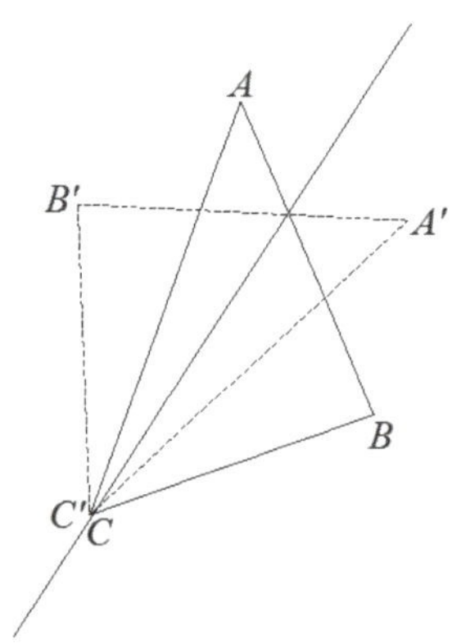

Lesson 15: Rotations, Reflections, and Symmetry

Student Outcomes

- Students learn the relationship between a reflection and a rotation.
- Students examine rotational symmetry within an individual figure.

Lesson Notes

In Lesson 15, students investigated how rotations and reflections can elicit the symmetries within a rectangle, parallelogram, trapezoid, and regular polygon. Students explore the differences between line symmetry and rotational symmetry and how to identify and apply each type of symmetry.

Classwork

Opening Exercise (5 minutes)

Opening Exercise

The original triangle, labeled *A*, has been reflected across the first line, resulting in the image labeled *B*. Reflect the image across the second line.

Carlos looked at the image of the reflection across the second line and said, "That's not the image of triangle *A* after two reflections; that's the image of triangle *A* after a rotation!" Do you agree? Why or why not?

While a rotation was not performed in this example, Carlos is correct that reflecting a figure twice over intersecting lines yields the same result as a rotation. The point R is the center of rotation.

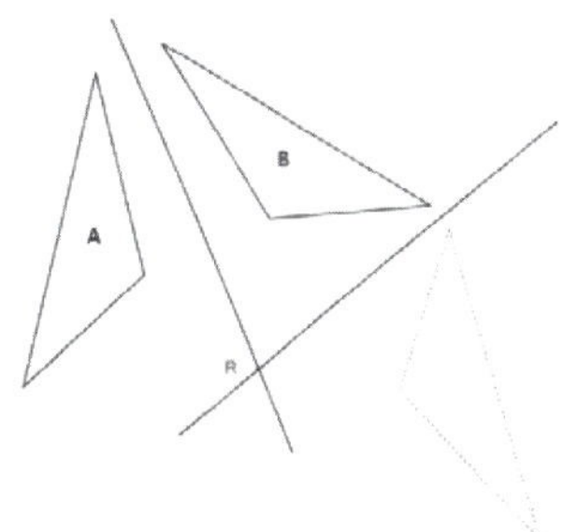

Discussion (2 minutes)

Discussion

When you reflect a figure across a line, the original figure and its image share a line of symmetry, which we have called the line of reflection. When you reflect a figure across a line and then reflect the image across a line that intersects the first line, your final image is a rotation of the original figure. The center of rotation is the point at which the two lines of reflection intersect. The angle of rotation is determined by connecting the center of rotation to a pair of corresponding vertices on the original figure and the final image. The figure above is a 210° rotation (or 150° clockwise rotation).

Exploratory Challenge (24 minutes)

Exploratory Challenge

LINE OF SYMMETRY OF A FIGURE: This is an isosceles triangle. By definition, an isosceles triangle has at least two congruent sides. A *line of symmetry* of the triangle can be drawn from the top vertex to the midpoint of the base, decomposing the original triangle into two congruent right triangles. This line of symmetry can be thought of as a reflection across itself that takes the isosceles triangle to itself. Every point of the triangle on one side of the line of symmetry has a corresponding point on the triangle on the other side of the line of symmetry, given by reflecting the point across the line. In particular, the line of symmetry is equidistant from all corresponding pairs of points. Another way of thinking about line symmetry is that a figure has line symmetry if there exists a line (or lines) such that the image of the figure when reflected over the line is itself.

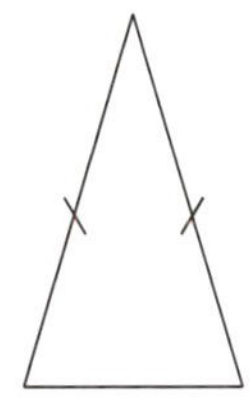

Does every figure have a line of symmetry?

No

Which of the following have multiple lines of symmetry?

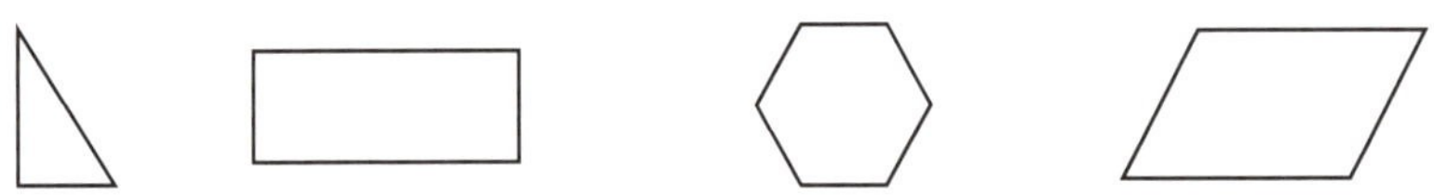

The rectangle and hexagon have multiple lines of symmetry.

Use your compass and straightedge to draw one line of symmetry on each figure above that has at least one line of symmetry. Then, sketch any remaining lines of symmetry that exist. What did you do to justify that the lines you constructed were, in fact, lines of symmetry? How can you be certain that you have found all lines of symmetry?

Students may have measured angles or used patty paper to prove the symmetry.

ROTATIONAL SYMMETRY OF A FIGURE: A nontrivial *rotational symmetry of a figure* is a rotation of the plane that maps the figure back to itself such that the rotation is greater than $0°$ but less than $360°$. Three of the four polygons above have a nontrivial rotational symmetry. Can you identify the polygon that does not have such symmetry?

The triangle does not have such symmetry.

When we studied rotations two lessons ago, we located both a center of rotation and an angle of rotation.

Identify the center of rotation in the equilateral triangle ABC below, and label it D. Follow the directions in the paragraph below to locate the center precisely.

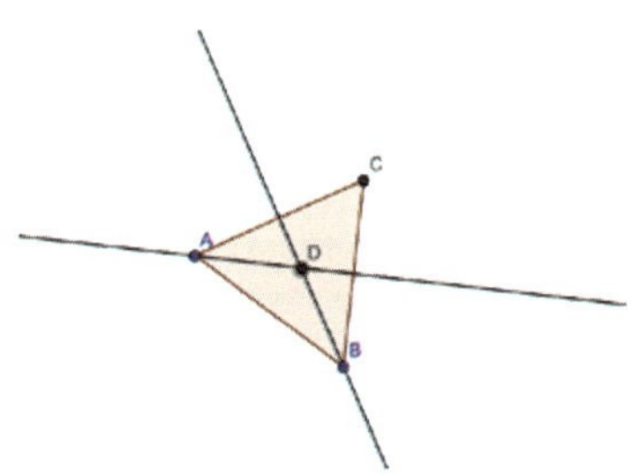

To identify the center of rotation in the equilateral triangle, the simplest method is finding the perpendicular bisector of at least two of the sides. The intersection of these two bisectors gives us the center of rotation. Hence, the center of rotation of an equilateral triangle is also the circumcenter of the triangle. In Lesson 5 of this module, you also located another special point of concurrency in triangles—the incenter. What do you notice about the incenter and circumcenter in the equilateral triangle?

They are the same point.

In any regular polygon, how do you determine the angle of rotation? Use the equilateral triangle above to determine the method for calculating the angle of rotation, and try it out on the rectangle, hexagon, and parallelogram above.

IDENTITY SYMMETRY: A symmetry of a figure is a basic rigid motion that maps the figure back onto itself. There is a special transformation that trivially maps any figure in the plane back to itself called the *identity transformation*. This transformation, like the function f defined on the real number line by the equation $f(x) = x$, maps each point in the plane back to the same point (in the same way that f maps 3 to 3, π to π, and so forth). It may seem strange to discuss the do-nothing *identity symmetry* (the symmetry of a figure under the identity transformation), but it is actually quite useful when listing all of the symmetries of a figure.

Let us look at an example to see why. The equilateral triangle ABC on the previous page has two nontrivial rotations about its circumcenter D, a rotation by $120°$ and a rotation by $240°$. Notice that performing two $120°$ rotations back-to-back is the same as performing one $240°$ rotation. We can write these two back-to-back rotations explicitly, as follows:

- First, rotate the triangle by $120°$ about D: $R_{D,120°}(\triangle ABC)$.
- Next, rotate the image of the first rotation by $120°$: $R_{D,120°}\left(R_{D,120°}(\triangle ABC)\right)$.

Rotating $\triangle ABC$ by $120°$ twice in a row is the same as rotating $\triangle ABC$ once by $120° + 120° = 240°$. Hence, rotating by $120°$ twice is equivalent to one rotation by $240°$:

$$R_{D,120°}\left(R_{D,120°}(\triangle ABC)\right) = R_{D,240°}(\triangle ABC).$$

In later lessons, we see that this can be written compactly as $R_{D,120°} \cdot R_{D,120°} = R_{D,240°}$. What if we rotated by $120°$ one more time? That is, what if we rotated $\triangle ABC$ by $120°$ three times in a row? That would be equivalent to rotating $\triangle ABC$ once by $120° + 120° + 120°$ or $360°$. But a rotation by $360°$ is equivalent to doing nothing (i.e., the identity transformation)! If we use I to denote the identity transformation ($I(P) = P$ for every point P in the plane), we can write this equivalency as follows:

$$R_{D,120°}\left(R_{D,120°}\left(R_{D,120°}(\triangle ABC)\right)\right) = I(\triangle ABC).$$

Continuing in this way, we see that rotating $\triangle ABC$ by $120°$ four times in a row is the same as rotating once by $120°$, rotating five times in a row is the same as $R_{D,240°}$, and so on. In fact, for a whole number n, rotating $\triangle ABC$ by $120°$ n times in a row is equivalent to performing one of the following three transformations:

$$\{R_{D,120°}, \quad R_{D,240°}, \quad I\}.$$

Hence, by including identity transformation I in our list of rotational symmetries, we can write any number of rotations of $\triangle ABC$ by $120°$ using only three transformations. For this reason, we include the identity transformation as a type of symmetry as well.

Exercises (10 minutes)

Exercises

Use Figure 1 to answer the questions below.

Figure 1

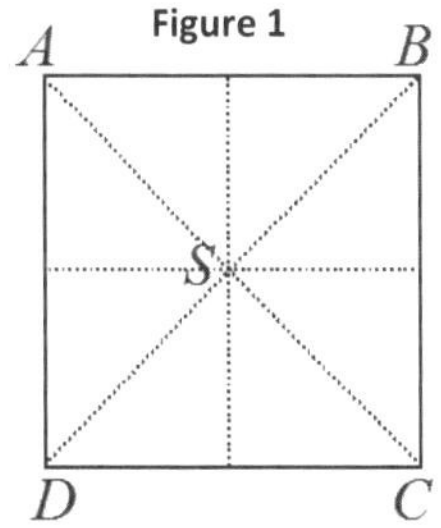

1. Draw all lines of symmetry. Locate the center of rotational symmetry.

2. Describe all symmetries explicitly.

   a. What kinds are there?

      Line symmetry, rotational symmetry, and the identity symmetry

   b. How many are rotations? (Include $360°$ rotational symmetry, i.e., the identity symmetry.)

      4 *($90°$, $180°$, $270°$, $360°$)*

   c. How many are reflections?

      4

3. Prove that you have found all possible symmetries.

   a. How many places can vertex A be moved to by some symmetry of the square that you have identified? (Note that the vertex to which you move A by some specific symmetry is known as the image of A under that symmetry. Did you remember the identity symmetry?)

      4—vertex A can be moved to A, B, C, or D.

   b. For a given symmetry, if you know the image of A, how many possibilities exist for the image of B?

      2

   c. Verify that there is symmetry for all possible images of A and B.

      $(A,B) \to (A,B)$ *is the identity*, $(A,B) \to (A,D)$ *is a reflection along* $\overline{AC}$, $(A,B) \to (D,A)$ *is a rotation by* $90°$, *etc.*

   d. Using part (b), count the number of possible images of A and B. This is the total number of symmetries of the square. Does your answer match up with the sum of the numbers from Exercise 2 parts (b) and (c)?

      8

Relevant Vocabulary

REGULAR POLYGON: A polygon is *regular* if all sides have equal length and all interior angles have equal measure.

Closing (1 minute)

- Describe line symmetry, rotational symmetry, and identity symmetry.
  - *Line symmetry is when there exists a line (or lines) such that the image of a figure when reflected over the line is itself. A nontrivial rotational symmetry of a figure is a rotation of the plane that maps the figure back to itself such that the rotation is greater than* $0°$ *but less than* $360°$. *Identity symmetry is a basic rigid motion that maps a figure back onto itself.*

Exit Ticket (3 minutes)

Name ______________________________ Date ______________

Lesson 15: Rotations, Reflections, and Symmetry

Exit Ticket

What is the relationship between a rotation and a reflection? Sketch a diagram that supports your explanation.

Exit Ticket Sample Solutions

What is the relationship between a rotation and a reflection? Sketch a diagram that supports your explanation.

Reflecting a figure twice over intersecting lines yields the same result as a rotation.

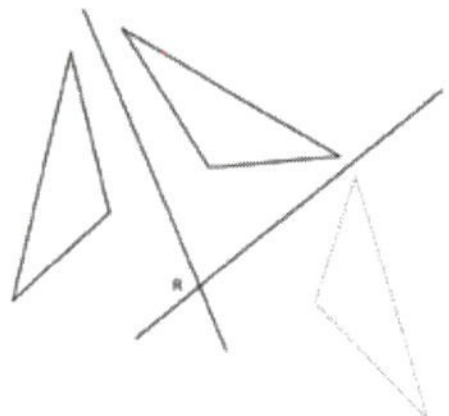

Problem Set Sample Solutions

Use Figure 1 to answer Problems 1–3.

1. Draw all lines of symmetry. Locate the center of rotational symmetry.

Figure 1

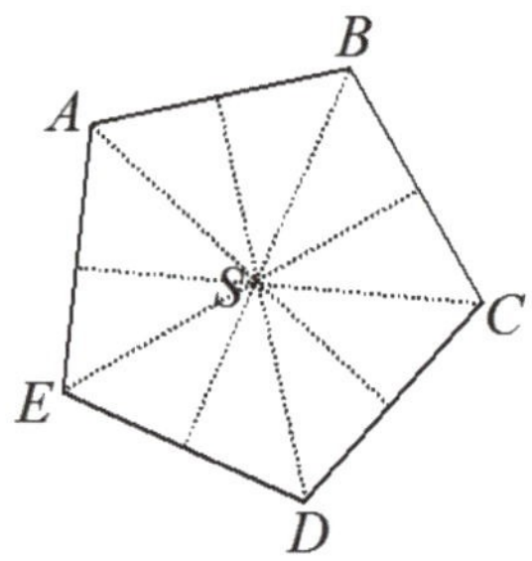

2. Describe all symmetries explicitly.
   a. What kinds are there?

      Rotational symmetry, line symmetry, and the identity symmetry

   b. How many are rotations (including the identity symmetry)?

      5

   c. How many are reflections?

      5

3. Now that you have found the symmetries of the pentagon, consider these questions:
   a. How many places can vertex A be moved to by some symmetry of the pentagon? (Note that the vertex to which you move A by some specific symmetry is known as the image of A under that symmetry. Did you remember the identity symmetry?)

      5*—vertex A can be moved to* A*,* B*,* C*,* D *or* E*.*

   b. For a given symmetry, if you know the image of A, how many possibilities exist for the image of B?

      $(A, B) \rightarrow (A, B)$ ***is the identity,*** $(A, B) \rightarrow (C, B)$ ***is a reflection along*** $\overline{SB}$***,*** $(A, B) \rightarrow (E, A)$ ***is a rotation by*** $72°$***, etc.***

   c. Verify that there is symmetry for all possible images of A and B.

   d. Using part (b), count the number of possible images of A and B. This is the total number of symmetries of the figure. Does your answer match up with the sum of the numbers from Problem 2 parts (b) and (c)?

      10

Use Figure 2 to answer Problem 4.

Figure 2

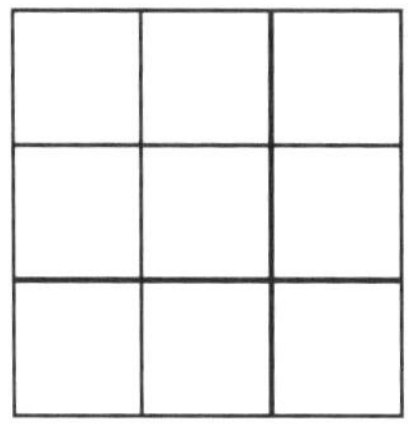

4. Shade exactly two of the nine smaller squares so that the resulting figure has
   a. Only one vertical and one horizontal line of symmetry.
   b. Only two lines of symmetry about the diagonals.
   c. Only one horizontal line of symmetry.
   d. Only one line of symmetry about a diagonal.
   e. No line of symmetry.

   Possible answers:

   a. 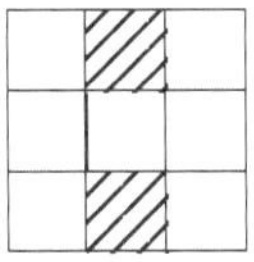 *b.* 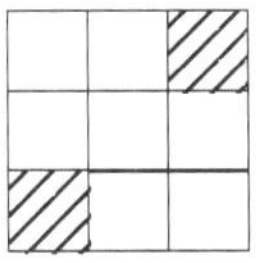 *c.* 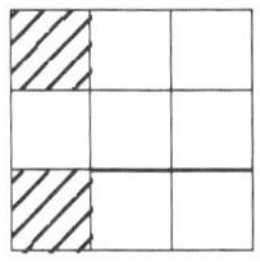 *d.* 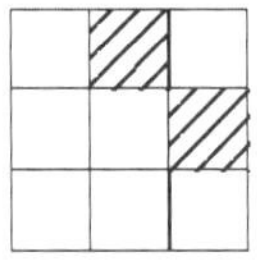 *e.* 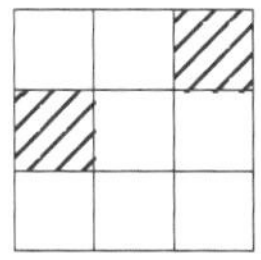

Use Figure 3 to answer Problem 5.

Figure 3

5. Describe all the symmetries explicitly.
   a. How many are rotations (including the identity symmetry)?

      6

   b. How many are reflections?

      12

   c. How could you shade the figure so that the resulting figure only has three possible rotational symmetries (including the identity symmetry)?

      See the figure for an example.

6. Decide whether each of the statements is true or false. Provide a counterexample if the answer is false.
   a. If a figure has exactly two lines of symmetry, it has exactly two rotational symmetries (including the identity symmetry).

      True

   b. If a figure has at least three lines of symmetry, it has at least three rotational symmetries (including the identity symmetry).

      True

   c. If a figure has exactly two rotational symmetries (including the identity symmetry), it has exactly two lines of symmetry.

      False; a parallelogram

   d. If a figure has at least three rotational symmetries (including the identity symmetry), it has at least three lines of symmetry.

      False; 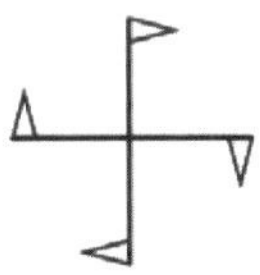

Lesson 16: Translations

Student Outcome

- Students learn the precise definition of a translation and perform a translation by construction.

Lesson Notes

In Lesson 16, students precisely define translations and use the construction of a parallelogram to demonstrate how to apply a translation to a figure. Students then use vectors to describe the translations. This may be the first time many students have seen vectors, so some additional explanation may be needed (a *vector* is a directed line segment that has both length and direction).

Refer to Grade 8 Module 2 Lesson 2 for supplementary materials on use of vectors, as well as on translations in general.

Classwork

Exploratory Challenge (5 minutes)

Exploratory Challenge

In Lesson 4, you completed a construction exercise that resulted in a pair of parallel lines (Problem 1 from the Problem Set). Now we examine an alternate construction.

Construct the line parallel to a given line AB through a given point P.

1. **Draw circle P: Center P, radius AB.**
2. **Draw circle B: Center B, radius AP.**
3. **Label the intersection of circle P and circle B as Q.**
4. **Draw $\overleftrightarrow{PQ}$.**

***Note:* Circles P and B intersect in two locations. Pick the intersection Q so that points A and Q are in opposite half-planes of line PB.**

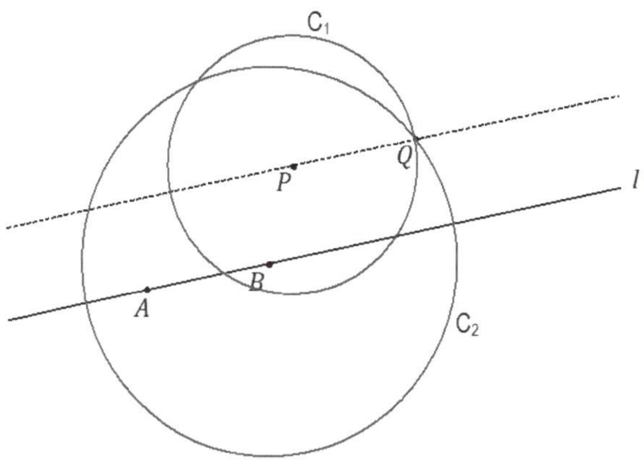

The construction shows that $m\angle ABP$ and $m\angle QPB$ are equal, alternate interior angles. Hence, by the alternate interior angles converse, $\overleftrightarrow{PQ} \parallel \overleftrightarrow{AB}$.

Discussion (10 minutes)

Discussion

To perform a translation, we need to use the previous construction. Let us investigate the definition of translation.

For vector $\overrightarrow{AB}$, the *translation along* $\overrightarrow{AB}$ is the transformation $T_{\overrightarrow{AB}}$ of the plane defined as follows:

1. **For any point P on the line AB, $T_{\overrightarrow{AB}}(P)$ is the point Q on $\overleftrightarrow{AB}$ so that $\overrightarrow{PQ}$ has the same length and the same direction as $\overrightarrow{AB}$, and**
2. **For any point P not on $\overleftrightarrow{AB}$, $T_{\overrightarrow{AB}}(P)$ is the point Q obtained as follows. Let l be the line passing through P and parallel to $\overleftrightarrow{AB}$. Let m be the line passing through B and parallel to line AP. The point Q is the intersection of l and m.**

Note: **The parallel line construction on the previous page shows a quick way to find the point Q in part 2 of the definition of translation.**

In the figure to the right, quadrilateral $ABCD$ has been translated the length and direction of vector $\overrightarrow{CC'}$. Notice that the distance and direction from each vertex to its corresponding vertex on the image are identical to that of $\overrightarrow{CC'}$.

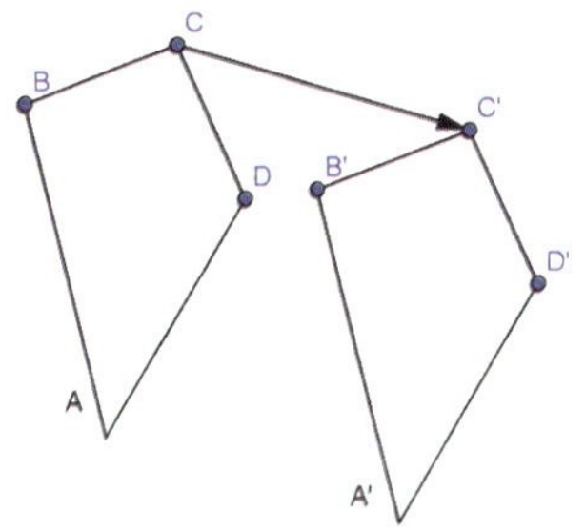

Example 1 (8 minutes)

Example 1

Draw the vector that defines each translation below.

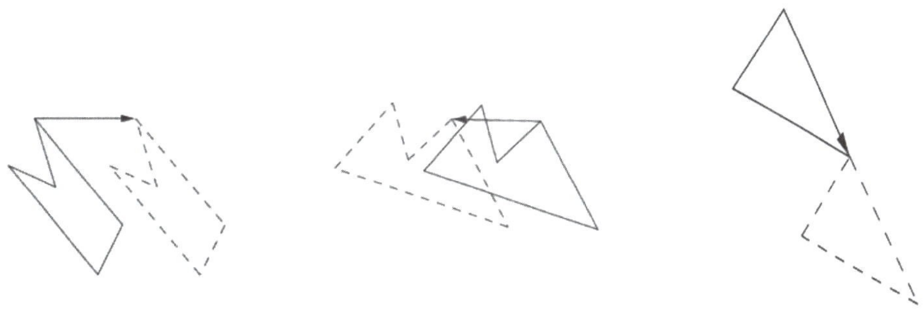

Finding the vector is relatively straightforward. Applying a vector to translate a figure is more challenging. To translate a figure, we must construct parallel lines to the vector through the vertices of the original figure and then find the points on those parallel lines that are the same direction and distance away as given by the vector.

Example 2 (8 minutes)

Example 2

Use your compass and straightedge to apply $T_{\overrightarrow{AB}}$ to segment P_1P_2.

Note: Use the steps from the Exploratory Challenge *twice* for this question, creating two lines parallel to $\overrightarrow{AB}$: one through P_1 and one through P_2.

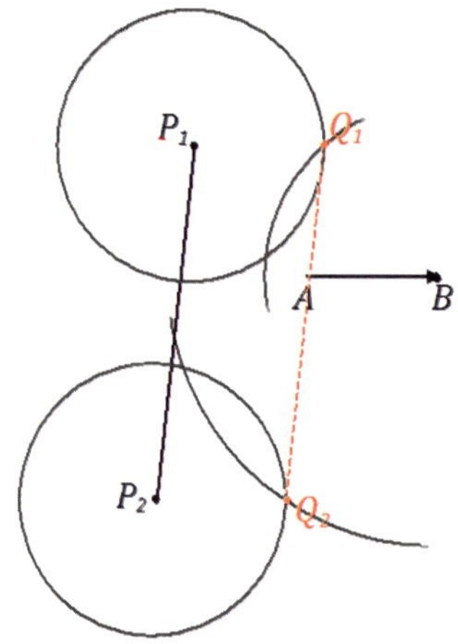

Example 3 (8 minutes)

Example 3

Use your compass and straightedge to apply $T_{\overrightarrow{AB}}$ to $\triangle P_1P_2P_3$.

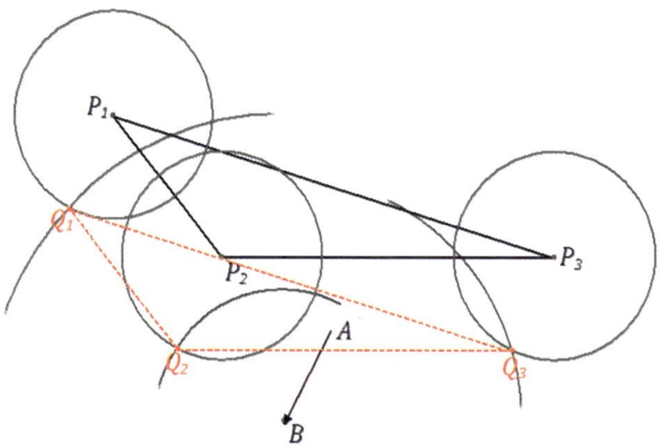

Relevant Vocabulary

PARALLEL: Two lines are *parallel* if they lie in the same plane and do not intersect. Two segments or rays are parallel if the lines containing them are parallel lines.

Closing (1 minute)

- How does translation T along vector AB affect the points in the plane?
    - *For any point P on the line AB, $T_{\overrightarrow{AB}}(P)$ is the point Q on $\overleftrightarrow{AB}$ so that $\overrightarrow{PQ}$ has the same length and the same direction as $\overrightarrow{AB}$. For any point P not on $\overleftrightarrow{AB}$, $T_{\overrightarrow{AB}}(P)$ is the point Q so that points P, Q, A, and B form a parallelogram.*

> Lesson Summary
>
> **A translation carries segments onto segments of equal length.**
>
> **A translation carries angles onto angles of equal measure.**

Exit Ticket (5 minutes)

Name ____ Date____________________

Lesson 16: Translations

Exit Ticket

Translate the figure one unit down and three units right. Draw the vector that defines the translation.

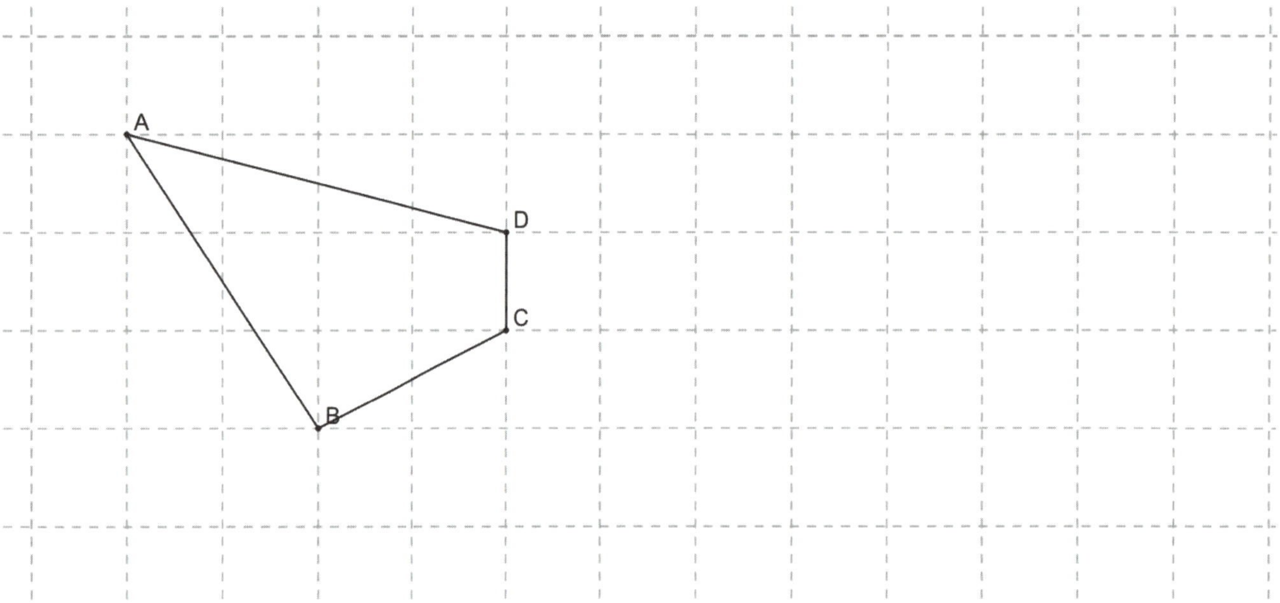

Exit Ticket Sample Solutions

Translate the figure one unit down and three units right. Draw the vector that defines the translation.

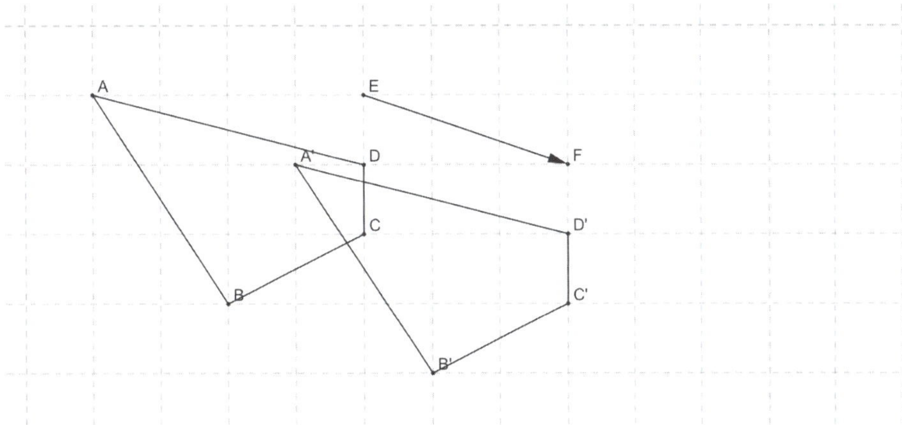

Problem Set Sample Solutions

Translate each figure according to the instructions provided.

1. 2 units down and 3 units left

   Draw the vector that defines the translation.

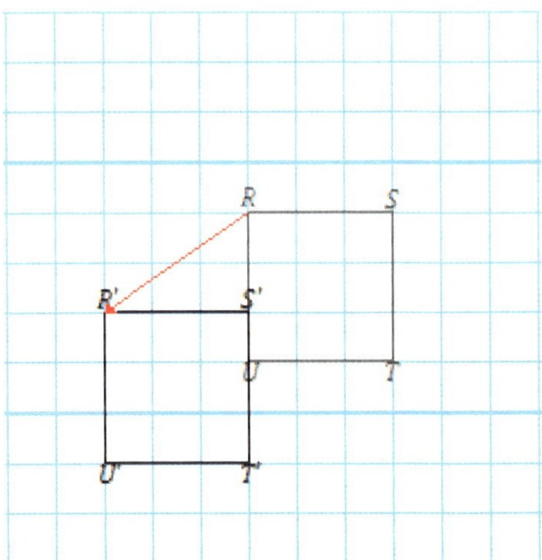

2. 1 unit up and 2 units right

   Draw the vector that defines the translation.

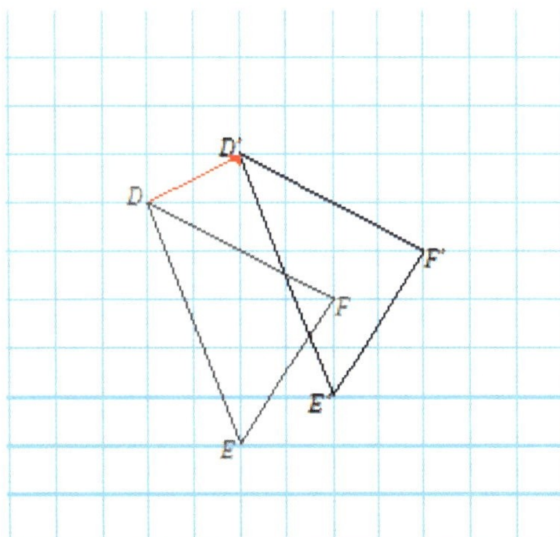

3. Use your compass and straightedge to apply $T_{\overrightarrow{AB}}$ to the circle below (center P_1, radius $\overline{P_1P_2}$).

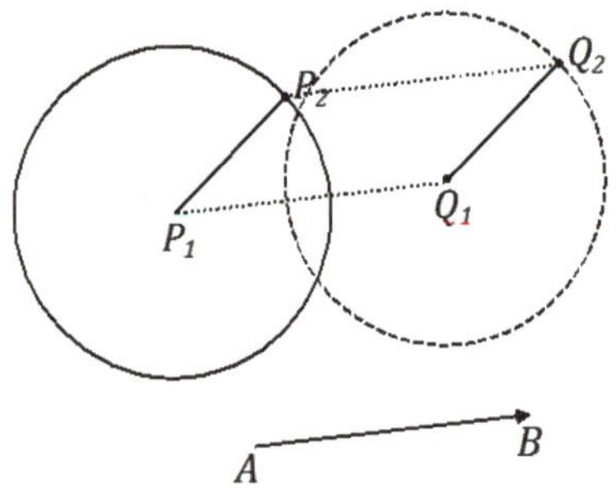

To translate the circle is to translate its radius.

4. Use your compass and straightedge to apply $T_{\overrightarrow{AB}}$ to the circle below.

   Hint: You need to first find the center of the circle. You can use what you learned in Lesson 4 to do this.

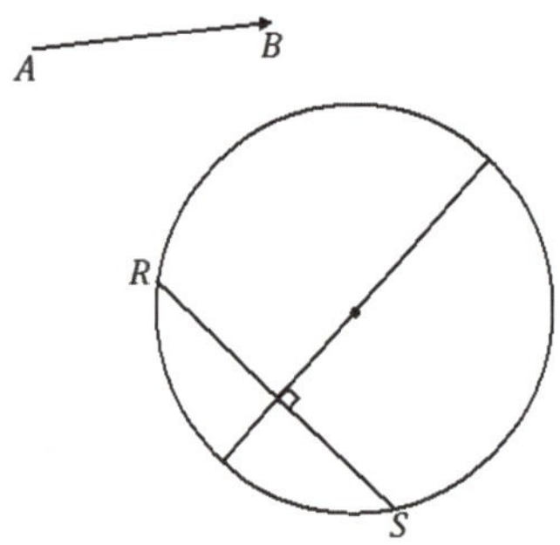

- *To find the center of the circle (and thereby also the radius of the circle), draw any chord within the circle. Construct the perpendicular bisector of the chord, and mark the diameter of the circle, which contains the center of the circle. Finally, use a perpendicular bisector to find the midpoint of the diameter.*
- *Once the center has been established, the problem is similar to Problem 3.*

Two classic toothpick puzzles appear below. Solve each puzzle.

5. Each segment on the fish represents a toothpick. Move (translate) exactly three toothpicks and the eye to make the fish swim in the opposite direction. Show the translation vectors needed to move each of the three toothpicks and the eye.

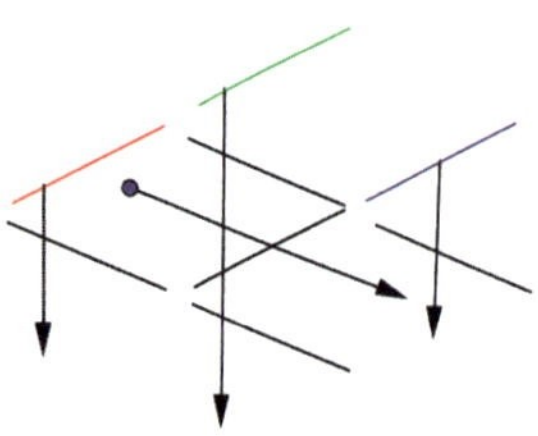

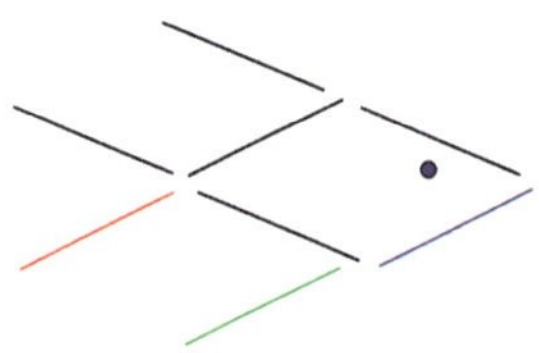

6. **Again, each segment represents a single toothpick. Move (translate) exactly three toothpicks to make the triangle point downward. Show the translation vectors needed to move each of the three toothpicks.**

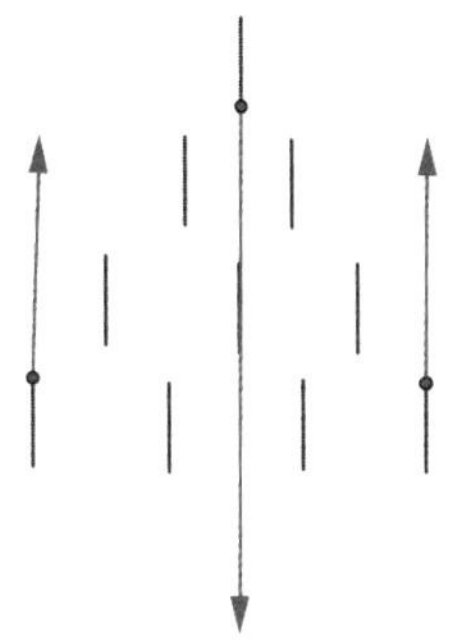

7. **Apply $T_{\overrightarrow{GH}}$ to translate $\triangle ABC$.**

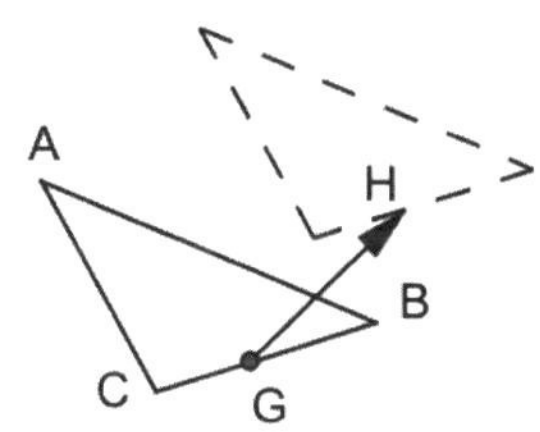

Lesson 17: Characterize Points on a Perpendicular Bisector

Student Outcomes

- Students understand that any point on a line of reflection is equidistant from any pair of pre-image and image points in a reflection.

Lesson Notes

In Lesson 17, students review the three types of rigid motions previously explored and examine their relationship to the construction of perpendicular lines and, more specifically, perpendicular bisectors. They further explore the *distance preservation* characteristic of rigid motions and see how this preservation applies in each type of transformation.

Classwork

Opening Exercise (4 minutes)

Opening Exercise

MP.5

In Lesson 3, you bisected angles, including straight angles. You related the bisection of straight angles in Lesson 3 to the construction of perpendicular bisectors in Lesson 4. Review the process of constructing a perpendicular bisector with the segment below. Then complete the definition of perpendicular lines below your construction.

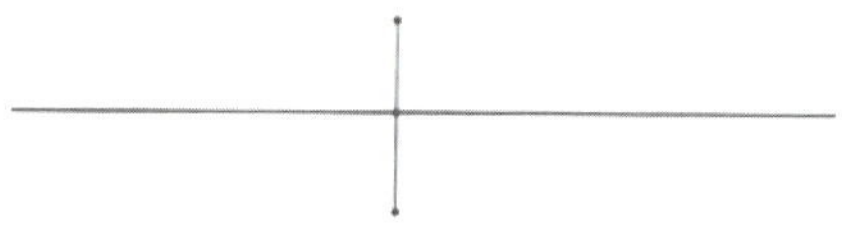

MP.6

Use the compass and straightedge construction from Lesson 4.

Two lines are perpendicular if they ***intersect*****, and if any of the angles formed by the intersection of the lines is a** ***right (or $90°$)*** **angle. Two segments are perpendicular if the lines containing them are** ***perpendicular*****.**

Ask students to discuss the definition of a perpendicular bisector and how having already learned the construction helps them to know and understand the definition.

Discussion/Examples 1–3 (15 minutes)

Discussion

The line you constructed in the Opening Exercise is called the perpendicular bisector of the segment. As you learned in Lesson 14, the perpendicular bisector is also known as the line of reflection of the segment. With a line of reflection, any point on one side of the line (pre-image) is the same distance from the line as its image on the opposite side of the line.

Example 1

Is it possible to find or construct a line of reflection that is *not* a perpendicular bisector of a segment connecting a point on the pre-image to its image? Try to locate a line of reflection between the two figures to the right without constructing any perpendicular bisectors.

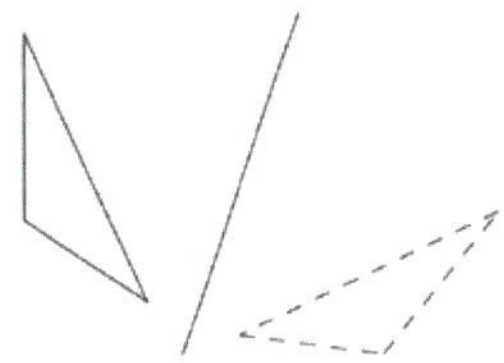

Discussion

Why were your attempts impossible? Look back at the definition of *reflection* from Lesson 14.

For a line l in the plane, a *reflection across* l is the transformation r_l of the plane defined as follows:

1. For any point P on the line l, $r_l(P) = P$, and
2. For any point P not on l, $r_l(P)$ is the point Q so that l is the perpendicular bisector of the segment PQ.

The key lies in the use of the term *perpendicular bisector*. For a point P not on l, explain how to construct the point Q so that l is the perpendicular bisector of the segment PQ.

Now, let's think about the problem from another perspective. We have determined that any point on the pre-image figure is the same distance from the line of reflection as its image. Therefore, the two points are equidistant from the point at which the line of reflection (perpendicular bisector) intersects the segment connecting the pre-image point to its image. What about other points on the perpendicular bisector? Are they also equidistant from the pre-image and image points? Let's investigate.

Example 2

Using the same figure from the previous investigation, but with the line of reflection, is it possible to conclude that any point on the perpendicular bisector is equidistant from any pair of pre-image and image points? For example, is $GP = HP$ in the figure? The point P is clearly *not* on the segment connecting the pre-image point G to its image H. How can you be certain that $GP = HP$? If r is the reflection, then $r(G) = H$ and $r(P) = P$. Since r preserves distances, $GP = HP$.

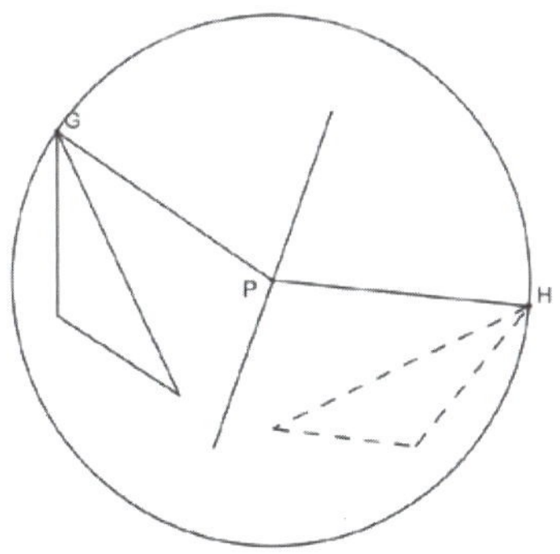

Discussion

We have explored perpendicular bisectors as they relate to reflections and have determined that they are essential to reflections. Are perpendicular lines, specifically, perpendicular bisectors, essential to the other two types of rigid motions: rotations and translations? Translations involve constructing parallel lines (which can certainly be done by constructing perpendiculars but are not essential to constructing parallels). However, perpendicular bisectors play an important role in rotations. In Lesson 13, we found that the intersection of the perpendicular bisectors of two segments connecting pairs of pre-image to image points determined the center of rotation.

Example 3

Find the center of rotation for the transformation below. How are perpendicular bisectors a major part of finding the center of rotation? Why are they essential?

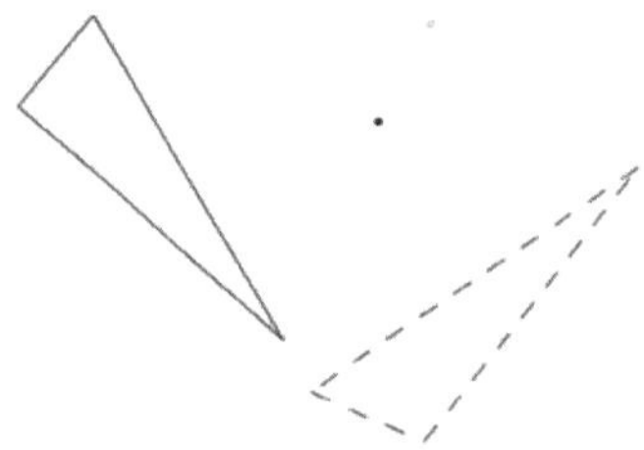

The center of rotation is the intersection of two perpendicular bisectors, each to a segment that joins a pair of corresponding points (between the figure and its image).

As you explore this figure, also note another feature of rotations. As with all rigid motions, rotations preserve distance. A transformation is said to be distance-preserving (or length-preserving) if the distance between the images of two points is always equal to the distance between the original two points. Which of the statements below is true of the distances in the figure? Justify your response.

1. $AB = A'B'$
2. $AA' = BB'$

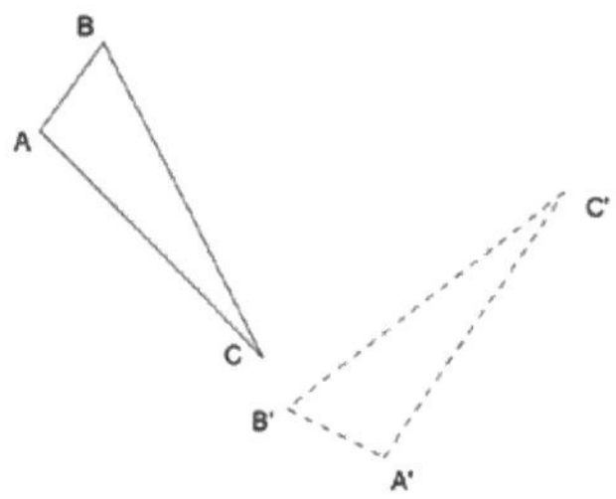

$AB = A'B'$ is a distance in the figure. The distance between A and B remains the same after the transformation. The second statement $AA' = BB'$ is incorrect because it involves distances between the original figure and the new figures. Rotations are distance-preserving; thus, we are only assured that distance between the images of two points is always equal to the distances between the original two points.

Exercises (20 minutes)

MP.7

Exercises

In each pre-image/image combination below, (a) identify the type of transformation; (b) state whether perpendicular bisectors play a role in constructing the transformation and, if so, what role; and (c) cite an illustration of the distance-preserving characteristic of the transformation (e.g., identify two congruent segments from the pre-image to the image). For the last requirement, you have to label vertices on the pre-image and image.

1.

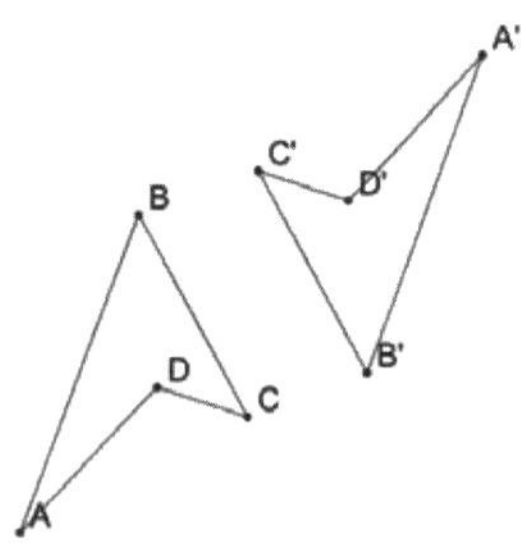

Transformation	Perpendicular Bisectors?	Examples of Distance Preservation
Rotation	*Yes, to construct the center of rotation*	$AB = A'B'$ $DC = D'C'$

2.

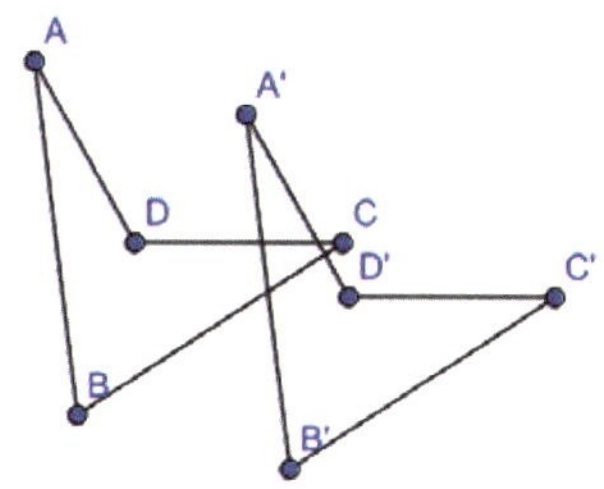

Transformation	Perpendicular Bisectors?	Examples of Distance Preservation
Translation	*No*	$AD = A'D'$ $DC = D'C'$

3.

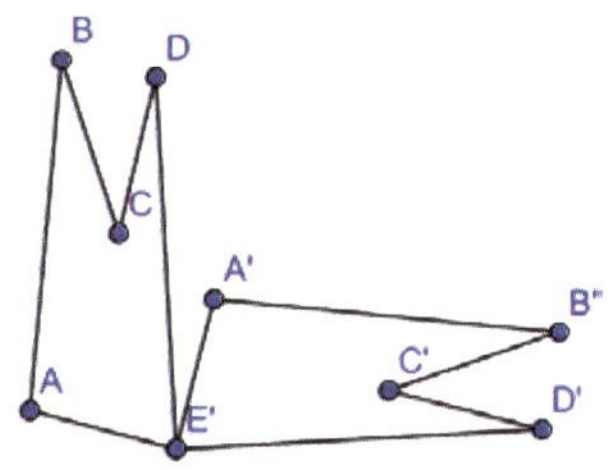

Transformation	Perpendicular Bisectors?	Examples of Distance Preservation
Rotation	*In general, yes; however, not required in this case*	$AB = A'B'$ $BC = B'C'$

4.

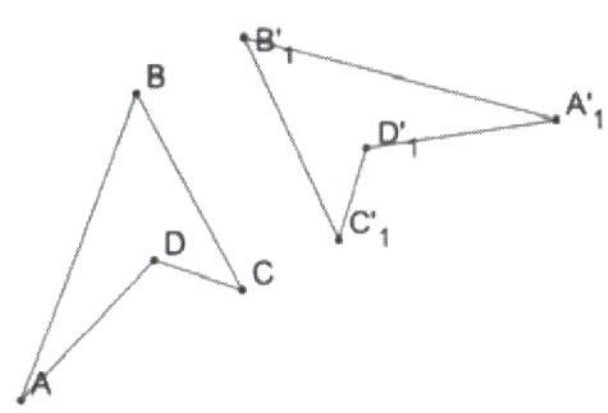

Transformation	Perpendicular Bisectors?	Examples of Distance Preservation
Reflection	*Yes, to construct the line of reflection*	$BC = B'_1C'_1$ $AD = A'_1D'_1$

5. In the figure to the right, $\overleftrightarrow{GH}$ is a segment of reflection. State and justify two conclusions about distances in this figure. At least one of your statements should refer to perpendicular bisectors.

Each point and its reflected image are the same distance from the line of reflection.

The line of reflection is the perpendicular bisector of the line connecting each point with its reflected image.

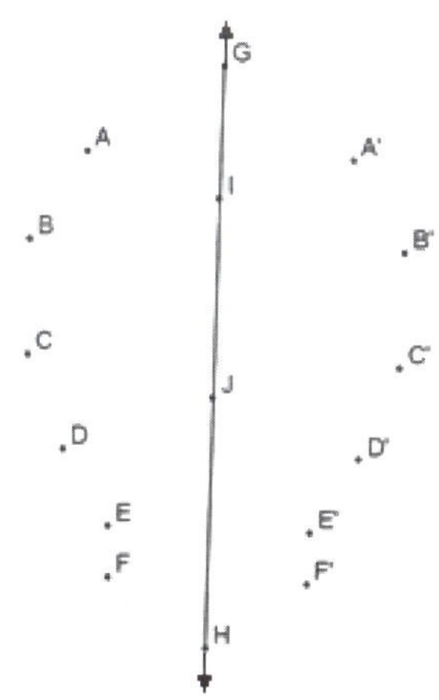

Closing (1 minute)

- Perpendicular bisectors are essential to both reflections and rotations. The perpendicular bisector is the line of reflection or line of symmetry in a reflection. In a rotation, the intersection of the perpendicular bisectors of two segments connecting pairs of pre-image to image points determine the center of rotation.

Exit Ticket (5 minutes)

Name ______________________________ Date__________________

Lesson 17: Characterize Points on a Perpendicular Bisector

Exit Ticket

Using your understanding of rigid motions, explain why *any* point on the perpendicular bisector is equidistant from *any* pair of pre-image and image points. Use your construction tools to create a figure that supports your explanation.

Exit Ticket Sample Solutions

Using your understanding of rigid motions, explain why *any* point on the perpendicular bisector is equidistant from *any* pair of pre-image and image points. Use your construction tools to create a figure that supports your explanation.

If r is the reflection, then $r(G) = H$ *and* $r(P) = P$. Since r preserves distances, $GP = HP$.

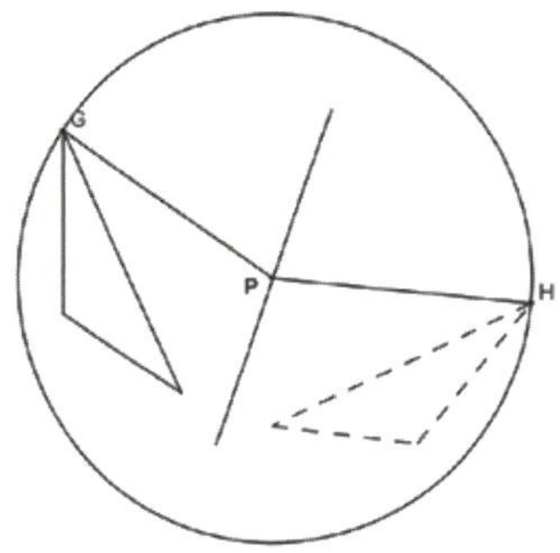

Problem Set Sample Solutions

Create/construct two problems involving transformations—one reflection and one rotation—that require the use of perpendicular bisectors. Your reflection problem may require locating the line of reflection or using the line of reflection to construct the image. Your rotation problem should require location of the point of rotation. (Why should your rotation problem *not* require construction of the rotated image?) Create the problems on one page, and construct the solutions on another. Another student will be solving your problems in the next class period.

Answers will vary.

Lesson 18: Looking More Carefully at Parallel Lines

Student Outcomes

- Students learn to construct a line parallel to a given line through a point not on that line using a rotation by $180°$. They learn how to prove the alternate interior angles theorem using the parallel postulate and the construction.

Lesson Notes

Lesson 18 is important. It may take two days to cover all of the information in this lesson. Hiding underneath its ideas are most of the reasons why reflection and translation can be defined based upon the geometry assumptions students learn in Lesson 33. (For example, only the geometry assumptions are used to construct a parallel line through a given point in Example 4.) While these ideas are hiding under the surface, do not focus on them. Instead, concentrate students' attention on the main geometry assumption of this lesson: the parallel postulate.

Parallel lines have already been defined, and in the missing angles problems (and perhaps discussed casually), the following two ideas have been used.

1. Suppose a transversal intersects a pair of lines. If a pair of alternate interior angles is equal in measure, then the pair of lines are parallel.
2. (A form of the parallel postulate) Suppose a transversal intersects a pair of lines. If the pair of lines are parallel, then the pair of alternate interior angles are equal in measure.

However, students have probably not made careful distinctions between these thoughts. More likely, students remember the following concept:

> *Suppose a transversal intersects a pair of lines. The pair of lines are parallel if and only if a pair of alternate interior angles are equal in measure.*

Perhaps students simply associate the ideas in this statement. When students see parallel lines cut by a transversal, they recognize the angle relations and then use those angle relations to recognize parallel lines.

This lesson is designed to help students carefully distinguish between these two ideas. In particular, students are shown why there is a need for the parallel postulate as one of our geometric assumptions: *Through a given external point there is at most one line parallel to a given line.*

Classwork

Opening Exercise (8 minutes)

Opening Exercise

Exchange Problem Sets from Lesson 17 with a classmate. Solve the problems posed by your classmate while he or she solves yours. Compare your solutions, and then discuss and resolve any discrepancies. Why were you asked only to locate the point of rotation rather than to rotate a pre-image to obtain the image? How did you use perpendicular bisectors in constructing your solutions?

Answers will vary.

Discussion

> **Discussion**
>
> We say that two lines are *parallel* if they lie in the same plane and do not intersect. Two segments or rays are parallel if the lines containing them are parallel.

Examples 1–7 (30 minutes)

Example 1

Why is the phrase *in the plane* critical to the definition of parallel lines? Explain and illustrate your reasoning.

Two lines in (3-dimensional space) are called skew lines if they do not lie in the same plane. In that case, they do not intersect (because if they did, they would share a plane together) and are not parallel. (If they were parallel, then they would both have to lie in the same plane.)

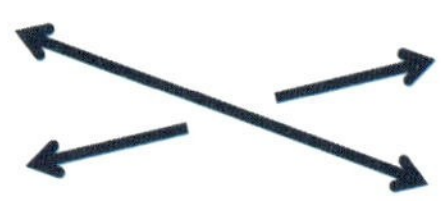

In Lesson 7, we recalled some basic facts learned in earlier grades about pairs of lines and angles created by a transversal to those lines. One of those basic facts is the following:

> *Suppose a transversal intersects a pair of lines. The lines are parallel if and only if a pair of alternate interior angles are equal in measure.*

Our goal in this lesson is to prove this theorem using basic rigid motions, geometry assumptions, and a geometry assumption we introduce in this lesson called the *parallel postulate.* Of all of the geometry assumptions we have given so far, the parallel postulate gets a special name because of the special role it played in the history of mathematics. (Euclid included a version of the parallel postulate in his books, and for $2,000$ years people tried to show that it was not a necessary assumption. Not only did it turn out that the assumption was necessary for Euclidean geometry, but study of the parallel postulate led to the creation of non-Euclidean geometries.)

The basic fact above really has two parts, which we prove separately:

1. Suppose a transversal intersects a pair of lines. If two alternate interior angles are equal in measure, then the pair of lines are parallel.
2. Suppose a transversal intersects a pair of lines. If the lines are parallel, then the pair of alternate interior angles are equal in measure.

The second part turns out to be an equivalent form of the parallel postulate. To build up to the theorem, first we need to do a construction.

Example 2

MP.7

Given a line l and a point P not on the line, follow the steps below to rotate l by $180°$ to a line l' that passes through P:

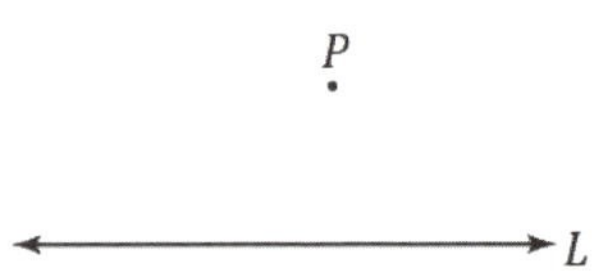

a. Label any point A on l.

b. Find the midpoint of segment AP using a ruler. (Measure the length of segment AP, and locate the point that is distance $\frac{AP}{2}$ from A between A and P.) Label the midpoint C.

MP.7

c. Perform a $180°$ rotation around center C. To quickly find the image of l under this rotation by hand:

   i. Pick another point B on l.

   ii. Draw $\overleftrightarrow{CB}$.

   iii. Draw circle: center C, radius CB.

   iv. Label the other point where the circle intersects $\overleftrightarrow{CB}$ by Q.

   v. Draw $\overleftrightarrow{PQ}$.

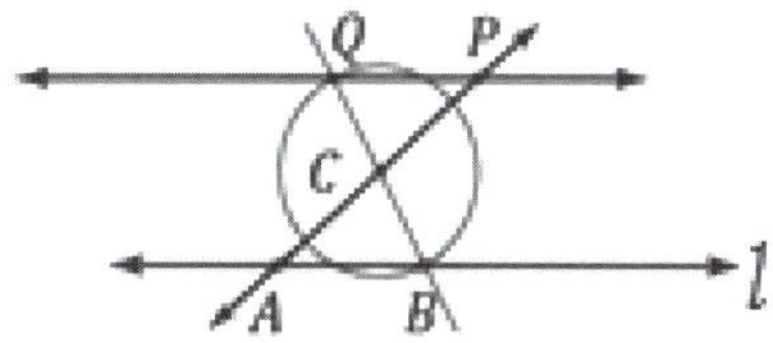

d. Label the image of the rotation by $180°$ of l by $l' = R_{C,180}(l)$.

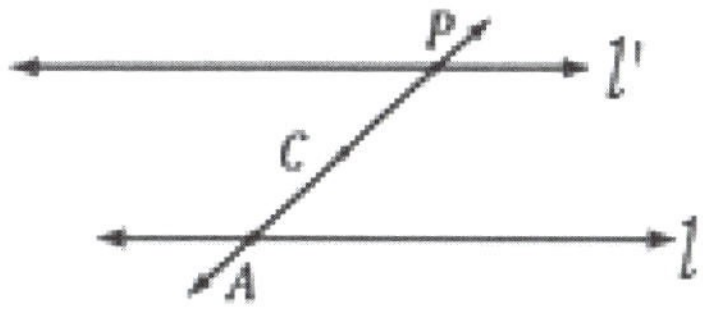

How does your construction relate to the geometry assumption stated above to rotations? Complete the statement below to clarify your observations:

$R_{C,180}$ is a $180°$ __*rotation*__ around C. Rotations preserve __*lines*__; therefore $R_{C,180}$, maps the line l to the line __*l'*__. What is $R_{C,180}(A)$? __*P*__

Example 3

The lines l and l' in the construction certainly look parallel, but we do not have to rely on looks.

Claim: In the construction, l is parallel to l'.

MP.7

PROOF: We show that assuming they are not parallel leads to a contradiction. If they are not parallel, then they must intersect somewhere. Call that point X. Since X is on l', it must be the image of some point S on l under the $R_{C,180}$ rotation (i.e., $R_{C,180}(S) = X$). Since $R_{C,180}$ is a $180°$ rotation, S and X must be the endpoints of a diameter of a circle that has center C. In particular, $\overleftrightarrow{SX}$ must contain C. Since S is a point on l, and X is a different point on l (it was the intersection of both lines), we have that $l = \overleftrightarrow{SX}$ because there is only one line through two points. But $\overleftrightarrow{SX}$ also contains C, which means that l contains C. However, C was constructed so that it was not on l. This is absurd.

There are only two possibilities for any two distinct lines l and l' in a plane: either the lines are parallel, or they are not parallel. Since assuming the lines were not parallel led to a false conclusion, the only possibility left is that l and l' were parallel to begin with.

Example 4

The construction and claim together implies the following theorem.

THEOREM: Given a line l and a point P not on the line, then there exists line l' that contains P and is parallel to l.

This is a theorem we have justified before using compass and straightedge constructions, but now we see it follows directly from basic rigid motions and our geometry assumptions.

MP.7

Example 5

We are now ready to prove the first part of the basic fact above. We have two lines, l and l', and all we know is that a transversal $\overleftrightarrow{AP}$ intersects l and l' such that a pair of alternate interior angles are equal in measure. (In the picture below, we are assuming $m\angle QPA = m\angle BAP$.)

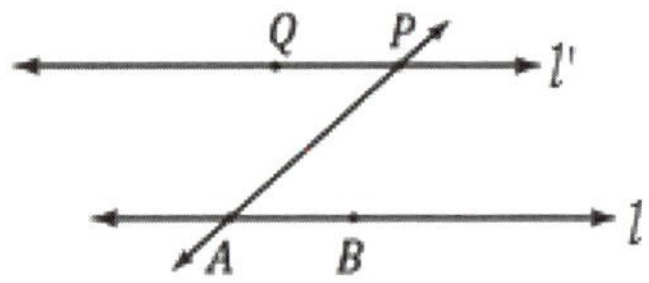

Let C be the midpoint of $\overline{AP}$. What happens if you rotate $180°$ around the center C? Is there enough information to show that $R_{C,180}(l) = l'$?

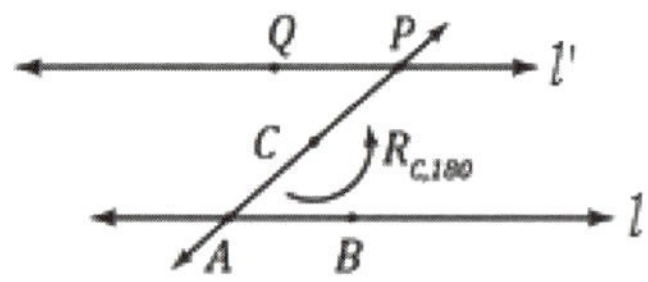

a. What is the image of the segment AP?

PA

b. In particular, what is the image of the point A?

P

c. Why are the points Q and $R_{C,180}(B)$ on the same side of $\overleftrightarrow{AP}$?

Sketch of the Answer: The rotation by $180°$ maps $\overleftrightarrow{AP}$ to itself because $\overleftrightarrow{AP}$ contains the center C. In particular, it can be shown that the rotation maps one half-plane of $\overleftrightarrow{AP}$ to the other half-plane and vice versa. Since Q and B are in opposite half-planes (by definition of alternate interior angles), and B and $R_{C,180}(B)$ are in opposite half-planes, Q and $R_{C,180}(B)$ must be in the same half-plane.

d. What is the image of $R_{C,180}(\angle BAP)$? Is it $\angle QPA$? Explain why.

Because under the rotation, the vertex of $\angle BAP$ maps to the vertex of $\angle QPA$, ray AP maps to ray PA, the point B goes to the same side as Q. Since $m\angle BAP = m\angle QPA$ (by assumption), ray AB must map to ray PQ. Thus, $\angle BAP$ maps to $\angle QPA$.

e. Why is $R_{C,180}(l) = l'$?

Since $\overrightarrow{AB}$ maps to $\overrightarrow{PQ}$, $R_{C,180}(l) = R_{C,180}(\overleftrightarrow{AB}) = \overleftrightarrow{PQ} = l'$.

We have just proven that a rotation by $180°$ takes l to l'. By the claim in Example 3, lines l and l' must be parallel, which is summarized below.

THEOREM: Suppose a transversal intersects a pair of lines. If a pair of alternate interior angles are equal in measure, then the pair of lines are parallel.

Discussion

In Example 5, suppose we had used a different rotation to construct a line parallel to l that contains P. Such constructions are certainly plentiful. For example, for every other point D on l, we can find the midpoint of segment PD and use the construction in Example 2 to construct a different $180°$ rotation around a different center such that the image of the line l is a parallel line through the point P. Are any of these parallel lines through P different? In other words,

Can we draw a line other than the line l' through P that never meets l?

The answer may surprise you; it stumped mathematicians and physicists for centuries. In nature, the answer is that it is sometimes possible and sometimes not. This is because there are places in the universe (near massive stars, for example) where the model geometry of space is not plane-like or flat but is actually quite curved. To rule out these other types of strange but beautiful geometries, we must assume that the answer to the previous question is only one line. That choice becomes one of our geometry assumptions:

(Parallel Postulate) Through a given external point there is at most one line parallel to a given line.

In other words, we assume that for any point P in the plane not lying on a line ℓ, every line in the plane that contains P intersects ℓ except at most one line—the one we call *parallel* to ℓ.

Example 6

We can use the parallel postulate to prove the second part of the basic fact.

THEOREM: Suppose a transversal intersects a pair of lines. If the pair of lines are parallel, then the pair of alternate interior angles are equal in measure.

PROOF: Suppose that a transversal $\overleftrightarrow{AP}$ intersects line l at A and l' at P, pick and label another point B on l, and choose a point Q on l' on the opposite side of $\overleftrightarrow{AP}$ as B. The picture might look like the figure below:

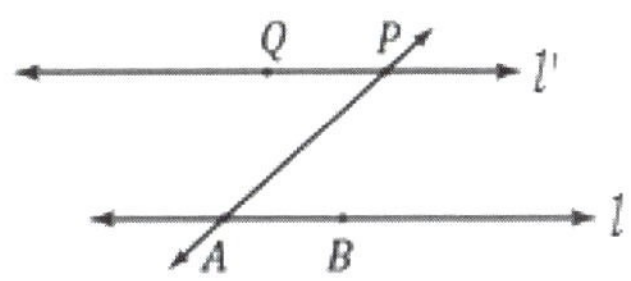

Let C be the midpoint of $\overline{AP}$, and apply a rotation by $180°$ around the center C. As in previous discussions, the image of l is the line $R_{C,180}(l)$, which is parallel to l and contains point P. Since l' and $R_{C,180}(l)$ are both parallel to l and contain P, by the parallel postulate, they must be the same line: $R_{C,180}(l) = l'$. In particular, $R_{C,180}(\angle BAP) = \angle QPA$. Since rotations preserve angle measures, $m\angle BAP = m\angle QPA$, which was what we needed to show.

Discussion

It is important to point out that, although we only proved the alternate interior angles theorem, the same sort of proofs can be done in the exact same way to prove the corresponding angles theorem and the interior angles theorem. Thus, all of the proofs we have done so far (in class and in the Problem Sets) that use these facts are really based, in part, on our assumptions about rigid motions.

Example 7

We end this lesson with a theorem that we just state but can be easily proved using the parallel postulate.

THEOREM: If three distinct lines l_1, l_2, and l_3 in the plane have the property that $l_1 \parallel l_2$ and $l_2 \parallel l_3$, then $l_1 \parallel l_3$. (In proofs, this can be written as, "If two lines are parallel to the same line, then they are parallel to each other.")

Note that students should at least remember that, in Euclidean Geometry, two lines are parallel if and only if alternate interior angles of any transversal are equal in measure and should be able to elaborate on what that means. This one statement includes both the parallel postulate and its converse. We can construct parallel lines without the parallel postulate, but in a geometry that does not satisfy the parallel postulate; there are many parallels to a given line through a point not on it. Without the parallel postulate, parallel lines are plentiful, and we cannot tell much about a line if all we know is that it passes through a point and is parallel to another line.

Relevant Vocabulary

PARALLEL: Two lines are *parallel* if they lie in the same plane and do not intersect. Two segments or rays are parallel if the lines containing them are parallel lines.

TRANSVERSAL: Given a pair of lines l and m in a plane, a third line t is a *transversal* if it intersects l at a single point and intersects m at a single but different point.

The definition of transversal rules out the possibility that any two of the lines l, m, and t are the same line.

ALTERNATE INTERIOR ANGLES: Let line t be a transversal to lines l and m such that t intersects l at point P and intersects m at point Q. Let R be a point on l and S be a point on m such that the points R and S lie in opposite half planes of t. Then the $\angle RPQ$ and the $\angle PQS$ are called *alternate interior angles* of the transversal t with respect to m and l.

CORRESPONDING ANGLES: Let line t be a transversal to lines l and m. If $\angle x$ and $\angle y$ are alternate interior angles, and $\angle y$ and $\angle z$ are vertical angles, then $\angle x$ and $\angle z$ are *corresponding angles*.

Closing (1 minute)

- Two lines are parallel if and only if alternate interior angles of any transversal are equal in measure. This can be separated into two statements:
  1. Suppose a transversal intersects a pair of lines. If two alternate interior angles are equal in measure, then the pair of lines are parallel.
  2. Suppose a transversal intersects a pair of lines. If the lines are parallel, then the pair of alternate interior angles are equal in measure.
- Statement 2 is an equivalent form of the parallel postulate: Through a given external point there is at most one line parallel to a given line.

Exit Ticket (6 minutes)

Name ______________________________ Date ____________

Lesson 18: Looking More Carefully at Parallel Lines

Exit Ticket

1. Construct a line through the point P below that is parallel to the line l by rotating l by $180°$ (using the steps outlined in Example 2).

2. Why is the parallel line you constructed the only line that contains P and is parallel to l?

Exit Ticket Sample Solutions

1. Construct a line through the point P below that is parallel to the line l by rotating l by $180°$ (using the steps outlined in Example 2).

   The construction should look like the steps in Example 2.

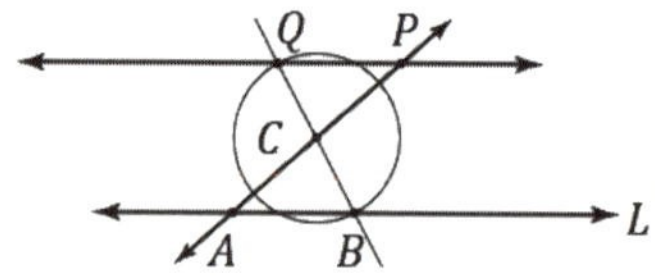

2. Why is the parallel line you constructed the only line that contains P and is parallel to l?

   The answer should reference the parallel postulate in a meaningful way.

Problem Set Sample Solutions

Notice that we are frequently asked two types of questions about parallel lines. If we are told that two lines are parallel, then we may be required to use this information to prove the congruence of two angles (corresponding, alternate interior, etc.). On the other hand, if we are given the fact that two angles are congruent (or perhaps supplementary), we may have to prove that two lines are parallel.

1. In the figure, $\overline{AL} \parallel \overline{BM}$, $\overline{AL} \perp \overline{CF}$, and $\overline{GK} \perp \overline{BM}$. Prove that $\overline{CF} \parallel \overline{GK}$.

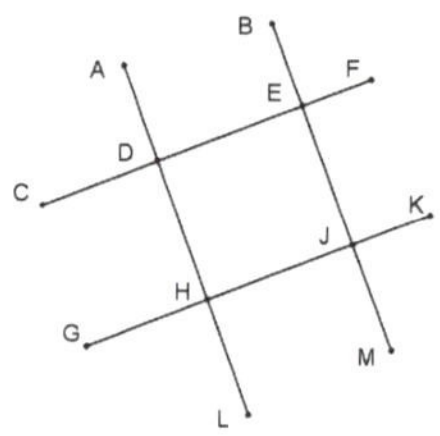

$\overline{AL} \parallel \overline{BM}$	*Given*
$\overline{AL} \perp \overline{CF}$	*Given*
$\overline{GK} \perp \overline{BM}$	*Given*
$m\angle ADC = 90°$	*Definition of perpendicular lines*
$m\angle BJH = 90°$	*Definition of perpendicular lines*
$m\angle BJH = m\angle AHG$	*If two parallel lines are cut by a transversal, then the corresponding angles are equal in measure.*
$m\angle ADC = m\angle AHG$	*Substitution property of equality*
$\overline{CF} \parallel \overline{GK}$	*If two lines are cut by a transversal such that the corresponding angles are equal in measure, then the lines are parallel.*

2. Given that $\angle B$ and $\angle C$ are supplementary and $m\angle A = m\angle C$, prove that $\overline{AD} \parallel \overline{BC}$.

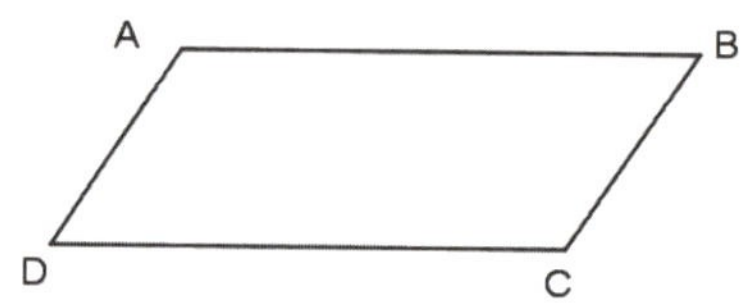

$\angle B$ and $\angle C$ are supplementary.	*Given*
$m\angle A = m\angle C$	*Given*
$m\angle B + m\angle C = 180$	*Definition of supplementary angles*
$m\angle B + m\angle A = 180$	*Substitution property of equality*
$\overline{AD} \parallel \overline{BC}$	*If a transversal intersects two lines such that the same side interior angles are supplementary, then the lines are parallel.*

3. Mathematicians state that *if a transversal to two parallel lines is perpendicular to one of the lines, then it is perpendicular to the other.* Prove this statement. (Include a labeled drawing with your proof.)

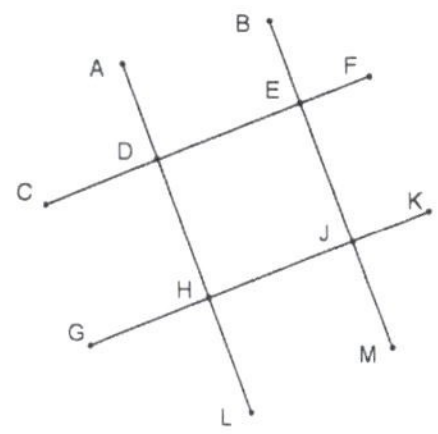

$\overline{AL} \parallel \overline{BM}$	*Given*
$\overline{AL} \perp \overline{CF}$	*Given*
$m\angle ADC = 90°$	*Definition of perpendicular lines*
$m\angle ADC = m\angle BED$	*If two parallel lines are cut by a transversal, then corresponding angles are equal in measure.*
$m\angle BED = 90°$	*Substitution*
$\overline{CF} \perp \overline{BM}$	*Definition of perpendicular lines*

4. In the figure, $\overline{AB} \parallel \overline{CD}$ and $\overline{EF} \parallel \overline{GH}$. Prove that $m\angle AFE = m\angle DKH$.

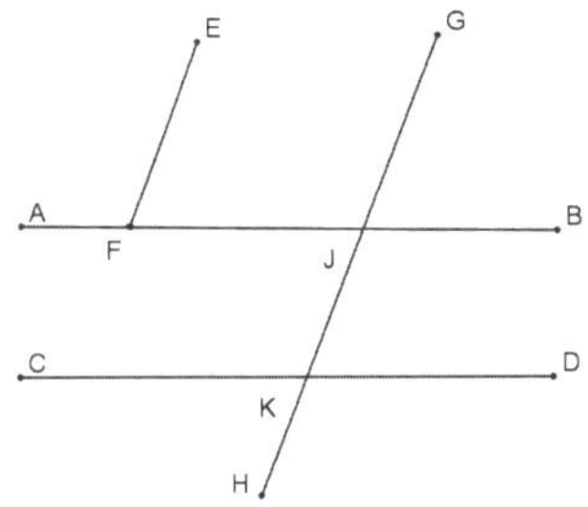

$\overline{AB} \parallel \overline{CD}$	*Given*
$\overline{EF} \parallel \overline{GH}$	*Given*
$m\angle AFE = m\angle GJF$	*If two parallel lines are cut by a transversal, then the corresponding angles are equal in measure.*
$m\angle JKC = m\angle GJF$	*If two parallel lines are cut by a transversal, then the corresponding angles are equal in measure.*
$m\angle JKC = m\angle DKH$	*Vertical angles are equal in measure.*
$m\angle AFE = m\angle DKH$	*Transitive property*

5. In the figure, $\angle E$ and $\angle AFE$ are complementary, and $\angle C$ and $\angle BDC$ are complementary. Prove that $\overline{AE} \parallel \overline{CB}$.

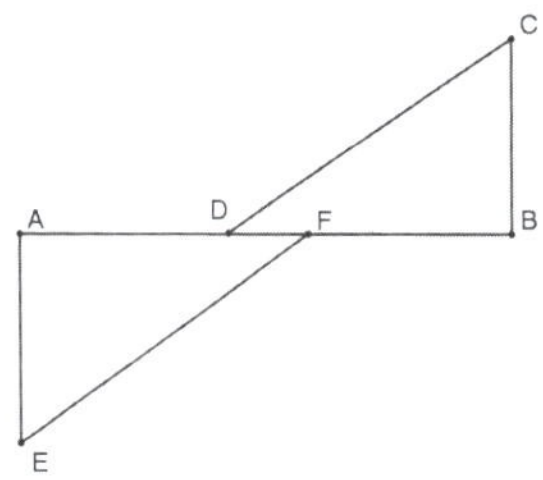

$\angle E$ *and* $\angle AFE$ *are complementary.*	*Given*
$\angle C$ *and* $\angle BDC$ *are complementary.*	*Given*
$m\angle E + m\angle AFE = 90°$	*Definition of complementary angles*
$m\angle C + m\angle BDC = 90°$	*Definition of complementary angles*
$m\angle E + m\angle AFE + m\angle A = 180°$	*Sum of the angle measures in a triangle is* $180°$.
$m\angle C + m\angle BDC + m\angle B = 180°$	*Sum of the angle measures in a triangle is* $180°$.
$m\angle A = 90°$; $m\angle = 90°$	*Subtraction property of equality*
$m\angle A = m\angle B$	*Substitution property of equality*
$\overline{AE} \parallel \overline{CB}$	*If two lines are cut by a transversal such that a pair of alternate interior angles are equal in measure, then the lines are parallel.*

6. Given a line l and a point P not on the line, the following directions can be used to draw a line m perpendicular to the line l through the point P based upon a rotation by $180°$:

   a. Pick and label a point A on the line l so that the circle (center P, radius AP) intersects l twice.

   b. Use a protractor to draw a perpendicular line n through the point A (by constructing a $90°$ angle).

   c. Use the directions in Example 2 to construct a parallel line m through the point P.

   Do the construction. Why is the line m perpendicular to the line l in the figure you drew? Why is the line m the only perpendicular line to l through P?

   Figures will vary but should follow the directions outlined.
   Students can use Problem 3 or reprove the fact that if a transversal to two parallel lines is perpendicular to one, then it is perpendicular to the other.
   Suppose line m' *contains* P *and is perpendicular to* l*; then by the interior angles theorem (or alternate interior angles or corresponding interior angles),* m' *is parallel to* n*. Since* m *and* m' *are both parallel to* n *and contain* P*, by the parallel postulate, they must be the same line (i.e.,* $m' = m$*).*
   A student might answer, "parallel postulate." If so, partial credit might be awarded (depending on your tastes for partial credit).

Problems 7–10 all refer to the figure to the right. The exercises are otherwise unrelated to each other.

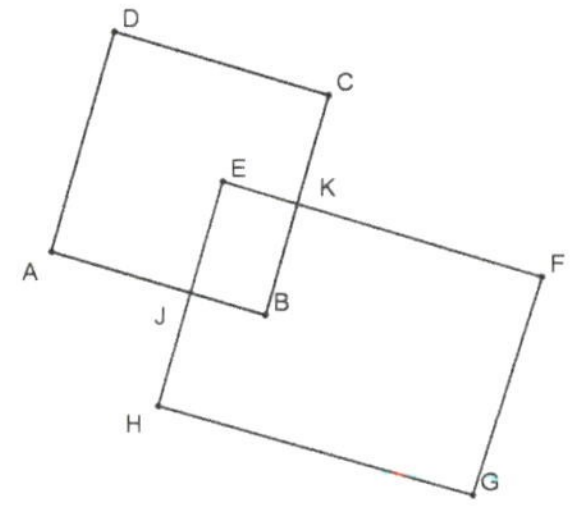

7. $\overline{AD} \parallel \overline{BC}$ and $\angle EJB$ is supplementary to $\angle JBK$. Prove that $\overline{AD} \parallel \overline{JE}$.

$\overline{AD} \parallel \overline{BC}$	*Given*
$\angle EJB$ *is supplementary to* $\angle JBK$.	*Given*
$\overline{EJ} \parallel \overline{BC}$	*If a transversal intersects two lines such that the same side interior angles are supplementary, then the lines are parallel.*
$\overline{AD} \parallel \overline{EJ}$	*If two segments are parallel to the same segment, then they are parallel to each other.*

8. $\overline{AD} \parallel \overline{FG}$ and $\overline{EJ} \parallel \overline{FG}$. Prove that $\angle DAJ$ and $\angle EJA$ are supplementary.

$\overline{AD} \parallel \overline{FG}$	*Given*
$\overline{EJ} \parallel \overline{FG}$	*Given*
$\overline{AD} \parallel \overline{EJ}$	*If two segments are parallel to the same segment, then they are parallel to each other.*
$m\angle DAJ + m\angle EJA = 180°$	*If a transversal intersects two parallel lines, then the interior angles on the same side of the transversal are supplementary.*

9. $m\angle C = m\angle G$ and $\angle B$ is supplementary to $\angle G$. Prove that $\overline{DC} \parallel \overline{AB}$.

$m\angle C = m\angle G$	*Given*
$\angle B$ *is supplementary to* $\angle G$	*Given*
$m\angle B + m\angle G = 180°$	*Definition of supplementary angles*
$m\angle B + m\angle C = 180°$	*Substitution property of equality*
$\overline{DC} \parallel \overline{AB}$	*If a transversal intersects two lines such that the same side interior angles are supplementary, then the lines are parallel.*

10. $\overline{AB} \parallel \overline{EF}$, $\overline{EF} \perp \overline{CB}$, and $\angle EKC$ is supplementary to $\angle KCD$. Prove that $\overline{AB} \parallel \overline{DC}$.

$\overline{AB} \parallel \overline{EF}$	*Given*
$\overline{EF} \perp \overline{CB}$	*Given*
$\angle EKC$ *is supplementary to* $\angle KCD$.	*Given*
$m\angle ABC + m\angle BKE = 180°$	*If parallel lines are cut by a transversal, then interior angles on the same side are supplementary.*
$m\angle EKC + m\angle KCD = 180°$	*Definition of supplementary angles*
$m\angle BKE = 90°$ *and* $m\angle EKC = 90°$	*Definition of right angles*
$m\angle ABC = 90°$ *and* $m\angle KCE = 90°$	*Subtraction property of equality*
$\angle ABC$ *and* $\angle KCE$ *are supplementary.*	*Definition of supplementary angles*
$\overline{AB} \parallel \overline{DC}$	*If two lines are cut by a transversal such that a pair of interior angles on the same side are supplementary, then the lines are parallel.*

Lesson 19: Construct and Apply a Sequence of Rigid Motions

Student Outcomes

- Students begin developing the capacity to speak and write articulately using the concept of congruence. This involves being able to repeat the definition of congruence and use it in an accurate and effective way.

Classwork

Opening (20 minutes)

> **Opening**
>
> We have been using the idea of congruence already (but in a casual and unsystematic way). In Grade 8, we introduced and experimented with concepts around congruence through *physical models, transparencies, or geometry software.* Specifically, we had to
>
> > *(1) Understand that a two-dimensional figure is congruent to another if the second can be obtained from the first by a sequence of rotations, reflections, and translations; and (2) describe a sequence that exhibits the congruence between two congruent figures.* (8.G.A.2)
>
> As with so many other concepts in high school Geometry, congruence is familiar, but we now study it with greater precision and focus on the language with which we discuss it.
>
> Let us recall some facts related to congruence that appeared previously in this unit.
>
> 1. We observed that rotations, translations, and reflections—and thus all rigid motions—preserve the lengths of segments and the measures of angles. We think of two segments (respectively, angles) as the *same* in an important respect if they have the same length (respectively, degree measure), and thus, sameness of these objects relating to measure is well characterized by the existence of a rigid motion mapping one thing to another. Defining *congruence* by means of rigid motions extends this notion of sameness to arbitrary figures, while clarifying the meaning in an articulate way.
> 2. We noted that a symmetry is a rigid motion that carries a figure to itself.
>
> So how do these facts about rigid motions and symmetry relate to congruence? We define two figures in the plane as congruent if there exists a finite composition of basic rigid motions that maps one figure onto the other.
>
> It might seem easy to equate two figures being congruent to having *same size and same shape*. The phrase *same size and same shape* has intuitive meaning and helps to paint a mental picture, but it is not a definition. As in a court of law, to establish guilt it is not enough to point out that the defendant looks like a sneaky, unsavory type. We need to point to exact pieces of evidence concerning the specific charges. It is also not enough that the defendant did something bad. It must be a violation of a specific law. Same size and same shape is on the level of, "He looks like a sneaky, bad guy who deserves to be in jail."
>
> It is also not enough to say that they are alike in all respects except position in the plane. We are saying that there is some particular rigid motion that carries one to another. Almost always, when we use congruence in an explanation or proof, we need to refer to the rigid motion. To show that two figures are congruent, we only need to show that there is a transformation that maps one directly onto the other. However, once we know that there is a transformation, then we know that there are actually many such transformations, and it can be useful to consider more than one. We see this when discussing the symmetries of a figure. A symmetry is nothing other than a congruence of an object with itself.

A figure may have many different rigid motions that map it onto itself. For example, there are six different rigid motions that take one equilateral triangle with side length 1 to another such triangle. Whenever this occurs, it is because of a symmetry in the objects being compared.

Lastly, we discuss the relationship between *congruence* and *correspondence*. A correspondence between two figures is a function from the parts of one figure to the parts of the other, with no requirements concerning same measure or existence of rigid motions. If we have rigid motion T that takes one figure to another, then we have a correspondence between the parts. For example, if the first figure contains segment AB, then the second includes a corresponding segment $T(A)T(B)$. But we do not need to have a congruence to have a correspondence. We might list the parts of one figure and pair them with the parts of another. With two triangles, we might match vertex to vertex. Then the sides and angles in the first have corresponding parts in the second. But being able to set up a correspondence like this does not mean that there is a rigid motion that produces it. The sides of the first might be paired with sides of different length in the second. Correspondence in this sense is important in triangle similarity.

Discussion/Examples (18 minutes)

Discussion

We now examine a figure being mapped onto another through a composition of rigid motions.

To map $\triangle PQR$ to $\triangle XYZ$ here, we first rotate $\triangle PQR$ $120°$ ($R_{D,120°}$) around the point, D. Then reflect the image ($r_{\overline{EF}}$) across $\overline{EF}$. Finally, translate the second image ($T_{\vec{v}}$) along the given vector to obtain $\triangle XYZ$. Since each transformation is a rigid motion, $\triangle PQR \cong \triangle XYZ$. We use function notation to describe the composition of the rotation, reflection, and translation:

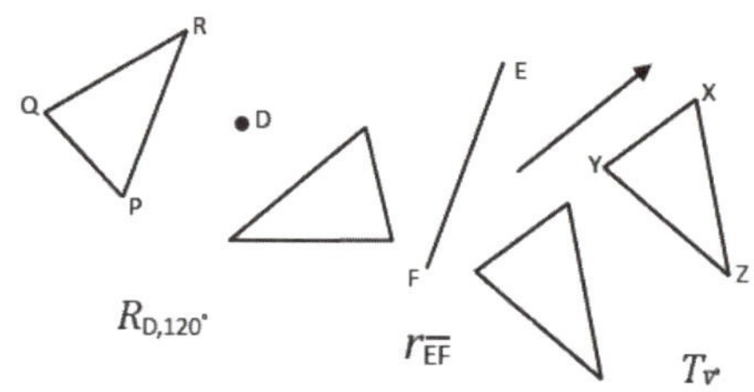

$$T_{\vec{v}}\left(r_{\overline{EF}}\left(RD, 120°(\triangle PQR)\right)\right) = \triangle XYZ.$$

Notice that (as with all composite functions) the innermost function/transformation (the rotation) is performed first, and the outermost (the translation) last.

Example 1

i. Draw and label a $\triangle PQR$ in the space below.

ii. Use your construction tools to apply one of each of the rigid motions we have studied to it in a sequence of your choice.

iii. Use function notation to describe your chosen composition here. Label the resulting image as $\triangle XYZ$:

____

iv. Complete the following sentences: (Some blanks are single words; others are phrases.)

$\triangle PQR$ is ___*congruent*___ to $\triangle XYZ$ because ___*rigid motions*___ map point P to point X, point Q to point Y, and point R to point Z. Rigid motions map segments onto ___*segments of equal length*___ and angles onto ___*angles of equal measure*___.

Example 2

On a separate piece of paper, trace the series of figures in your composition but do *NOT* include the center of rotation, the line of reflection, or the vector of the applied translation.

Swap papers with a partner, and determine the composition of transformations your partner used. Use function notation to show the composition of transformations that renders $\triangle PQR \cong \triangle XYZ$.

Closing (1 minute)

- Two figures in the plane are said to be congruent if there exists a finite composition of basic rigid motions that maps one figure onto the other.
- To notate composite functions, the innermost function/transformation is performed first and the outermost last.

Exit Ticket (6 minutes)

Name ______________________________ Date__________________

Lesson 19: Construct and Apply a Sequence of Rigid Motions

Exit Ticket

Assume that the following figures are drawn to scale. Use your understanding of congruence to explain why square $ABCD$ and rhombus $GHIJ$ are not congruent.

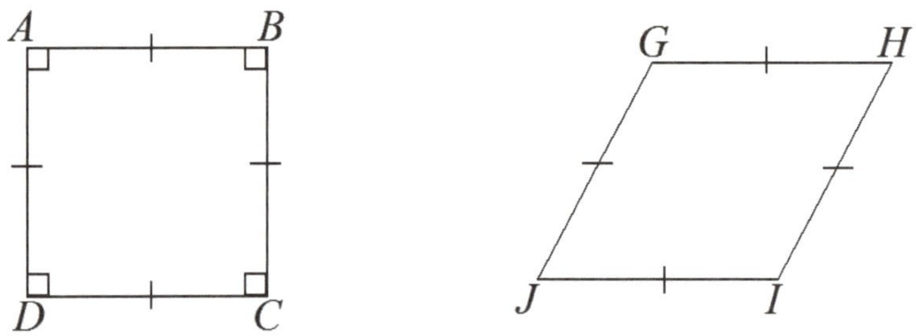

Exit Ticket Sample Solutions

Assume that the following figures are drawn to scale. Use your understanding of congruence to explain why square $ABCD$ and rhombus $GHIJ$ are not congruent.

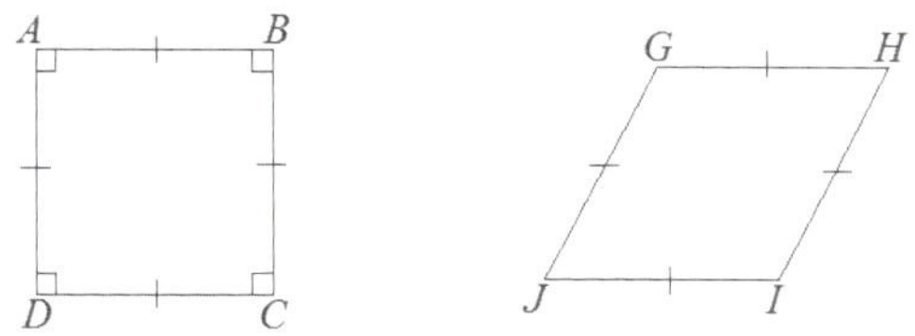

Rigid motions map angles onto angles of equal measure, and the measures of the angles of square $ABCD$ are all $90°$, whereas the angles of rhombus $GHIJ$ are not. Therefore, there is no rigid motion that maps square $ABCD$ onto rhombus $GHIJ$.

Problem Set Sample Solutions

1. Use your understanding of congruence to explain why a triangle cannot be congruent to a quadrilateral.

   a. Why can't a triangle be congruent to a quadrilateral?

   A triangle cannot be congruent to a quadrilateral because there is no rigid motion that takes a figure with three vertices to a figure with four vertices.

   b. Why can't an isosceles triangle be congruent to a triangle that is not isosceles?

   An isosceles triangle cannot be congruent to a triangle that is not isosceles because rigid motions map segments onto segments of equal length, and the lengths of an isosceles triangle differ from those of a triangle that is not isosceles.

2. Use the figures below to answer each question:

   a. $\triangle ABD \cong \triangle CDB$. What rigid motion(s) maps $\overline{CD}$ onto $\overline{AB}$? Find two possible solutions.

   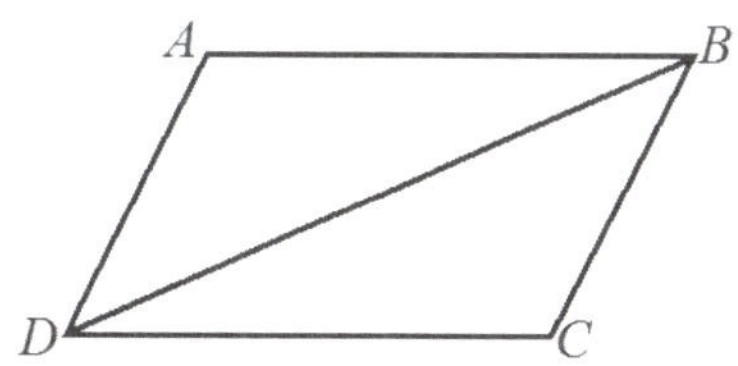

   A $180°$ rotation about the midpoint of $\overline{DB}$

   A reflection over the line that joins the midpoints of $\overline{AD}$ and $\overline{BC}$, followed by another reflection over the line that joins the midpoints of $\overline{AB}$ and $\overline{DC}$

   b. All of the smaller triangles are congruent to each other. What rigid motion(s) map $\overline{ZB}$ onto $\overline{AZ}$? Find two possible solutions.

   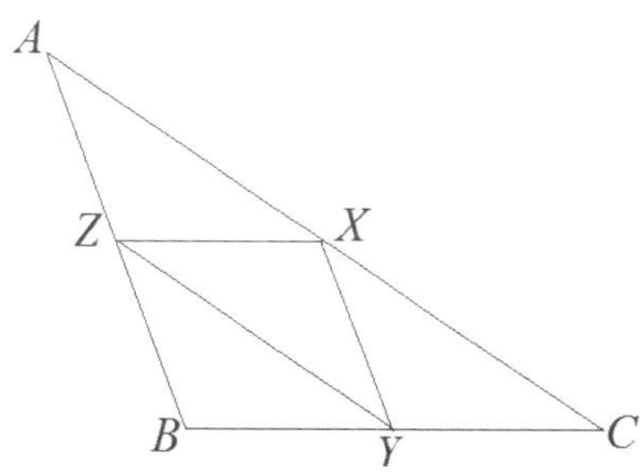

   A translation $T_{\vec{ZA}}$

   A $180°$ rotation about the midpoint of $\overline{ZY}$, followed by a $180°$ rotation about the midpoint of $\overline{ZX}$

Lesson 20: Applications of Congruence in Terms of Rigid Motions

Student Outcomes

- Students understand that a congruence between figures gives rise to a correspondence between parts such that corresponding parts are congruent, and they are able to state the correspondence that arises from a given congruence.
- Students recognize that correspondences may be set up even in cases where no congruence is present. They know how to describe and notate all the possible correspondences between two triangles or two quadrilaterals, and they know how to state a correspondence between two polygons.

Classwork

Opening (18 minutes)

Opening

Every congruence gives rise to a correspondence.

Under our definition of congruence, when we say that one figure is congruent to another, we mean that there is a rigid motion that maps the first onto the second. That rigid motion is called a congruence.

Recall the Grade 7 definition: *A correspondence between two triangles is a pairing of each vertex of one triangle with one and only one vertex of the other triangle.* When reasoning about figures, it is useful to be able to refer to corresponding parts (e.g., sides and angles) of the two figures. We look at one part of the first figure and compare it to the corresponding part of the other. Where does a correspondence come from? We might be told by someone how to make the vertices correspond. Conversely, we might make our own correspondence by matching the parts of one triangle with the parts of another triangle based on appearance. Finally, if we have a congruence between two figures, the congruence gives rise to a correspondence.

A rigid motion F always produces a one-to-one correspondence between the points in a figure (the *pre-image*) and points in its image. If P is a point in the figure, then the corresponding point in the image is $F(P)$. A rigid motion also maps each part of the figure to a corresponding part of the image. As a result, *corresponding parts of congruent figures are congruent* since the very same rigid motion that makes a congruence between the figures also makes a congruence between each part of the figure and the corresponding part of the image.

In proofs, we frequently refer to the fact that corresponding angles, sides, or parts of congruent triangles are congruent. This is simply a repetition of the definition of congruence. If $\triangle ABC$ is congruent to $\triangle DEG$ because there is a rigid motion F such that $F(A) = D$, $F(B) = E$, and $F(C) = G$, then $\overline{AB}$ is congruent to $\overline{DE}$, $\triangle ABC$ is congruent to $\triangle DEG$, and so forth because the rigid motion F takes $\overline{AB}$ to $\overline{DE}$ and $\angle BAC$ to $\angle EDF$.

There are correspondences that do not come from congruences.

The sides (and angles) of two figures might be compared even when the figures are not congruent. For example, a carpenter might want to know if two windows in an old house are the same, so the screen for one could be interchanged with the screen for the other. He might list the parts of the first window and the analogous parts of the second, thus making a correspondence between the parts of the two windows. Checking part by part, he might find that the angles in the frame of one window are slightly different from the angles in the frame of the other, possibly because the house has tilted slightly as it aged. He has used a correspondence to help describe the differences between the windows not to describe a congruence.

In general, given any two triangles, one could make a table with two columns and three rows and then list the vertices of the first triangle in the first column and the vertices of the second triangle in the second column in a random way. This would create a correspondence between the triangles, though generally not a very useful one. No one would expect a random correspondence to be very useful, but it is a correspondence nonetheless.

Later, when we study similarity, we find that it is very useful to be able to set up correspondences between triangles despite the fact that the triangles are not congruent. Correspondences help us to keep track of which part of one figure we are comparing to that of another. It makes the rules for associating part to part explicit and systematic so that other people can plainly see what parts go together.

Discussion (10 minutes)

Discussion

Let's review function notation for rigid motions.

a. To name a translation, we use the symbol $T_{\overrightarrow{AB}}$. We use the letter T to signify that we are referring to a translation and the letters A and B to indicate the translation that moves each point in the direction from A to B along a line parallel to line AB by distance AB. The image of a point P is denoted $T_{\overrightarrow{AB}}(P)$. Specifically, $T_{\overrightarrow{AB}}(A) = B$.

b. To name a reflection, we use the symbol r_l, where l is the line of reflection. The image of a point P is denoted $r_l(P)$. In particular, if A is a point on l, $r_l(A) = A$. For any point P, line l is the perpendicular bisector of segment $Pr_l(P)$.

c. To name a rotation, we use the symbol R_{C,x° to remind us of the word *rotation*. C is the center point of the rotation, and x represents the degree of the rotation counterclockwise around the center point. Note that a positive degree measure refers to a counterclockwise rotation, while a negative degree measure refers to a clockwise rotation.

Examples 1–3 (10 minutes)

Example 1

In each figure below, the triangle on the left has been mapped to the one on the right by a 240° rotation about P. Identify all six pairs of corresponding parts (vertices and sides).

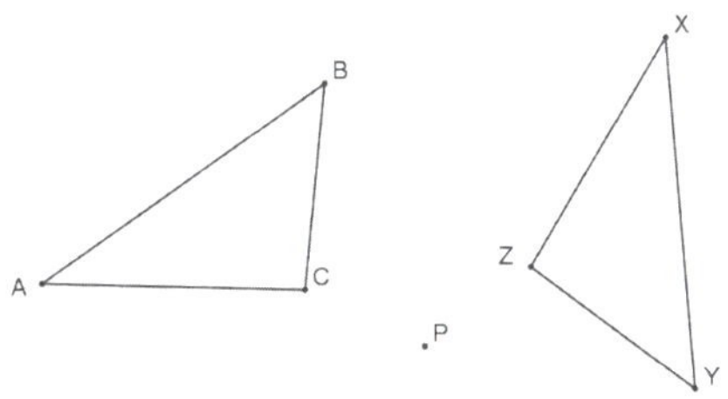

Corresponding Vertices	Corresponding Sides
$A \rightarrow X$	$\overline{AB} \rightarrow \overline{XY}$
$B \rightarrow Y$	$\overline{AC} \rightarrow \overline{XZ}$
$C \rightarrow Z$	$\overline{BC} \rightarrow \overline{YZ}$

What rigid motion mapped $\triangle ABC$ onto $\triangle XYZ$? Write the transformation in function notation.

$R_{P,240^\circ}(\triangle ABC) \rightarrow \triangle XYZ$

Example 2

Given a triangle with vertices A, B, and C, list all the possible correspondences of the triangle with itself.

1. $A \to A$ $B \to B$ $C \to C$
2. $A \to A$ $B \to C$ $C \to B$
3. $A \to B$ $B \to A$ $C \to C$
4. $A \to B$ $B \to C$ $C \to A$
5. $A \to C$ $B \to A$ $C \to B$
6. $A \to C$ $B \to B$ $C \to A$

Example 3

Give an example of two quadrilaterals and a correspondence between their vertices such that (a) corresponding sides are congruent, but (b) corresponding angles are not congruent.

$A \to J$

$B \to K$

$C \to L$

$D \to M$

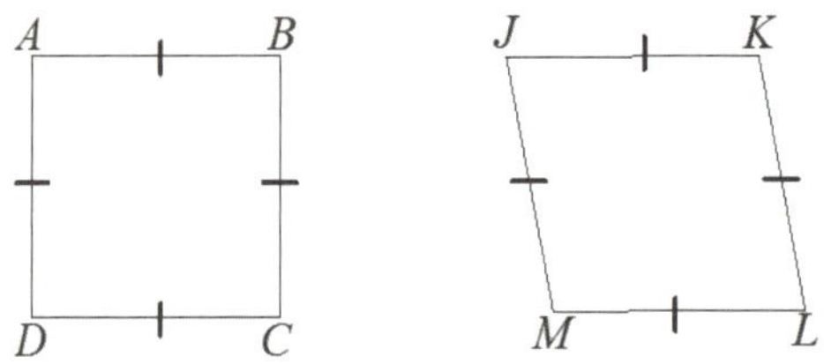

Closing (2 minute)

- A natural correspondence exists between congruent figures, but figures need not be congruent for there to be a correspondence. A correspondence can be assigned.
- Correspondences allow for a systematic way of keeping track of which part of one figure we are comparing to that of another.

Exit Ticket (5 minutes)

Name ______________________________ Date ______________

Lesson 20: Applications of Congruence in Terms of Rigid Motions

Exit Ticket

1. What is a correspondence? Why does a congruence naturally yield a correspondence?

2. Each side of $\triangle XYZ$ is twice the length of each side of $\triangle ABC$. Fill in the blanks below so that each relationship between lengths of sides is true.

_________ $\times 2 =$ _________

_________ $\times 2 =$ _________

_________ $\times 2 =$ _________

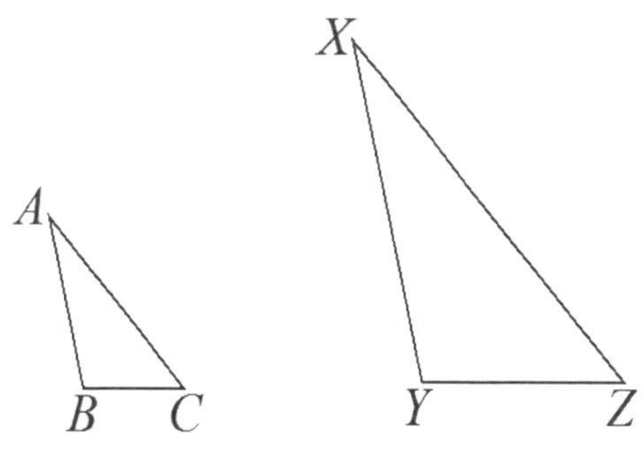

Exit Ticket Sample Solutions

1. What is a correspondence? Why does a congruence naturally yield a correspondence?

   A correspondence between two triangles is a pairing of each vertex of one triangle with one and only one vertex of the other triangle. This pairing can be expanded to figures other than triangles. A congruence naturally yields a correspondence since a rigid motion maps each part of a figure to what we call a corresponding part of the image.

2. Each side of $\triangle XYZ$ is twice the length of each side of $\triangle ABC$. Fill in the blanks below so that each relationship between lengths of sides is true.

   $\underline{AB} \times 2 = \underline{XY}$

   $\underline{BC} \times 2 = \underline{YZ}$

   $\underline{CA} \times 2 = \underline{ZX}$

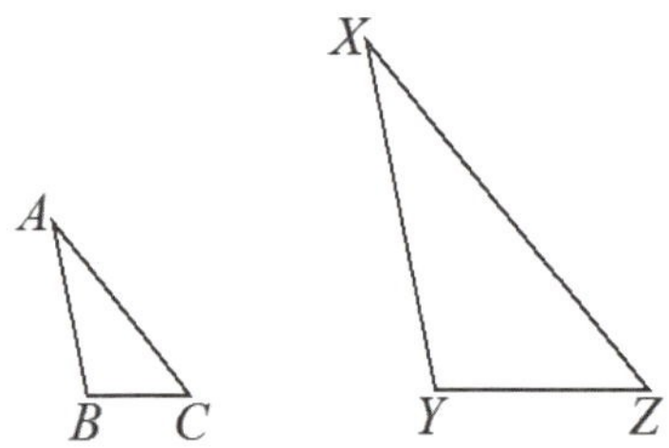

Problem Set Sample Solutions

1. Given two triangles, one with vertices A, B, and C, and the other with vertices X, Y, and Z, there are six different correspondences of the first with the second.

   a. One such correspondence is the following:

   $$A \to Z$$
   $$B \to X$$
   $$C \to Y$$

   Write the other five correspondences.

$A \to X$	$A \to Y$	$A \to X$	$A \to Y$	$A \to Z$
$B \to Z$	$B \to Z$	$B \to Y$	$B \to X$	$B \to Y$
$C \to Y$	$C \to X$	$C \to Z$	$C \to Z$	$C \to X$

   b. If all six of these correspondences come from congruences, then what can you say about $\triangle ABC$?

   It must be equilateral.

   c. If two of the correspondences come from congruences, but the others do not, then what can you say about $\triangle ABC$?

   It must be isosceles and cannot be equilateral.

   d. Why can there be no two triangles where three of the correspondences come from congruences, but the others do not?

   By part (c), if two correspondences come from congruences, then the triangle must be isosceles. A third correspondence implies that the triangles must be equilateral. But then all six correspondences would be congruences, contradicting that the others are not.

EUREKA MATH

2. Give an example of two triangles and a correspondence between them such that (a) all three corresponding angles are congruent, but (b) corresponding sides are not congruent.

$A \to E$

$B \to D$

$C \to F$

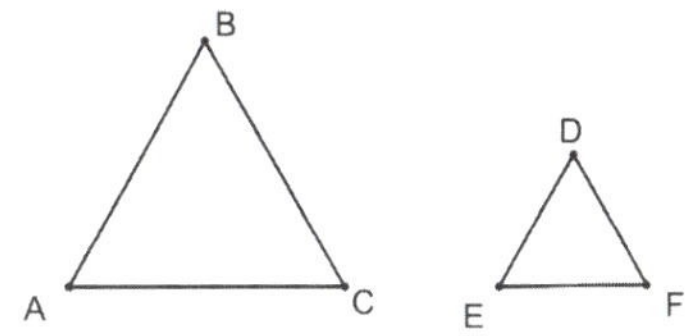

3. Give an example of two triangles and a correspondence between their vertices such that (a) one angle in the first is congruent to the corresponding angle in the second, and (b) two sides of the first are congruent to the corresponding sides of the second, but (c) the triangles themselves are not congruent.

$A \to X$

$B \to Y$

$C \to Z$

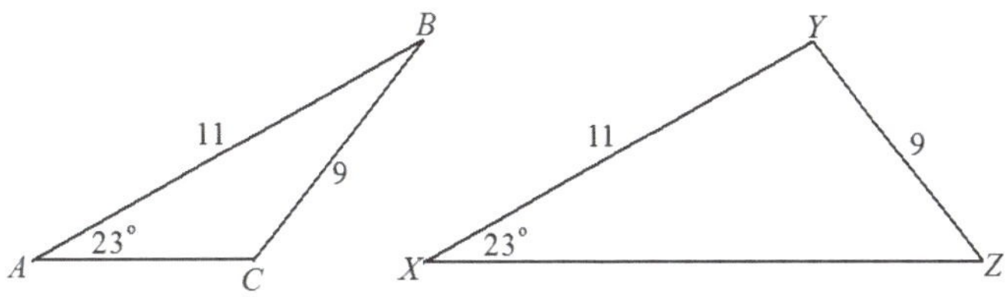

4. Give an example of two quadrilaterals and a correspondence between their vertices such that (a) all four corresponding angles are congruent, and (b) two sides of the first are congruent to two sides of the second, but (c) the two quadrilaterals are not congruent.

$A \to J$

$B \to K$

$C \to L$

$D \to M$

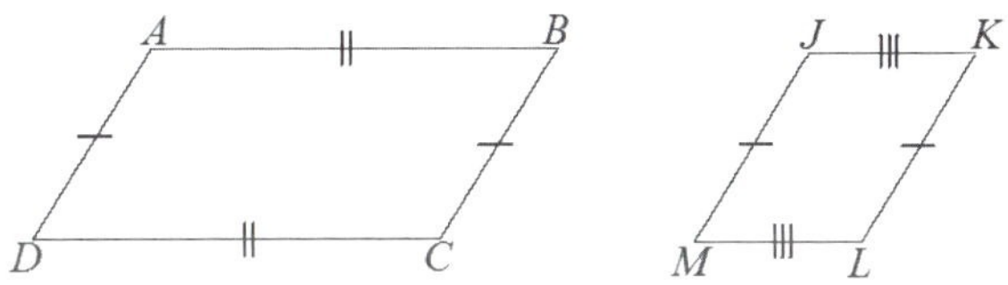

5. A particular rigid motion, M, takes point P as input and gives point P' as output. That is, $M(P) = P'$. The same rigid motion maps point Q to point Q'. Since rigid motions preserve distance, is it reasonable to state that $PP' = QQ'$? Does it matter which type of rigid motion M is? Justify your response for each of the three types of rigid motion. Be specific. If it is indeed the case, for some class of transformations, that $PP' = QQ'$ is true for all P and Q, explain why. If not, offer a counterexample.

This is not always true. A rotation around a vertex does not move each point the same distance. In a rigid motion, the distance that is preserved is within the figure distance. A reflection also does not satisfy $PP' = QQ'$ for all P and Q. Reflecting a figure over one of its sides does not move the points on the line of reflection, and other points are moved by a distance proportionate to their distance from the reflection line. A translation, however, does satisfy the condition that $PP' = QQ'$ for all P and Q.

Lesson 21: Correspondence and Transformations

Student Outcomes

- Students practice applying a sequence of rigid motions from one figure onto another figure in order to demonstrate that the figures are congruent.

Lesson Notes

In Lesson 21, students consolidate their understanding of congruence in terms of rigid motions with knowledge of corresponding vertices and sides of triangles. They identify specific sides and angles in the pre-image of a triangle that map onto specific angles and sides of the image. If a rigid motion results in every side and every angle of the pre-image mapping onto every corresponding side and angle of the image, they assert that the triangles are congruent.

Classwork

Opening Exercise (6 minutes)

Opening Exercise

The figure to the right represents a rotation of $\triangle ABC$ 80° around vertex C. Name the triangle formed by the image of $\triangle ABC$. Write the rotation in function notation, and name all corresponding angles and sides.

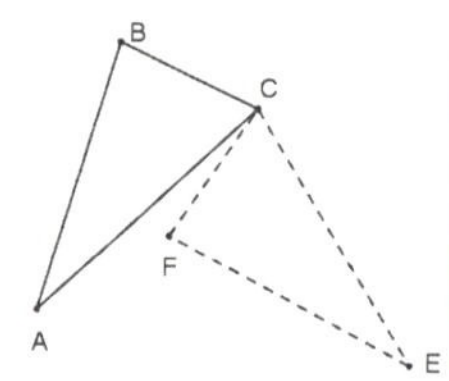

Corresponding Angles	*Corresponding Sides*
$\angle A \rightarrow \angle E$ $\angle B \rightarrow \angle F$ $\angle C \rightarrow \angle C$	$\overline{AB} \rightarrow \overline{EF}$ $\overline{BC} \rightarrow \overline{FC}$ $\overline{AC} \rightarrow \overline{EC}$

$\triangle EFC$

$R_{D,80^\circ}(\triangle EFC)$

Discussion (5 minutes)

Discussion

In the Opening Exercise, we explicitly showed a single rigid motion, which mapped every side and every angle of $\triangle ABC$ onto $\triangle EFC$. Each corresponding pair of sides and each corresponding pair of angles was congruent. When each side and each angle on the pre-image maps onto its corresponding side or angle on the image, the two triangles are congruent. Conversely, if two triangles are congruent, then each side and angle on the pre-image is congruent to its corresponding side or angle on the image.

Example (8 minutes)

Example

$ABCD$ is a square, and $\overline{AC}$ is one diagonal of the square. $\triangle ABC$ is a reflection of $\triangle ADC$ across segment AC. Complete the table below, identifying the missing corresponding angles and sides.

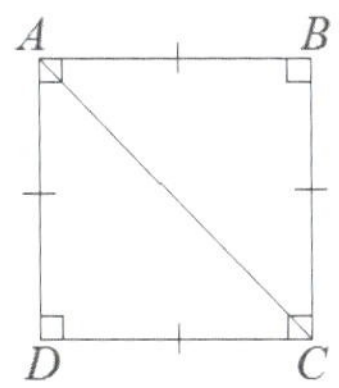

Corresponding Angles	Corresponding Sides
$\angle BAC \rightarrow \angle DAC$ $\angle ABC \rightarrow \angle ADC$ $\angle BCA \rightarrow \angle DCA$	$\overline{AB} \rightarrow \overline{AD}$ $\overline{BC} \rightarrow \overline{DC}$ $\overline{AC} \rightarrow \overline{AC}$

a. Are the corresponding sides and angles congruent? Justify your response.

Since a reflection is a rigid transformation, all angles and sides maintain their size.

b. Is $\triangle ABC \cong \triangle ADC$? Justify your response.

Yes, since $\triangle ADC$ is a reflection of $\triangle ABC$, they must be congruent.

Exercises (20 minutes)

Exercises

Each exercise below shows a sequence of rigid motions that map a pre-image onto a final image. Identify each rigid motion in the sequence, writing the composition using function notation. Trace the congruence of each set of corresponding sides and angles through all steps in the sequence, proving that the pre-image is congruent to the final image by showing that every side and every angle in the pre-image maps onto its corresponding side and angle in the image. Finally, make a statement about the congruence of the pre-image and final image.

1.

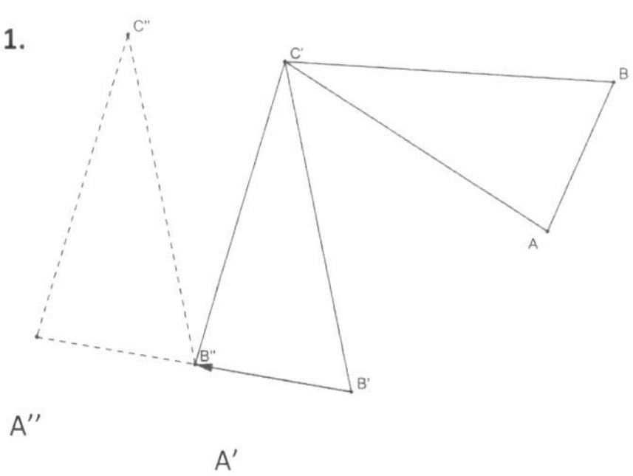

Sequence of Rigid Motions (2)	*rotation, translation*
Composition in Function Notation	$T_{\overrightarrow{B'B''}}\left(R_{c,90^\circ}(\triangle ABC)\right)$
Sequence of Corresponding Sides	$\overline{AB} \rightarrow \overline{A''B''}$ $\overline{BC} \rightarrow \overline{B''C''}$ $\overline{AC} \rightarrow \overline{A''C''}$
Sequence of Corresponding Angles	$A \rightarrow A''$ $B \rightarrow B''$ $C \rightarrow C''$
Triangle Congruence Statement	$\triangle ABC \cong \triangle A''B''C''$

2.

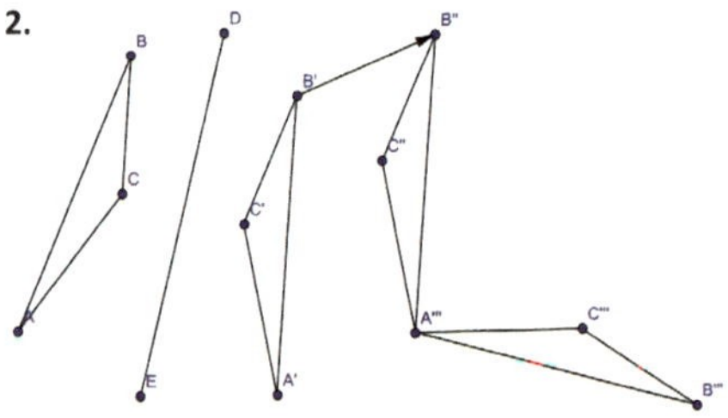

Sequence of Rigid Motions (3)	*reflection, translation, rotation*
Composition in Function Notation	$R_{A'',100^\circ}\left(T_{\overline{B'B''}}\left(r_{\overline{DE}}(\triangle ABC)\right)\right)$
Sequence of Corresponding Sides	$\overline{AB} \rightarrow \overline{A'''B'''}$ $\overline{BC} \rightarrow \overline{B'''C'''}$ $\overline{AC} \rightarrow \overline{A'''C'''}$
Sequence of Corresponding Angles	$A \rightarrow A'''$ $B \rightarrow B'''$ $C \rightarrow C'''$
Triangle Congruence Statement	$\triangle ABC \cong \triangle A'''B'''C'''$

3.

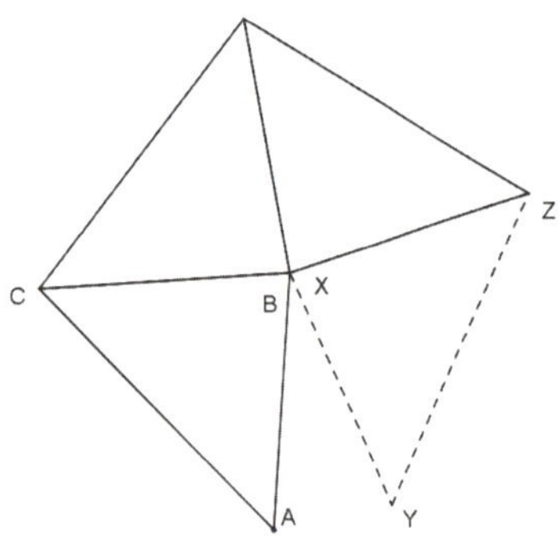

Sequence of Rigid Motions (3)	*reflections*
Composition in Function Notation	$r_{\overline{XZ}}\left(r_{\overline{BA'}}\left(r_{\overline{BC}}(\triangle ABC)\right)\right)$
Sequence of Corresponding Sides	$\overline{AB} \rightarrow \overline{YX}$ $\overline{AC} \rightarrow \overline{YZ}$ $\overline{BC} \rightarrow \overline{XZ}$
Sequence of Corresponding Angles	$A \rightarrow Y$ $B \rightarrow X$ $C \rightarrow Z$
Triangle Congruence Statement	$\triangle ABC \cong \triangle YXZ$

Closing (1 minute)

- When each side and each angle of a triangle on the pre-image maps onto its corresponding side or angle on the image, the two triangles are congruent. Conversely, if two triangles are congruent, then each side and angle on the pre-image is congruent to its corresponding side or angle on the image.

Exit Ticket (5 minutes)

Name ______________________________ Date ______________

Lesson 21: Correspondence and Transformations

Exit Ticket

Complete the table based on the series of rigid motions performed on $\triangle ABC$ below.

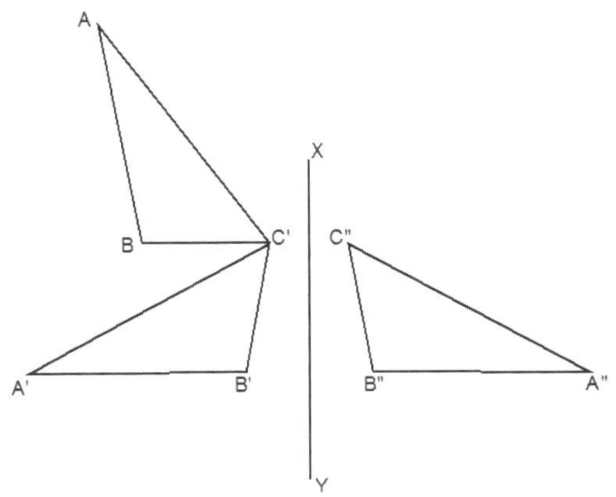

Sequence of Rigid Motions (2)	
Composition in Function Notation	
Sequence of Corresponding Sides	
Sequence of Corresponding Angles	
Triangle Congruence Statement	

Exit Ticket Sample Solutions

Complete the table based on the series of rigid motions performed on $\triangle ABC$ below.

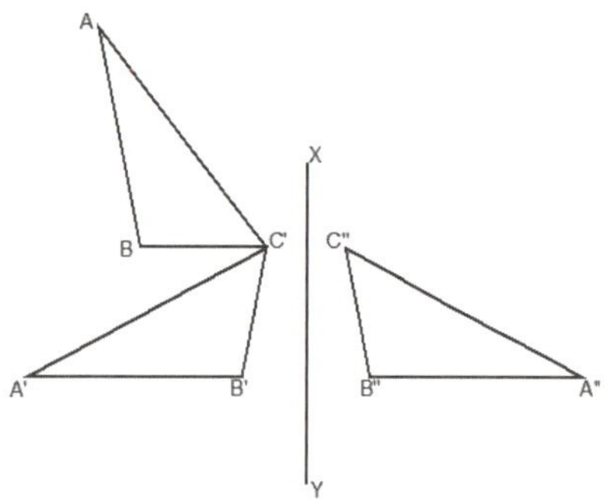

Sequence of Rigid Motions (2)	*rotation, reflection*
Composition in Function Notation	$\left(r_{\overline{XY}}\left(R_{C,90^\circ}\right)\right)$
Sequence of Corresponding Sides	$\overline{AB} \rightarrow \overline{A''B''}$ $\overline{BC} \rightarrow \overline{B''C''}$ $\overline{AC} \rightarrow \overline{A''C''}$
Sequence of Corresponding Angles	$\angle A \rightarrow \angle A'$ $\angle B \rightarrow \angle B'$ $\angle C \rightarrow \angle C'$
Triangle Congruence Statement	$\triangle ABC \cong \triangle A''B''C''$

Problem Set Sample Solutions

1. Exercise 3 mapped $\triangle ABC$ onto $\triangle YXZ$ in three steps. Construct a fourth step that would map $\triangle YXZ$ back onto $\triangle ABC$.

   Construct an angle bisector for the $\angle ABY$, and reflect $\triangle YXZ$ over that line.

2. Explain triangle congruence in terms of rigid motions. Use the terms *corresponding sides* and *corresponding angles* in your explanation.

   Triangle congruence can be found using a series of rigid motions in which you map an original or pre-image of a figure onto itself. By doing so, all the corresponding sides and angles of the figure map onto their matching corresponding sides or angles, which proves the figures are congruent.

Name ______________________________ Date ______________

1. State precise definitions of angle, circle, perpendicular, parallel, and line segment based on the undefined notions of point, line, distance along a line, and distance around a circular arc.

   Angle:

   Circle:

   Perpendicular:

   Parallel:

   Line segment:

2. A rigid motion, J, of the plane takes a point, A, as input and gives C as output (i.e., $J(A) = C$). Similarly, $J(B) = D$ for input point B and output point D.

   Jerry claims that knowing nothing else about J, we can be sure that $\overline{AC} \cong \overline{BD}$ because rigid motions preserve distance.

   a. Show that Jerry's claim is incorrect by giving a counterexample (hint: a counterexample would be a specific rigid motion and four points A, B, C, and D in the plane such that the motion takes A to C and B to D, yet $\overline{AC} \ncong \overline{BD}$).

   b. There is a type of rigid motion for which Jerry's claim is always true. Which type below is it?

   Rotation Reflection Translation

   c. Suppose Jerry claimed that $\overline{AB} \cong \overline{CD}$. Would this be true for any rigid motion that satisfies the conditions described in the first paragraph? Why or why not?

3.

a. In the diagram below, l is a line, A is a point on the line, and B is a point not on the line. C is the midpoint of $\overline{AB}$. Show how to create a line parallel to l that passes through B by using a rotation about C.

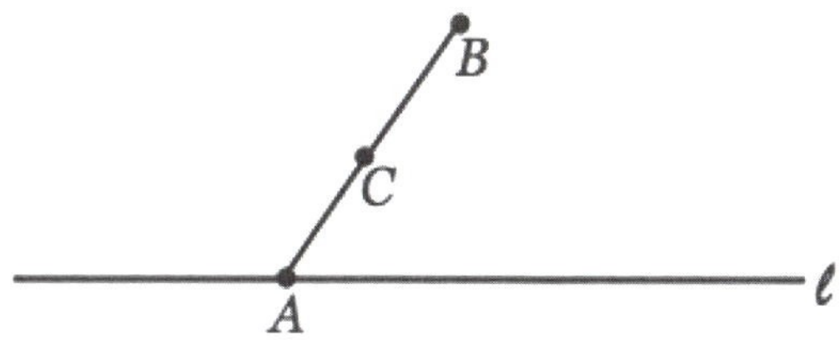

b. Suppose that four lines in a given plane, l_1, l_2, m_1, and m_2 are given, with the conditions (also given) that $l_1 \parallel l_2$, $m_1 \parallel m_2$, and l_1 is neither parallel nor perpendicular to m_1.

i. Sketch (freehand) a diagram of l_1, l_2, m_1, and m_2 to illustrate the given conditions.

ii. In any diagram that illustrates the given conditions, how many distinct angles are formed? Count only angles that measure less than $180°$, and count two angles as the same only if they have the same vertex and the same edges. Among these angles, how many different angle measures are formed? Justify your answer.

4. In the figure below, there is a reflection that transforms $\triangle ABC$ to $\triangle A'B'C'$.

   Use a straightedge and compass to construct the line of reflection, and list the steps of the construction.

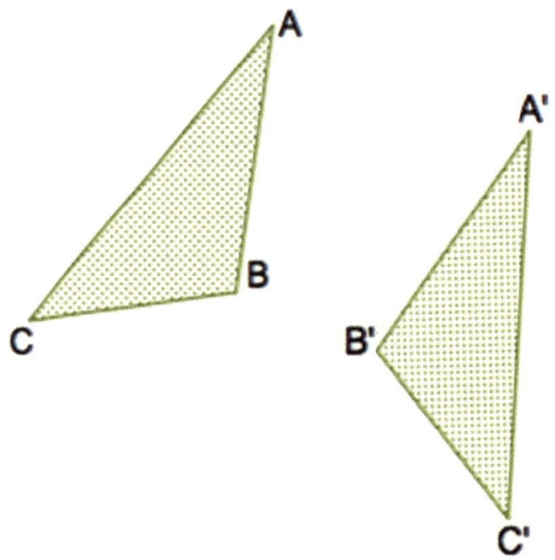

5. Precisely define each of the three rigid motion transformations identified.

a. $T_{\overrightarrow{AB}}(P)$ ____

____

____

____

____

b. $r_l(P)$ ____

____

____

____

____

c. $R_{C,30^\circ}(P)$ ____

____

____

____

____

6. Given in the figure below, line l is the perpendicular bisector of $\overline{AB}$ and of $\overline{CD}$.

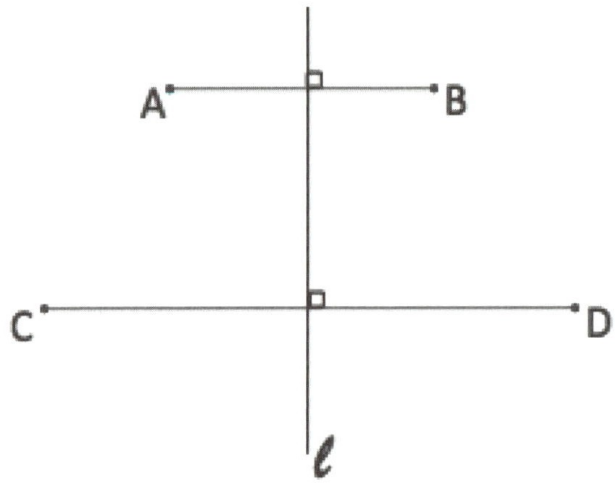

a. Show $\overline{AC} \cong \overline{BD}$ using rigid motions.

b. Show $\angle ACD \cong \angle BDC$.

c. Show $\overline{AB} \parallel \overline{CD}$.

A Progression Toward Mastery					
Assessment Task Item		**STEP 1** **Missing or incorrect answer and little evidence of reasoning or application of mathematics to solve the problem.**	**STEP 2** **Missing or incorrect answer but evidence of some reasoning or application of mathematics to solve the problem.**	**STEP 3** **A correct answer with some evidence of reasoning or application of mathematics to solve the problem, OR an incorrect answer with substantial evidence of solid reasoning or application of mathematics to solve the problem.**	**STEP 4** **A correct answer supported by substantial evidence of solid reasoning or application of mathematics to solve the problem.**
1	G-CO.A.1	Student accurately and precisely articulates the definitions of two or fewer of the five terms.	Student accurately and precisely articulates the definitions of at least three of the five terms.	Student accurately and precisely articulates the definitions of at least four of the five terms.	Student accurately and precisely articulates the definitions of all five terms.
2	a–c G-CO.A.2	Student circles "translation" in part (b) but does not provide a correct response in parts (a) and (b) or provides a response that does not show clear understanding of the application of rigid motions.	Student provides a response that includes a counterexample in part (a) OR presents an idea to prove that $\overline{AB} \cong \overline{CD}$ in part (c). However, whichever is presented is less than perfectly clear in stating the solutions. Student circles "translation" in part (b).	Student provides a counterexample in part (a) and presents an idea to prove that $\overline{AB} \cong \overline{CD}$ in part (c), but both are less than perfectly clear in stating the solutions. Student circles "translation" in part (b).	Student provides a correctly reasoned counterexample in part (a), circles "translation" in part (b), and justifies the claim that $\overline{AB} \cong \overline{CD}$ for any rigid motion in part (c).
3	a–b G-CO.A.1 G-CO.C.9 G-CO.D.12	Student provides an incomplete or irrelevant response in parts (a) and (b.ii) but provides an appropriate, clearly labeled sketch in part (b.i), or student provides correct responses in parts (a) and (b.ii) but includes incomplete sketch or no sketch in part (b.i).	Student provides an incomplete description of the rotation of line l about C in part (a) or an incomplete sketch for part (b.i), and an incorrect number of angles formed or an incorrect set of angle measures in part (b.ii).	Student provides an incomplete description of the rotation of line l about C in part (a), an appropriate, clearly labeled sketch for part (b.i), and a justification for why there are 16 relevant angles and 2 different angle measures in part (b.ii).	Student provides a correct description of the rotation of line l about C in part (a), an appropriate, clearly labeled sketch for part (b.i), and a justification for why there are 16 relevant angles and 2 different angle measures in part (b.ii).

4	G-CO.A.5 G-CO.D.12	Student provides a drawing that is not an appropriate construction and an underdeveloped list of steps.	Student provides appropriate construction marks but makes more than one error in the construction or the steps; the line of reflection is drawn.	Student provides appropriate construction marks but makes one error in the construction or the steps; the line of reflection is drawn.	Student draws a correct construction showing all appropriate marks, including the line of reflection, and the accompanying list of steps is also correct.
5	a–c G-CO.A.4	Student provides inaccurate definitions for the three rigid motions.	Student provides definitions that lack the precise language of an exemplary response and does not address the points that are unchanged (i.e., does not mention that the rotation of the center remains fixed).	Student provides definitions that lack the precise language of an exemplary response.	Student provides precise definitions for each rigid motion with correct usage of notation.
6	a–c G-CO.B.6 G-CO.C.9	Student provides an incorrect response or a response that shows little evidence of understanding the properties of reflections.	Student provides an incorrect response, but the response shows evidence of the beginning of understanding of the properties of reflections.	Student provides a response that lacks the precision of an exemplary response, but the response shows an understanding of the properties of reflections.	Student provides a correct response for each of the three parts that demonstrates a clear understanding of the properties of reflections.

Name ______________________________ Date ______________

1. State precise definitions of angle, circle, perpendicular, parallel, and line segment based on the undefined notions of point, line, distance along a line, and distance around a circular arc.

Angle:

An angle is formed by two rays that share a common vertex. An angle is proper if the two rays do not lie on the same line.

Circle:

A circle is the set of all points in a plane that are equidistant from the center point. The circle in plane P with center A and radius AB is the set of all points in P whose distance from A is the same as the distance from A to B.

Perpendicular:

Two lines are perpendicular if they have one point in common and if the four angles formed by the intersection are all right angles.

Parallel:

Two lines are parallel if they lie in the same plane and have no points in common.

Line segment:

A line segment is the set of point A and B and all points on line AB between A and B.

2. A rigid motion, J, of the plane takes a point, A, as input and gives C as output (i.e., $J(A) = C$). Similarly, $J(B) = D$ for input point B and output point D.

Jerry claims that knowing nothing else about J, we can be sure that $\overline{AC} \cong \overline{BD}$ because rigid motions preserve distance.

a. Show that Jerry's claim is incorrect by giving a counterexample (hint: a counterexample would be a specific rigid motion and four points A, B, C, and D in the plane such that the motion takes A to C and B to D, yet $\overline{AC} \ncong \overline{BD}$).

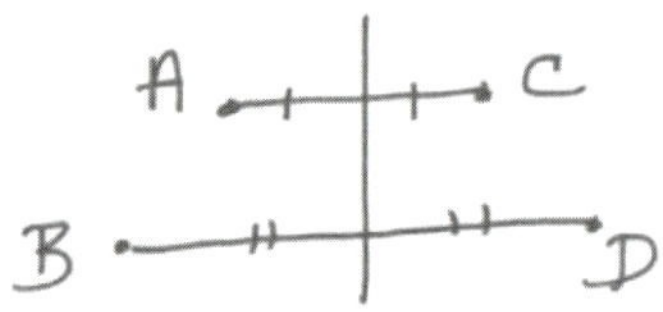

Here, J is a reflection across a vertical line. The distance from A to the line is different from the distance from B to the line. Therefore, the distance from A to its image (C) is different from the distance from B to its image (D).

b. There is a type of rigid motion for which Jerry's claim is always true. Which type below is it?

Rotation Reflection Translation (circled)

c. Suppose Jerry claimed that $\overline{AB} \cong \overline{CD}$. Would this be true for any rigid motion that satisfies the conditions described in the first paragraph? Why or why not?

Yes, because rigid motions always preserve distance.

3.

a. In the diagram below, l is a line, A is a point on the line, and B is a point not on the line. C is the midpoint of $\overline{AB}$. Show how to create a line parallel to l that passes through B by using a rotation about C.

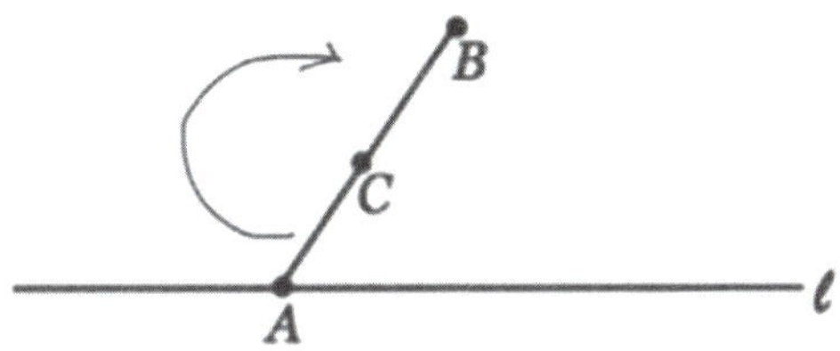

Take a 180° rotation about C. Line m is the image of ℓ under this rotation. Line m passes through B and is parallel to ℓ.

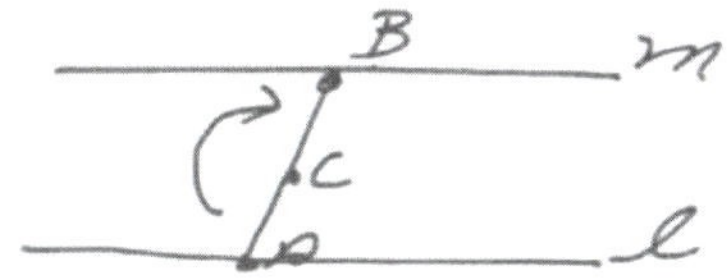

b. Suppose that four lines in a given plane, l_1, l_2, m_1, and m_2 are given, with the conditions (also given) that $l_1 \parallel l_2$, $m_1 \parallel m_2$, and l_1 is neither parallel nor perpendicular to m_1.

i. Sketch (freehand) a diagram of l_1, l_2, m_1, and m_2 to illustrate the given conditions.

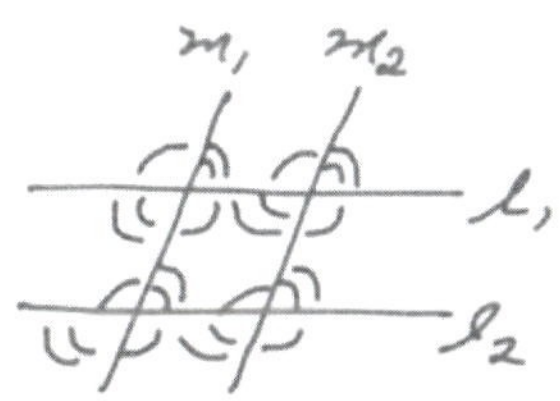

ii. In any diagram that illustrates the given conditions, how many distinct angles are formed? Count only angles that measure less than $180°$, and count two angles as the same only if they have the same vertex and the same edges. Among these angles, how many different angle measures are formed? Justify your answer.

There are 16 distinct angles with two different angle measures because alternate interior/exterior angles are congruent and corresponding angles are congruent.

4. In the figure below, there is a reflection that transforms $\triangle ABC$ to $\triangle A'B'C'$.

   Use a straightedge and compass to construct the line of reflection and list the steps of the construction.

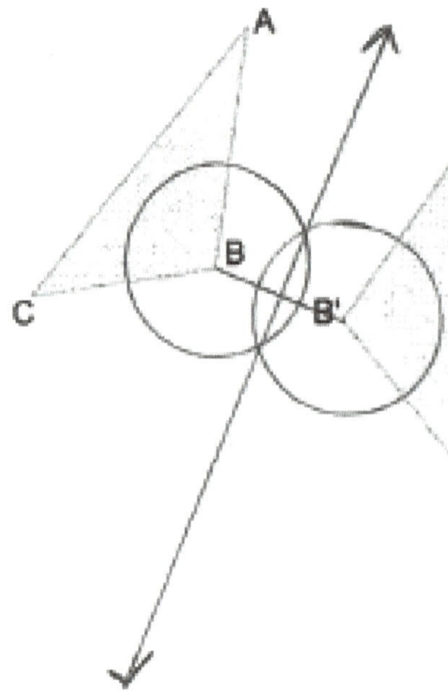

1. Draw segment BB'
2. Construct circle B with radius $\overline{BB'}$
3. Construct circle B' with radius $\overline{BB'}$
4. Connect the two intersections of circles B and B'
5. This forms the line of reflection between $\triangle ABC$ and $\triangle A'B'C'$

5. Precisely define each of the three rigid motion transformations identified.

a. $T_{\overrightarrow{AB}}(P)$ For vector $\overrightarrow{AB}$, the translation along $\overrightarrow{AB}$ is a translation of the plane: (1) For every point P on line AB, $T_{\overrightarrow{AB}}(P)$ is the point Q on $\overleftrightarrow{AB}$ so that $\overrightarrow{PQ}$ has the same length and direction as $\overrightarrow{AB}$ and (2) For P not on $\overleftrightarrow{AB}$, let l_1 be the line through P parallel to $\overleftrightarrow{AB}$ and l_2 be the line through B parallel to $\overleftrightarrow{AB}$. Q is the intersection of l_1 and l_2.

b. $r_l(P)$ For a line l, a reflection across l is the transformation r_l of the plane defined as follows: (1) For any point P on l, $r_l(P) = P$, and (2) For any point P not on l, $r_l(P)$ is the point Q so that l is the perpendicular bisector of $\overline{PQ}$

c. $R_{C,30°}(P)$ The rotation of 30° around center C is defined as follows: (1) For the center point C, $R_{C,30°}(C) = C$, and (2) For any other point P, $R_{C,30°}$ is the point Q that lies in the counterclockwise half-plane of ray $\overrightarrow{CP}$, such that $CQ = CP$ and $\angle PCQ = 30°$.

6. Given in the figure below, line l is the perpendicular bisector of $\overline{AB}$ and of $\overline{CD}$.

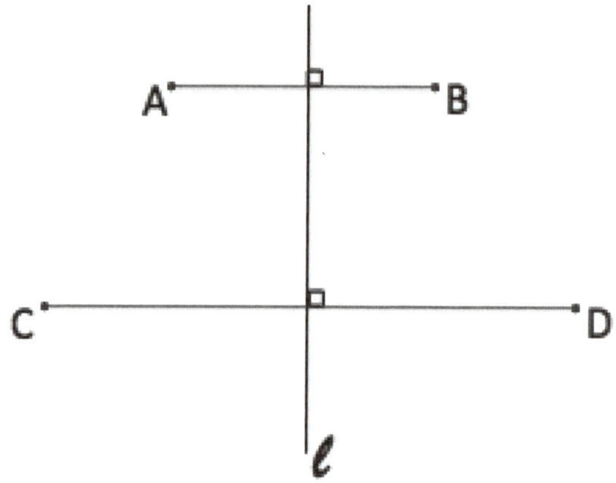

a. Show $\overline{AC} \cong \overline{BD}$ using rigid motions.

Since l is the perpendicular bisector of $\overline{AB}$ and $\overline{CD}$, the reflection through line l brings A to B and C to D. Because reflections take line segments to congruent line segments, $\overline{AC}$ is congruent to $\overline{BD}$.

b. Show $\angle ACD \cong \angle BDC$.

The reflection through line l brings A to B and C to D and D to C. Therefore ray $\overrightarrow{CA}$ goes to ray $\overrightarrow{DB}$. Ray $\overrightarrow{CD}$ goes to ray $\overrightarrow{DC}$. The image of $\angle ACD$ is therefore congruent to $\angle BDC$.

c. Show $\overline{AB} \parallel \overline{CD}$.

$\overline{AB} \parallel \overline{CD}$ because the perpendicular bisector intersects the two lines creating congruent corresponding angles.

Topic D

Congruence

G-CO.B.7, G-CO.B.8

Focus Standards:	G-CO.B.7	Use the definition of congruence in terms of rigid motions to show that two triangles are congruent if and only if corresponding pairs of sides and corresponding pairs of angles are congruent.
	G-CO.B.8	Explain how the criteria for triangle congruence (ASA, SAS, and SSS) follow from the definition of congruence in terms of rigid motions.
Instructional Days:	6	
Lesson 22:	Congruence Criteria for Triangles—SAS (P)[1]	
Lesson 23:	Base Angles of Isosceles Triangles (E)	
Lesson 24:	Congruence Criteria for Triangles—ASA and SSS (P)	
Lesson 25:	Congruence Criteria for Triangles—AAS and HL (E)	
Lessons 26–27:	Triangle Congruency Proofs (P, P)	

In Topic D, students use the knowledge of rigid motions developed in Topic C to determine and prove triangle congruence. At this point, students have a well developed definition of congruence supported by empirical investigation. They can now develop an understanding of traditional congruence criteria for triangles, such as SAS, ASA, and SSS, and devise formal methods of proof by direct use of transformations. As students prove congruence using the three criteria, they investigate why AAS also leads toward a viable proof of congruence and why they cannot use SSA to establish congruence. Examining and establishing these methods of proving congruency leads to analysis and application of specific properties of lines, angles, and polygons in Topic E.

[1]Lesson Structure Key: **P**-Problem Set Lesson, **M**-Modeling Cycle Lesson, **E**-Exploration Lesson, **S**-Socratic Lesson

Lesson 22: Congruence Criteria for Triangles—SAS

Student Outcomes

- Students learn why any two triangles that satisfy the SAS congruence criterion must be congruent.

Lesson Notes

Lesson 22 begins to investigate criteria, or the indicators, of triangle congruence. Students are introduced to the concept in Grade 8 but justified the criteria of triangle congruence (i.e., ASA, SAS, and SSS) in a more hands-on manner, manipulating physical forms of triangles through rigid motions to determine whether or not a pair of triangles is congruent. In this lesson, students formally prove the triangle congruency criteria.

Note that in the exercises that follow, proofs may employ *both* statements of equality of measure of angles and lengths of segments *and* statements of congruence of angles and segments. While not introduced formally, it is intuitively clear that two segments are congruent if and only if they are equal in length; similarly, two angles are equal in measure if and only if they are congruent. That is, a segment can be mapped onto another if and only if they are equal in length, and an angle can be mapped onto another if and only if they are equal in measure. Another implication is that some of our key facts and discoveries may also be stated in terms of congruence, such as "Vertical angles are *congruent.*" Or, "If two lines are cut by a transversal such that a pair of alternate interior angles are *congruent*, then the lines are parallel." Discuss these results with your students. Exercise 4 of this lesson should help students understand the logical equivalency of these statements.

Classwork

Opening Exercise (4 minutes)

Opening Exercise

Answer the following question. Then discuss your answer with a partner.

Do you think it is possible to know whether there is a rigid motion that takes one triangle to another without actually showing the particular rigid motion? Why or why not?

Answers may vary. Some students may think it is not possible because it is necessary to show the transformation as proof of its existence. Others may think it is possible by examining the triangles carefully.

It is common for curricula to take indicators of triangle congruence such as SAS and ASA as axiomatic, but this curriculum defines congruence in terms of rigid motions (as defined in the **G-CO** domain). However, it can be shown that these commonly used statements (SAS, ASA, etc.) *follow from* this definition of congruence and the properties of basic rigid motions (**G-CO.B.8**). Thus, these statements are indicators of whether rigid motions exist to take one triangle to the other. In other words, we have agreed to use the word *congruent* to mean *there exists a composition of basic rigid motion of the plane that maps one figure to the other.* We see that SAS, ASA, and SSS imply the existence of the rigid motion needed, but precision demands that we explain how and why.

While there are multiple proofs that show that SAS follows from the definition of congruence in terms of rigid motions and the properties of basic rigid motions, the one that appears in this lesson is one of the versions most accessible for students.

Discussion (20 minutes)

Discussion

It is true that we do not need to show the rigid motion to be able to know that there is one. We are going to show that there are criteria that refer to a few parts of the two triangles and a correspondence between them that guarantee congruency (i.e., existence of rigid motion). We start with the Side-Angle-Side (SAS) criteria.

SIDE-ANGLE-SIDE TRIANGLE CONGRUENCE CRITERIA (SAS): Given two triangles $\triangle ABC$ and $\triangle A'B'C'$ so that $AB = A'B'$ (Side), $m\angle A = m\angle A'$ (Angle), and $AC = A'C'$ (Side). Then the triangles are congruent.

The steps below show the most general case for determining a congruence between two triangles that satisfy the SAS criteria. Note that not all steps are needed for every pair of triangles. For example, sometimes the triangles already share a vertex. Sometimes a reflection is needed, sometimes not. It is important to understand that we can always use some or all of the steps below to determine a congruence between the two triangles that satisfies the SAS criteria.

PROOF: Provided the two distinct triangles below, assume $AB = A'B'$ (Side), $m\angle A = m\angle A'$ (Angle), and $AC = A'C'$ (Side).

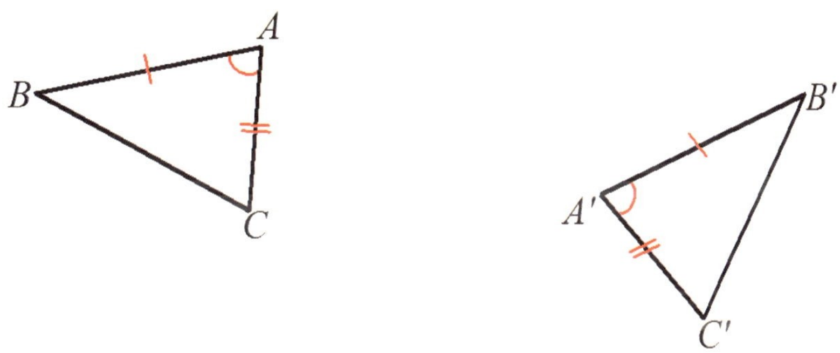

By our definition of congruence, we have to find a composition of rigid motions that maps $\triangle A'B'C'$ to $\triangle ABC$. We must find a congruence F so that $(\triangle A'B'C') = \triangle ABC$. First, use a translation T to map a common vertex.

Which two points determine the appropriate vector?

A', A

Can any other pair of points be used? Why or why not?

No. We use A' and A because only these angles are congruent by assumption.

State the vector in the picture below that can be used to translate $\triangle A'B'C'$.

$\overrightarrow{A'A}$

Using a dotted line, draw an intermediate position of $\triangle A'B'C'$ as it moves along the vector:

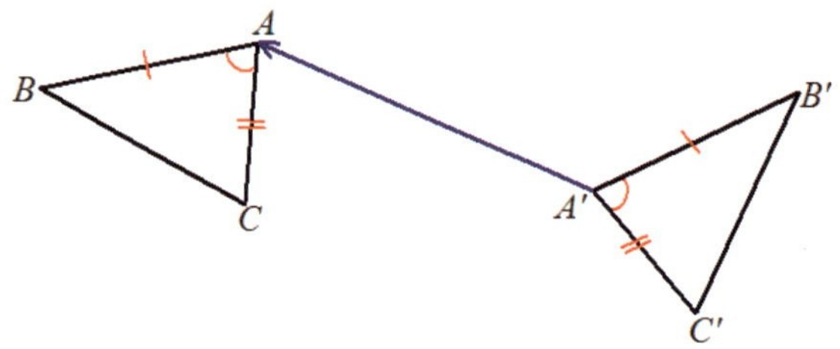

After the translation (below), $T_{\overrightarrow{A'A}}(\triangle A'B'C')$ shares one vertex with $\triangle ABC$, A. In fact, we can say $T_{\overrightarrow{A'A}}(\triangle A'B'C') = \triangle AB''C''$.

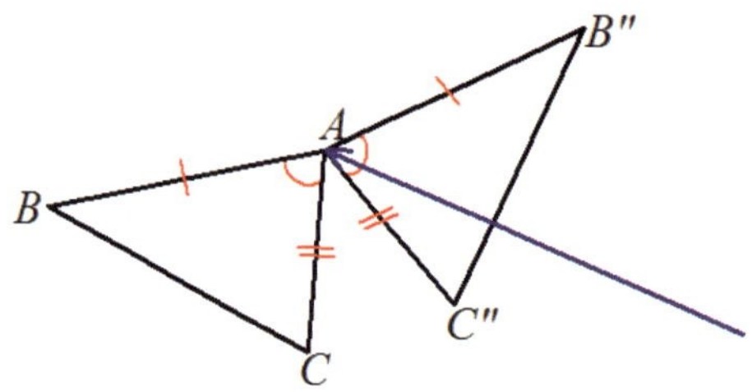

Next, use a clockwise rotation $R_{\angle CAC''}$ to bring the side $\overline{AC''}$ to $\overline{AC}$ (or a counterclockwise rotation to bring $\overline{AB''}$ to $\overline{AB}$).

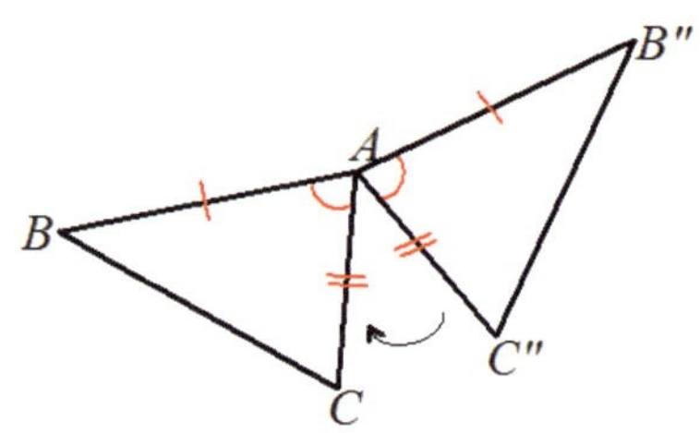

A rotation of appropriate measure maps $\overrightarrow{AC''}$ to $\overrightarrow{AC}$, but how can we be sure that vertex C'' maps to C? Recall that part of our assumption is that the lengths of sides in question are equal, ensuring that the rotation maps C'' to C. ($AC = AC''$; the translation performed is a rigid motion, and thereby did not alter the length when $\overline{AC'}$ became $\overline{AC''}$.)

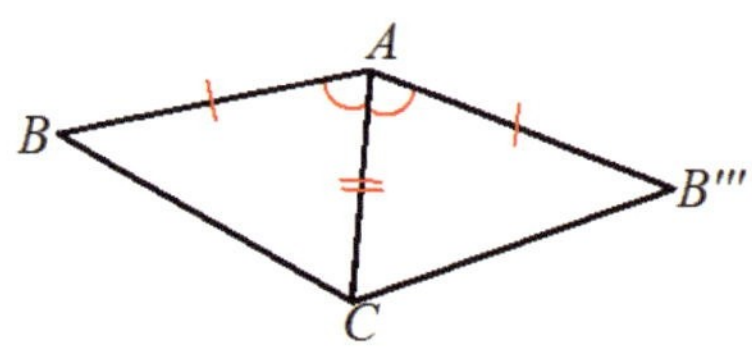

After the rotation $R_{\angle CAC''}(\triangle AB''C'')$, a total of two vertices are shared with $\triangle ABC$, A and C. Therefore, $R_{\angle CAC''}(\triangle AB''C'') = \triangle AB'''C$.

Finally, if B''' and B are on opposite sides of the line that joins AC, a reflection $r_{\overline{AC}}$ brings B''' to the same side as B.

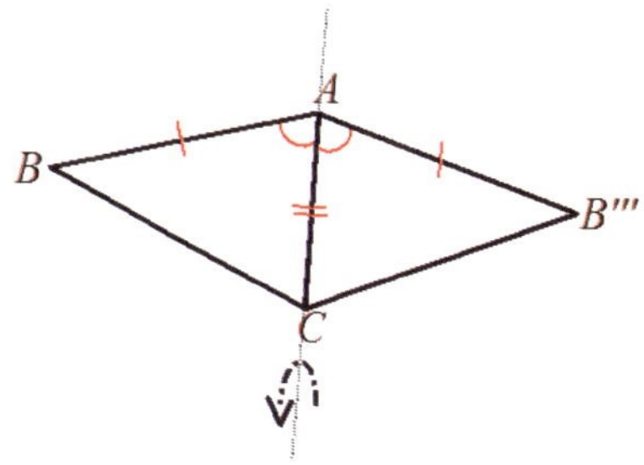

Since a reflection is a rigid motion and it preserves angle measures, we know that $m\angle B'''AC = m\angle BAC$ and so $\overrightarrow{AB'''}$ maps to $\overrightarrow{AB}$. If, however, $\overrightarrow{AB'''}$ coincides with $\overrightarrow{AB}$, can we be certain that B''' actually maps to B? We can, because not only are we certain that the rays coincide but also by our assumption that $AB = AB'''$. (Our assumption began as $AB = A'B'$, but the translation and rotation have preserved this length now as AB'''.) Taken together, these two pieces of information ensure that the reflection over $\overline{AC}$ brings B''' to B.

Another way to visually confirm this is to draw the marks of the ___*perpendicular bisector*___ construction for $\overline{AC}$.

Write the transformations used to correctly notate the congruence (the composition of transformations) that take $\triangle A'B'C' \cong \triangle ABC$:

F ___*Translation*___

G ___*Rotation*___

H ___*Reflection*___

$H\left(G\left(F(\triangle A'B'C')\right)\right) = \triangle ABC$

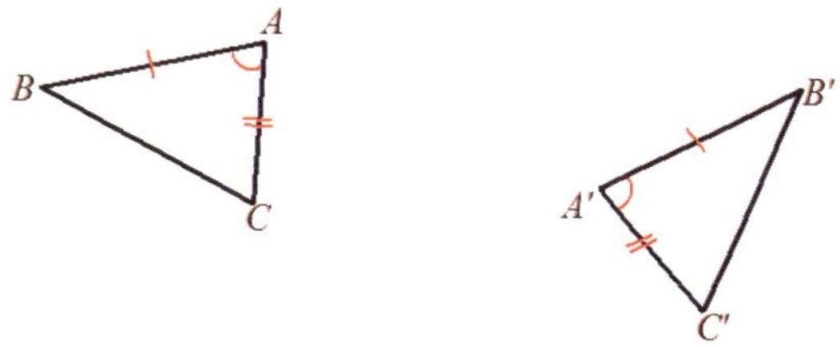

We have now shown a sequence of rigid motions that takes $\triangle A'B'C'$ to $\triangle ABC$ with the use of just three criteria from each triangle: two sides and an included angle. Given any two distinct triangles, we could perform a similar proof. There is another situation when the triangles are not distinct, where a modified proof is needed to show that the triangles map onto each other. Examine these below. Note that when using the Side-Angle-Side triangle congruence criteria as a reason in a proof, you need only state the congruence and *SAS*.

Example (5 minutes)

Students try an example based on the Discussion.

Example

What if we had the SAS criteria for two triangles that were not distinct? Consider the following two cases. How would the transformations needed to demonstrate congruence change?

Case	Diagram	Transformations Needed
Shared Side	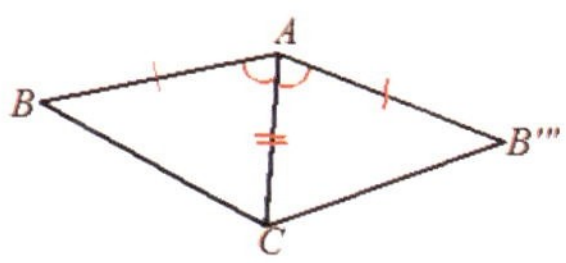	*reflection*
Shared Vertex	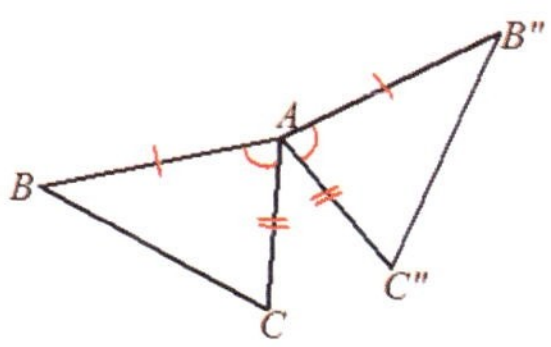	*rotation, reflection*

Exercises (7 minutes)

Exercises

1. Given: Triangles with a pair of corresponding sides of equal length and a pair of included angles of equal measure. Sketch and label three phases of the sequence of rigid motions that prove the two triangles to be congruent.

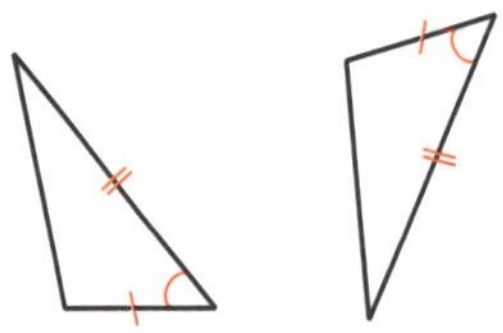

Translation	*Rotation*	*Reflection*
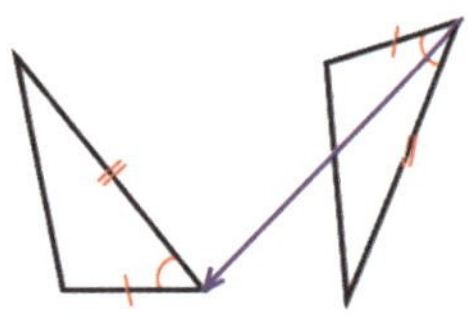		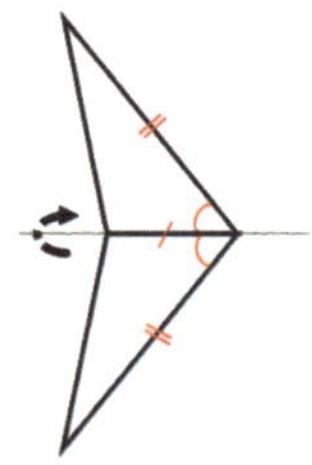

Justify whether the triangles meet the SAS congruence criteria; explicitly state which pairs of sides or angles are congruent and why. If the triangles do meet the SAS congruence criteria, describe the rigid motion(s) that would map one triangle onto the other.

2. Given: $m\angle LNM = m\angle LNO$, $MN = ON$

   Do $\triangle LNM$ and $\triangle LNO$ meet the SAS criteria?

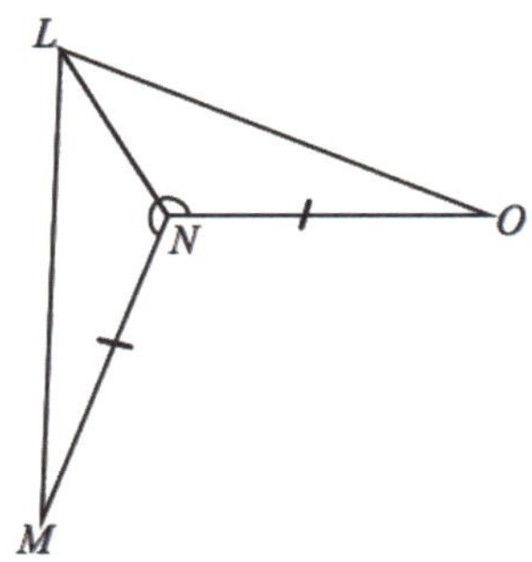

$m\angle LNM = m\angle LNO$	*Given*
$MN = ON$	*Given*
$LN = LN$	*Reflexive property*
$\triangle LNM \cong \triangle LNO$	*SAS*

The triangles map onto one another with a reflection over $\overline{LN}$.

3. Given: $m\angle HGI = m\angle JIG$, $HG = JI$

   Do $\triangle HGI$ and $\triangle JIG$ meet the SAS criteria?

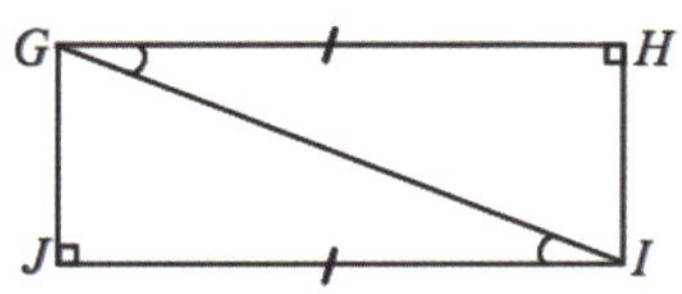

$m\angle HGI = m\angle JIG$	*Given*
$HG = JI$	*Given*
$GI = IG$	*Reflexive property*
$\triangle HGI \cong \triangle JIG$	*SAS*

The triangles map onto one another with a $180°$ rotation about the midpoint of the diagonal.

> **4. Is it true that we could also have proved $\triangle HGI$ and $\triangle JIG$ meet the SAS criteria if we had been given that $\angle HGI \cong \angle JIG$ and $\overline{HG} \cong \overline{JI}$? Explain why or why not.**
>
> ***Yes, this is true. Whenever angles are equal, they can also be described as congruent. Since rigid motions preserve angle measure, for two angles of equal measure, there always exists a sequence of rigid motions that will carry one onto the other. Additionally, since rigid motions preserve distance, for two segments of equal length, there always exists a sequence of rigid motions that carries one onto the other.***

Closing (1 minute)

- Two triangles, $\triangle ABC$ and $\triangle A'B'C'$, meet the Side-Angle-Side criteria when $AB = A'B'$ (Side), $m\angle A = m\angle A'$ (Angle), and $AC = A'C'$ (Side). The SAS criteria implies the existence of a congruence that maps one triangle onto the other.

Exit Ticket (8 minutes)

Name ______________________________ Date ____________

Lesson 22: Congruence Criteria for Triangles—SAS

Exit Ticket

If two triangles satisfy the SAS criteria, describe the rigid motion(s) that would map one onto the other in the following cases.

1. The two triangles share a single common vertex.

2. The two triangles are distinct from each other.

3. The two triangles share a common side.

Exit Ticket Sample Solutions

If two triangles satisfy the SAS criteria, describe the rigid motion(s) that would map one onto the other in the following cases.

1. The two triangles share a single common vertex.

   Rotation, reflection

2. The two triangles are distinct from each other.

   Translation, rotation, reflection

3. The two triangles share a common side.

   Reflection

Problem Set Sample Solutions

Justify whether the triangles meet the SAS congruence criteria; explicitly state which pairs of sides or angles are congruent and why. If the triangles do meet the SAS congruence criteria, describe the rigid motion(s) that would map one triangle onto the other.

1. Given: $\overline{AB} \parallel \overline{CD}$, and $AB = CD$

   Do $\triangle ABD$ and $\triangle CDB$ meet the SAS criteria?

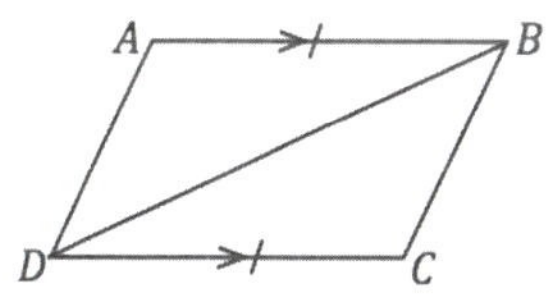

$AB = CD, \overline{AB} \parallel \overline{CD}$	*Given*
$BD = DB$	*Reflexive property*
$m\angle ABC = m\angle CDB$	*If a transversal intersects two parallel lines, then the measures of the alternate interior angles are equal in measure.*
$\triangle ABD \cong \triangle CDB$	*SAS*

   The triangles map onto one another with a $180°$ rotation about the midpoint of the diagonal.

2. Given: $m\angle R = 25°$, $RT = 7"$, $SU = 5"$, and $ST = 5"$

   Do $\triangle RSU$ and $\triangle RST$ meet the SAS criteria?

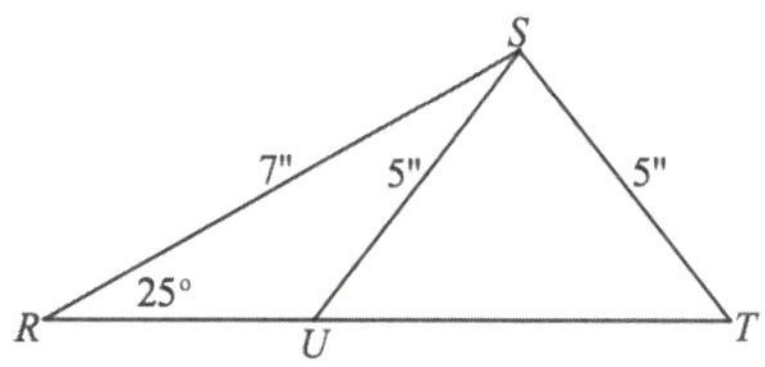

   There is not enough information given.

3. Given: $\overline{KM}$ and $\overline{JN}$ bisect each other

   Do $\triangle JKL$ and $\triangle NML$ meet the SAS criteria?

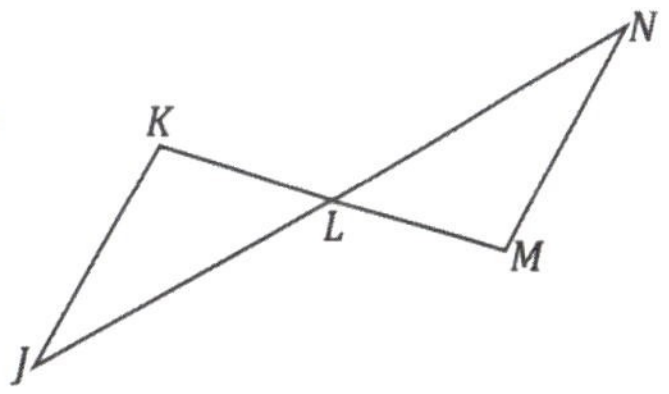

$\overline{KM}$ and $\overline{JN}$ bisect each other	*Given*
$m\angle KLJ = m\angle MLN$	*Vertical angles are equal in measure*
$KL = ML$	*Definition of a segment bisector*
$JL = NL$	*Definition of a segment bisector*
$\triangle JKL \cong \triangle NML$	*SAS*

   The triangles map onto one another with a $180°$ rotation about L.

4. Given: $m\angle 1 = m\angle 2$, and $BC = DC$

Do $\triangle ABC$ and $\triangle ADC$ meet the SAS criteria?

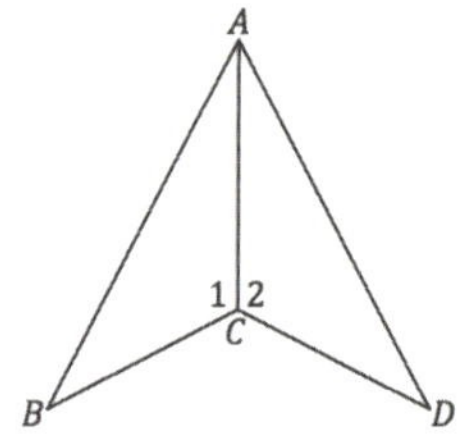

$m\angle 1 = m\angle 2$	*Given*
$BC = DC$	*Given*
$AC = AC$	*Reflexive property*
$\triangle ABC \cong \triangle ADC$	*SAS*

The triangles map onto one another with a reflection over $\overleftrightarrow{AC}$.

5. Given: $\overline{AE}$ bisects angle $\angle BCD$, and $BC = DC$

Do $\triangle CAB$ and $\triangle CAD$ meet the SAS criteria?

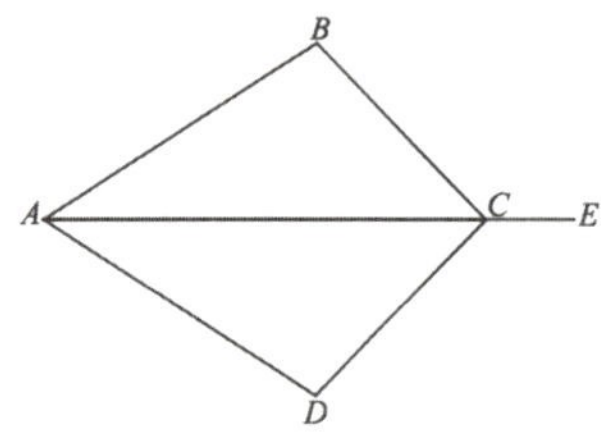

$\overline{AE}$ *bisects angle* $\angle BCD$	*Given*
$m\angle BCA = m\angle DCA$	*Definition of an angle bisector*
$BC = DC$	*Given*
$AC = AC$	*Reflexive property*
$\triangle CAD \cong \triangle CAB$	*SAS*

The triangles map onto one another with a reflection over $\overleftrightarrow{AC}$.

6. Given: $\overline{SU}$ and $\overline{RT}$ bisect each other

Do $\triangle SVR$ and $\triangle UVT$ meet the SAS criteria?

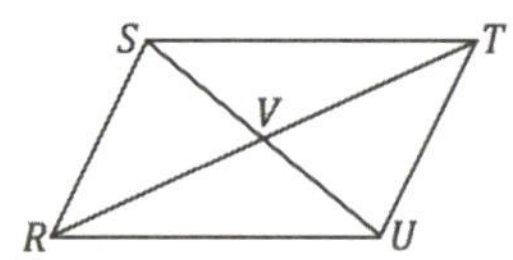

$\overline{SU}$ *and* $\overline{RT}$ *bisect each other*	*Given*
$SV = UV$	*Definition of a segment bisector*
$RV = VT$	*Definition of a segment bisector*
$m\angle SVR = m\angle UVT$	*Vertical angles are equal in measure*
$\triangle SVR \cong \triangle UVT$	*SAS*

The triangles map onto one another with a $180°$ *rotation about* V.

7. Given: $JM = KL$, $\overline{JM} \perp \overline{ML}$, and $\overline{KL} \perp \overline{ML}$

Do $\triangle JML$ and $\triangle KLM$ meet the SAS criteria?

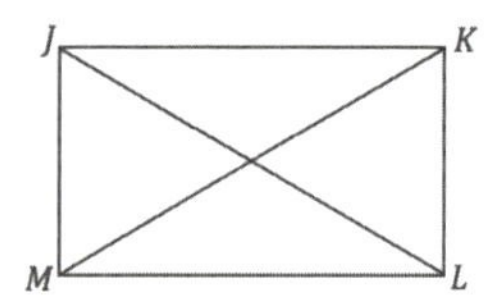

$JM = KL$	*Given*
$\overline{JM} \perp \overline{ML}$, $\overline{KM} \perp \overline{ML}$	*Given*
$m\angle JML = 90°$, $m\angle KLM = 90°$	*Definition of perpendicular lines*
$m\angle JML = m\angle KLM$	*Transitive property*
$ML = LM$	*Reflexive property*
$\triangle JML \cong \triangle KLM$	*SAS*

The triangles map onto one another with a reflection over the perpendicular bisector of $\overline{ML}$.

8. Given: $\overline{BF} \perp \overline{AC}$, and $\overline{CE} \perp \overline{AB}$

Do $\triangle BED$ and $\triangle CFD$ meet the SAS criteria?

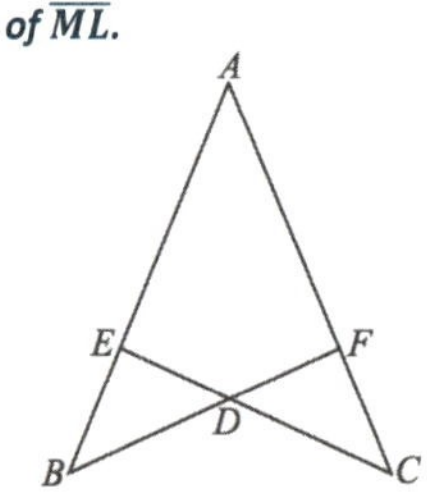

There is not enough information given.

9. Given: $m\angle VXY = m\angle VYX$

Do $\triangle VXW$ and $\triangle VYZ$ meet the SAS criteria?

There is not enough information given.

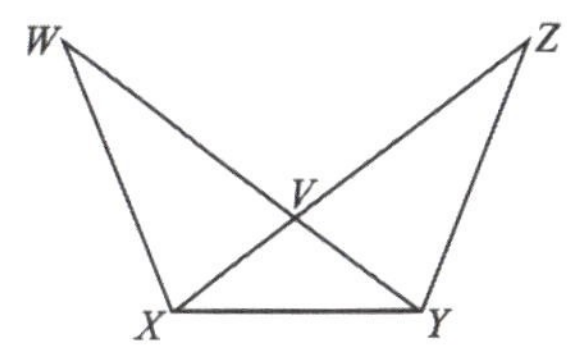

10. Given: $\triangle RST$ is isosceles, with $RS = RT$, and $SY = TZ$

Do $\triangle RSY$ and $\triangle RTZ$ meet the SAS criteria?

$\triangle RST$ *is isosceles with* $RS = RT$	*Given*
$m\angle S = m\angle T$	*Base angles of an isosceles triangle are equal in measure.*
$SY = TZ$	*Given*
$\triangle RSY \cong \triangle RTZ$	*SAS*

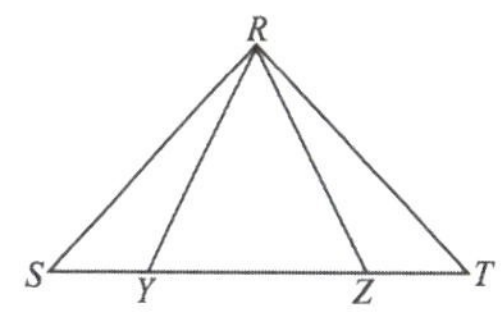

Lesson 23: Base Angles of Isosceles Triangles

Student Outcomes

- Students examine two different proof techniques via a familiar theorem.
- Students complete proofs involving properties of an isosceles triangle.

Lesson Notes

In Lesson 23, students study two proofs to demonstrate why the base angles of isosceles triangles are congruent. The first proof uses transformations, while the second uses the recently acquired understanding of the SAS triangle congruency. The demonstration of both proofs highlight the utility of the SAS criteria. Encourage students to articulate why the SAS criteria is so useful.

The goal of this lesson is to compare two different proof techniques by investigating a familiar theorem. Be careful not to suggest that proving an established fact about isosceles triangles is somehow the significant aspect of this lesson. However, if necessary, you can help motivate the lesson by appealing to history. Point out that Euclid used SAS and that the first application he made of it was in proving that base angles of an isosceles triangle are congruent. This is a famous part of the Elements. Today, proofs using SAS and proofs using rigid motions are valued equally.

Classwork

Opening Exercise (6 minutes)

Opening Exercise

Describe the additional piece of information needed for each pair of triangles to satisfy the SAS triangle congruence criteria.

a. **Given:** $AB = DC$

$\overline{AB} \perp \overline{BC}$, $\overline{DC} \perp \overline{CB}$, or $m\angle B = m\angle C$

Prove: $\triangle ABC \cong \triangle DCB$

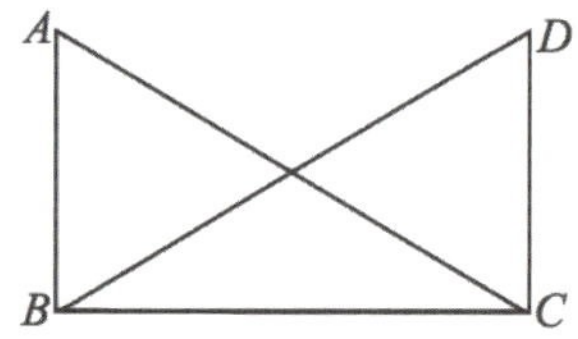

b. **Given:** $AB = RS$

$\overline{AB} \parallel \overline{RS}$

$CB = TS$, or $CT = BS$

Prove: $\triangle ABC \cong \triangle RST$

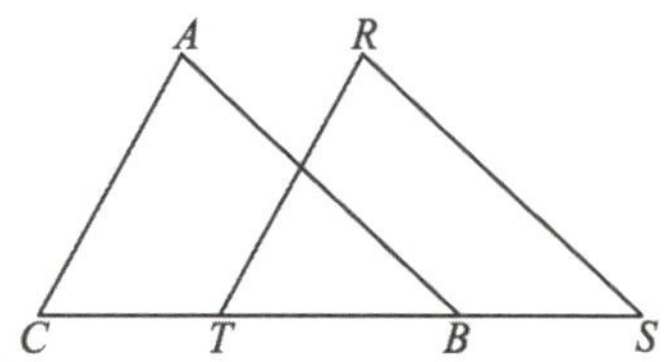

Exploratory Challenge (25 minutes)

Exploratory Challenge

Today we examine a geometry fact that we already accept to be true. We are going to prove this known fact in two ways: (1) by using transformations and (2) by using SAS triangle congruence criteria.

Here is isosceles triangle ABC. We accept that an isosceles triangle, which has (at least) two congruent sides, also has congruent base angles.

Label the congruent angles in the figure.

Now we prove that the base angles of an isosceles triangle are always congruent.

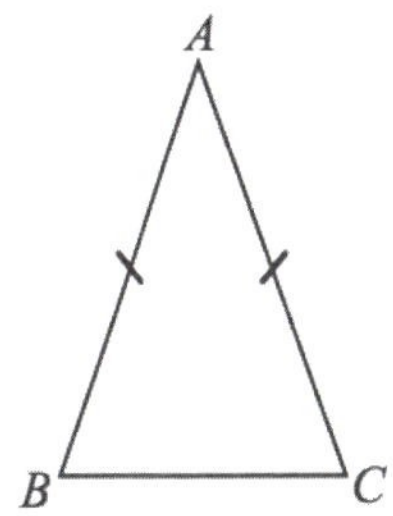

Prove Base Angles of an Isosceles are Congruent: Transformations

While simpler proofs do exist, this version is included to reinforce the idea of congruence as defined by rigid motions. Allow 15 minutes for the first proof.

Prove Base Angles of an Isosceles are Congruent: Transformations

Given: **Isosceles $\triangle ABC$, with $AB = AC$**

Prove: $m\angle B = m\angle C$

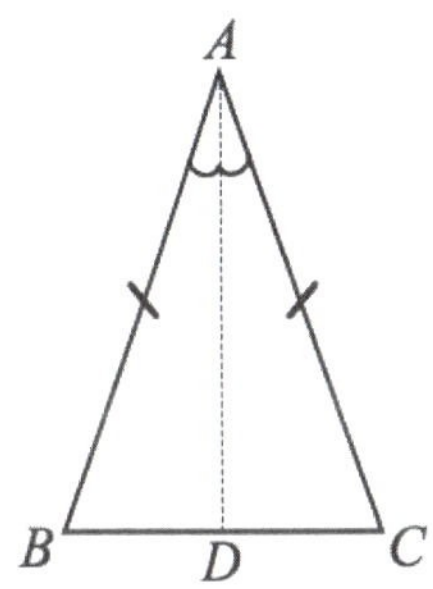

***Construction*: Draw the angle bisector $\overrightarrow{AD}$ of $\angle A$, where D is the intersection of the bisector and $\overline{BC}$. We need to show that rigid motions maps point B to point C and point C to point B.**

Let r be the reflection through $\overleftrightarrow{AD}$. Through the reflection, we want to demonstrate two pieces of information that map B to point C and vice versa: (1) $\overrightarrow{AB}$ maps to $\overrightarrow{AC}$, and (2) $AB = AC$.

Since A is on the line of reflection, $\overleftrightarrow{AD}$, $r(A) = A$. Reflections preserve angle measures, so the measure of the reflected angle $r(\angle BAD)$ equals the measure of $\angle CAD$; therefore, $r(\overrightarrow{AB}) = \overrightarrow{AC}$. Reflections also preserve lengths of segments; therefore, the reflection of $\overline{AB}$ still has the same length as $\overline{AB}$. By hypothesis, $AB = AC$, so the length of the reflection is also equal to AC. Then $r(B) = C$. Using similar reasoning, we can show that $r(C) = B$.

Reflections map rays to rays, so $r(\overrightarrow{BA}) = \overrightarrow{CA}$ and $r(\overrightarrow{BC}) = \overrightarrow{CB}$. Again, since reflections preserve angle measures, the measure of $r(\angle ABC)$ is equal to the measure of $\angle ACB$.

We conclude that $m\angle B = m\angle C$. Equivalently, we can state that $\angle B \cong \angle C$. In proofs, we can state that "base angles of an isosceles triangle are equal in measure" or that "base angles of an isosceles triangle are congruent."

Prove Base Angles of an Isosceles are Congruent: SAS

Allow 10 minutes for the second proof.

> **Prove Base Angles of an Isosceles are Congruent: SAS**
>
> **Given:** Isosceles $\triangle ABC$, with $AB = AC$
>
> **Prove:** $\angle B \cong \angle C$
>
> 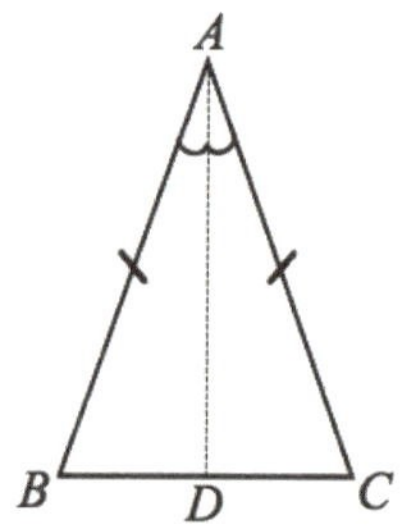
>
>
> *Construction:* Draw the angle bisector $\overrightarrow{AD}$ of $\angle A$, where D is the intersection of the bisector and $\overline{BC}$. We are going to use this auxiliary line towards our SAS criteria.

Allow students five minutes to attempt the proof on their own.

$AB = AC$	*Given*
$AD = AD$	*Reflexive property*
$m\angle BAD = m\angle CAD$	*Definition of an angle bisector*
$\triangle ABD \cong \triangle ACD$	*SAS*
$\angle B \cong \angle C$	*Corresponding angles of congruent triangles are congruent.*

Exercises (10 minutes)

Note that Exercise 4 asks students to use transformations in the proof. This is intended to reinforce the notion that congruence is defined in terms of rigid motions.

> **Exercises**
>
> 1. Given: $JK = JL$; $\overline{JR}$ bisects $\overline{KL}$
>   Prove: $\overline{JR} \perp \overline{KL}$
>
> 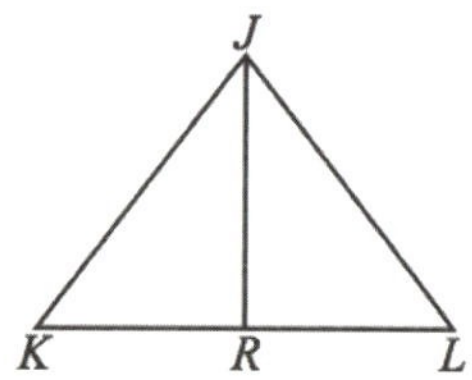
>

$JK = JL$	*Given*
$\angle K \cong \angle L$	*Base angles of an isosceles triangle are congruent.*
$KR = RL$	*Definition of a segment bisector*
$\therefore \triangle JRK \cong \triangle JRL$	*SAS*
$\angle JRK \cong \angle JRL$	*Corresponding angles of congruent triangles are congruent.*
$m\angle JRK + m\angle JRL = 180°$	*Linear pairs form supplementary angles.*
$2(m\angle JRK) = 180°$	*Substitution property of equality*
$m\angle JRK = 90°$	*Division property of equality*
$\therefore \overline{JR} \perp \overline{KL}$	*Definition of perpendicular lines*

2. Given: $AB = AC,\ XB = XC$

Prove: $\overline{AX}$ bisects $\angle BAC$

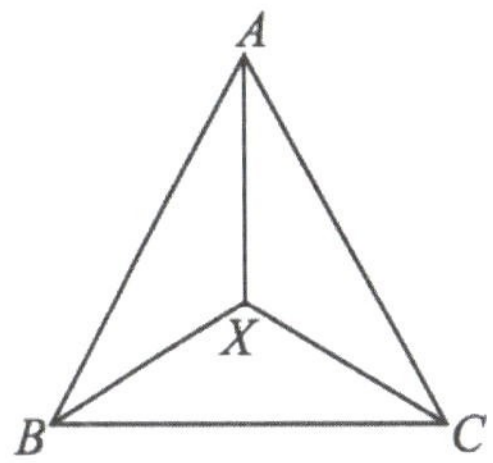

$AB = AC$	*Given*
$m\angle ABC = m\angle ACB$	*Base angles of an isosceles triangle are equal in measure.*
$XB = XC$	*Given*
$m\angle XBC = m\angle XCB$	*Base angles of an isosceles triangle are equal in measure.*
$m\angle ABC = m\angle ABX + m\angle XBC$,	
$m\angle ACB = m\angle ACX + m\angle XCB$	*Angle addition postulate*
$m\angle ABX = m\angle ABC - m\angle XBC$,	
$m\angle ACX = m\angle ACB - m\angle XCB$	*Subtraction property of equality*
$m\angle ABX = m\angle ACX$	*Substitution property of equality*
$\therefore \triangle ABX \cong \triangle ACX$	*SAS*
$\angle BAX \cong \angle CAX$	*Corresponding parts of congruent triangles are congruent.*
$\therefore\ \overline{AX}$ bisects $\angle BAC$.	*Definition of an angle bisector*

3. Given: $JX = JY, KX = LY$

Prove: $\triangle JKL$ is isosceles

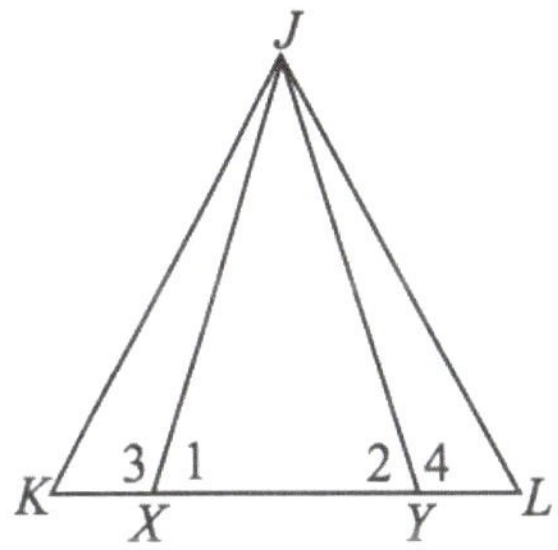

$JX = JY$	*Given*
$KX = LY$	*Given*
$m\angle 1 = m\angle 2$	*Base angles of an isosceles triangle are equal in measure.*
$m\angle 1 + m\angle 3 = 180°$	*Linear pairs form supplementary angles.*
$m\angle 2 + m\angle 4 = 180°$	*Linear pairs form supplementary angles.*
$m\angle 3 = 180° - m\angle 1$	*Subtraction property of equality*
$m\angle 4 = 180° - m\angle 3$	*Subtraction property of equality*
$m\angle 3 = m\angle 4$	*Substitution property of equality*
$\therefore \triangle JKX \cong \triangle JYL$	*SAS*
$\overline{JK} \cong \overline{JL}$	*Corresponding parts of congruent triangles are congruent.*
$\therefore \triangle JKL$ is isosceles.	*Definition of an isosceles triangle*

4. Given: $\triangle ABC$, with $m\angle CBA = m\angle BCA$

Prove: $BA = CA$

(Converse of base angles of isosceles triangle)

Hint: Use a transformation.

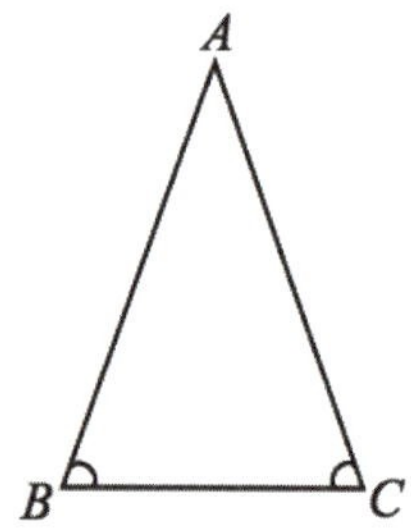

PROOF:

We can prove that $AB = AC$ by using rigid motions. Construct the perpendicular bisector l to $\overline{BC}$. Note that we do not know whether the point A is on l. If we did, we would know immediately that $AB = AC$, since all points on the perpendicular bisector are equidistant from B and C. We need to show that A is on l, or equivalently, that the reflection across l takes A to A.

Let r_l be the transformation that reflects the $\triangle ABC$ across l. Since l is the perpendicular bisector, $r_l(B) = C$ and $r_l(C) = B$. We do not know where the transformation takes A; for now, let us say that $r_l(A) = A'$.

Since $\angle CBA \cong \angle BCA$ and rigid motions preserve angle measures, after applying r_l to $\angle BCA$, we get that $\angle CBA \cong \angle CBA'$. $\angle CBA$ and $\angle CBA'$ share a ray BC, are of equal measure, and A and A' are both in the same half-plane with respect to line BC. Hence, $\overrightarrow{BA}$ and $\overrightarrow{BA'}$ are the same ray. In particular, A' is a point somewhere on $\overrightarrow{BA}$.

Likewise, applying r_l to $\angle CBA$ gives $\angle BCA \cong \angle BCA'$, and for the same reasons in the previous paragraph, A' must also be a point somewhere on $\overrightarrow{CA}$. Therefore, A' is common to both $\overrightarrow{BA}$ and $\overrightarrow{CA}$.

The only point common to both $\overrightarrow{BA}$ and $\overrightarrow{CA}$ is point A; thus, A and A' must be the same point, i.e., $A = A'$.

Hence, the reflection takes A to A, which means A is on the line l and $r_l(\overline{BA}) = \overline{CA'} = \overline{CA}$, or $BA = CA$.

5. Given: $\triangle ABC$, with $\overline{XY}$ is the angle bisector of $\angle BYA$, and $\overline{BC} \parallel \overline{XY}$

Prove: $YB = YC$

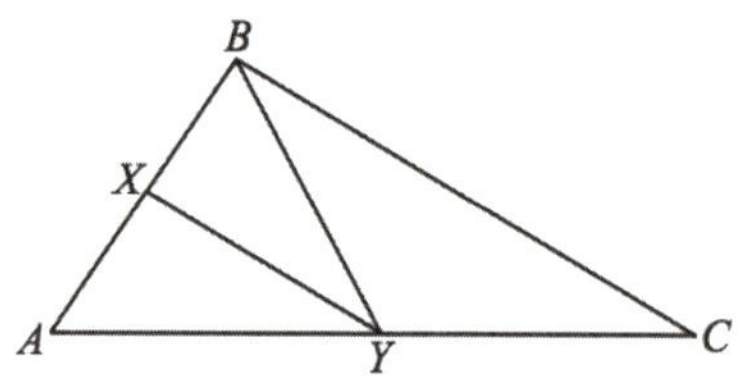

$\overline{BC} \parallel \overline{XY}$	*Given*
$m\angle XYB = m\angle CBY$	*When two parallel lines are cut by a transversal, the alternate interior angles are equal.*
$m\angle XYA = m\angle BCY$	*When two parallel lines are cut by a transversal, the corresponding angles are equal.*
$m\angle XYA = m\angle XYB$	*Definition of an angle bisector*
$m\angle CBY = m\angle BCY$	*Substitution property of equality*
$YB = YC$	*When the base angles of a triangle are equal in measure, the triangle is isosceles.*

Closing (1 minute)

- Would you prefer to prove that the base angles of an isosceles triangle are equal in measure using transformations or by using SAS? Why?

Exit Ticket (3 minutes)

Name ______________________________ Date ______________

Lesson 23: Base Angles of Isosceles Triangles

Exit Ticket

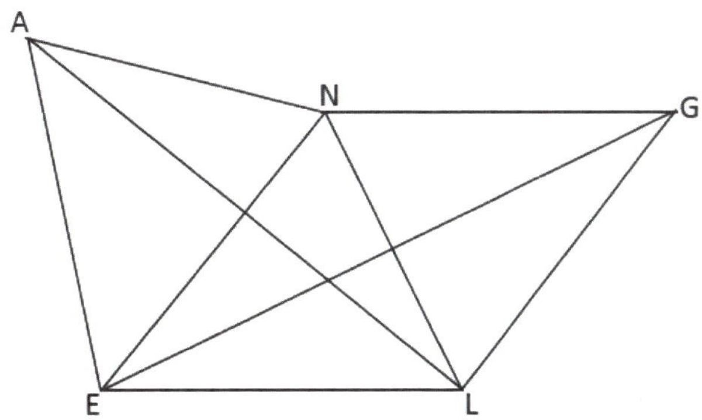

For each of the following, if the given congruence exists, name the isosceles triangle and the pair of congruent angles for the triangle based on the image above.

1. $\overline{AE} \cong \overline{LE}$

2. $\overline{LE} \cong \overline{LG}$

3. $\overline{AN} \cong \overline{LN}$

4. $\overline{EN} \cong \overline{GN}$

5. $\overline{NG} \cong \overline{LG}$

6. $\overline{AE} \cong \overline{NE}$

Exit Ticket Sample Solutions

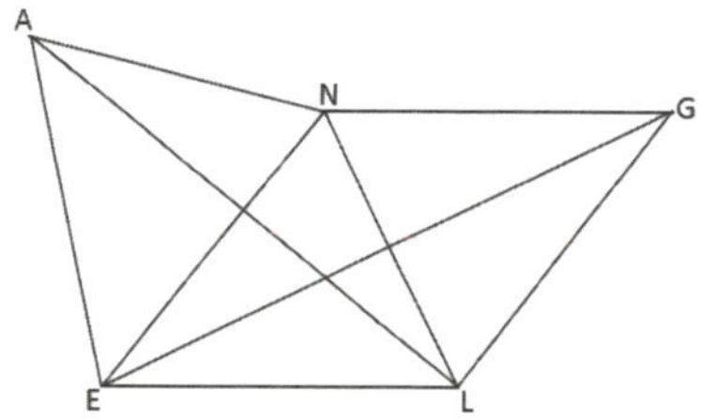

For each of the following, if the given congruence exists, name the isosceles triangle and the pair of congruent angles for the triangle based on the image above.

1. $\overline{AE} \cong \overline{LE}$

   $\triangle AEL,\ \angle EAL \cong \angle ELA$

2. $\overline{LE} \cong \overline{LG}$

   $\triangle LEG,\ \angle LEG \cong \angle LGE$

3. $\overline{AN} \cong \overline{LN}$

   $\triangle NLA,\ \angle NLA \cong \angle NAL$

4. $\overline{EN} \cong \overline{GN}$

   $\triangle NGE,\ \angle NGE \cong \angle NEG$

5. $\overline{NG} \cong \overline{LG}$

   $\triangle GLN,\ \angle GNL \cong \angle GLN$

6. $\overline{AE} \cong \overline{NE}$

   $\triangle AEN,\ \angle ENA \cong \angle EAN$

Problem Set Sample Solutions

1. Given: $AB = BC, AD = DC$

   Prove: $\triangle ADB$ and $\triangle CDB$ are right triangles

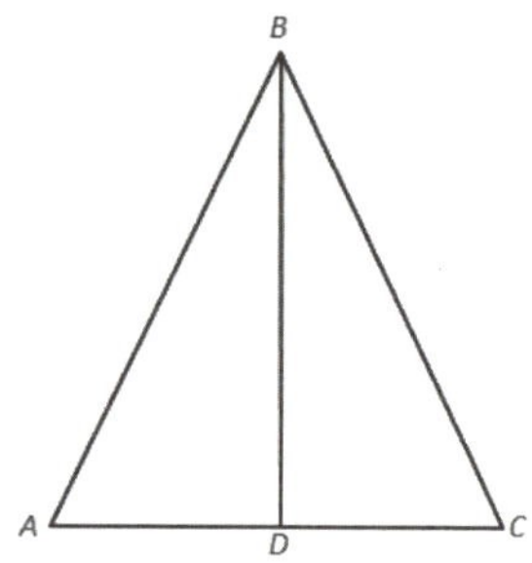

$AB = BC$	*Given*
$\triangle ABC$ *is isosceles.*	*Definition of isosceles triangle*
$m\angle A = m\angle C$	*Base angles of an isosceles triangle are equal in measure.*
$AD = DC$	*Given*
$\triangle ADB \cong \triangle CDB$	*SAS*
$m\angle ADB = m\angle CDB$	*Corresponding angles of congruent triangles are equal in measure.*
$m\angle ADB + m\angle CDB = 180^\circ$	*Linear pairs form supplementary angles.*
$2(m\angle ADB) = 180^\circ$	*Substitution property of equality*
$m\angle ADB = 90^\circ$	*Division property of equality*
$m\angle CDB = 90^\circ$	*Transitive property*
$\triangle ADB$ *and* $\triangle CDB$ *are right triangles.*	*Definition of a right triangle*

2. Given: $AC = AE$ and $\overline{BF} \parallel \overline{CE}$

Prove: $AB = AF$

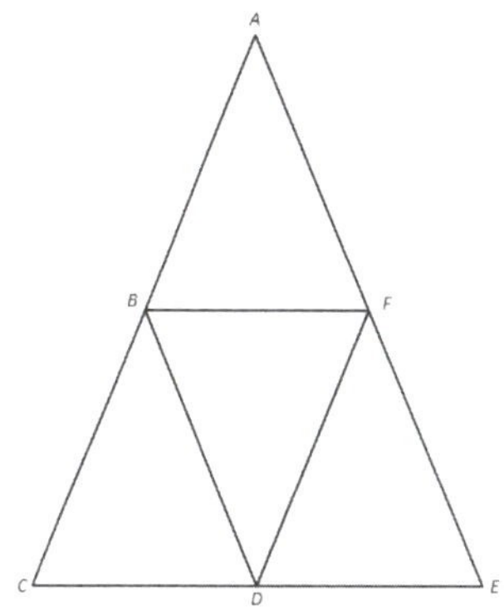

$AC = AE$	*Given*
$\triangle ACE$ *is isosceles.*	*Definition of isosceles triangle*
$m\angle ACE = m\angle AEC$	*Base angles of an isosceles triangle are equal in measure.*
$\overline{BF} \parallel \overline{CE}$	*Given*
$m\angle ACE = m\angle ABF$	*If parallel lines are cut by a transversal, then corresponding angles are equal in measure.*
$m\angle AFB = m\angle AEC$	*If parallel lines are cut by a transversal, then corresponding angles are equal in measure.*
$m\angle AFB = m\angle ABF$	*Transitive property*
$\triangle ABF$ *is isosceles.*	*If the base angles of a triangle are equal in measure, the triangle is isosceles.*
$AB = AF$	*Definition of an isosceles triangle*

3. In the diagram, $\triangle ABC$ is isosceles with $\overline{AC} \cong \overline{AB}$. In your own words, describe how transformations and the properties of rigid motions can be used to show that $\angle C \cong \angle B$.

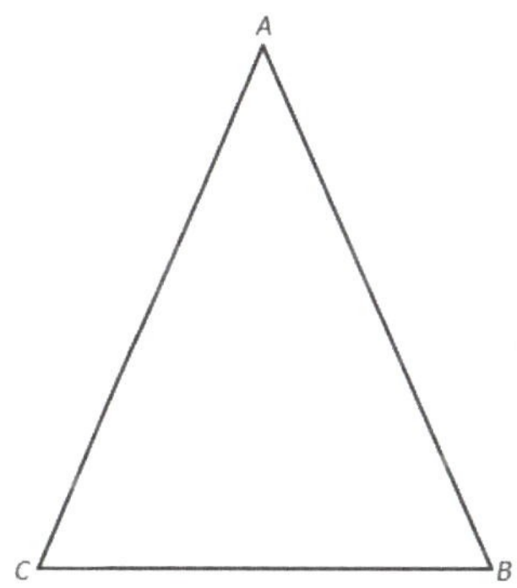

Answers will vary. A possible response would summarize the key elements of the proof from the lesson, such as the response below.

It can be shown that $\angle C \cong \angle B$ *using a reflection across the bisector of* $\angle A$. *The two angles formed by the bisector of* $\angle A$ *would be mapped onto one another since they share a ray, and rigid motions preserve angle measure.*

Since segment length is also preserved, B *is mapped onto* C *and vice versa.*

Finally, since rays are taken to rays and rigid motions preserve angle measure, $\overrightarrow{CA}$ *is mapped onto* $\overrightarrow{BA}$, $\overrightarrow{CB}$ *is mapped onto* $\overrightarrow{BC}$, *and* $\angle C$ *is mapped onto* $\angle B$. *From this, we see that* $\angle C \cong \angle B$.

Lesson 24: Congruence Criteria for Triangles—ASA and SSS

Student Outcomes

- Students learn why any two triangles that satisfy the ASA or SSS congruence criteria must be congruent.

Lesson Notes

This is the third lesson in the congruency topic. So far, students have studied the SAS triangle congruence criteria and how to prove base angles of an isosceles triangle are congruent. Students examine two more triangle congruence criteria in this lesson: ASA and SSS. Each proof assumes the initial steps from the proof of SAS; ask students to refer to their notes on SAS to recall these steps before proceeding with the rest of the proof. Exercises require using all three triangle congruence criteria.

Classwork

Opening Exercise (7 minutes)

Opening Exercise

Use the provided 30° angle as one base angle of an isosceles triangle. Use a compass and straight edge to construct an appropriate isosceles triangle around it.

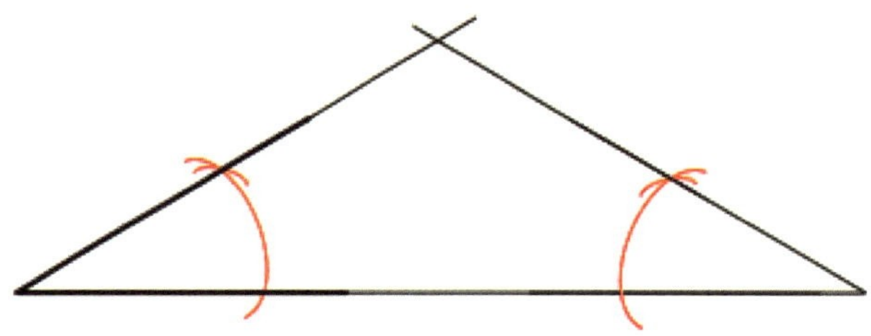

Compare your constructed isosceles triangle with a neighbor's. Does using a given angle measure guarantee that all the triangles constructed in class have corresponding sides of equal lengths?

No, side lengths may vary.

Discussion (24 minutes)

There are a variety of proofs of the ASA and SSS criteria. These follow from the SAS criteria already proved in Lesson 22.

Discussion

Today we are going to examine two more triangle congruence criteria, Angle-Side-Angle (ASA) and Side-Side-Side (SSS), to add to the SAS criteria we have already learned. We begin with the ASA criteria.

ANGLE-SIDE-ANGLE TRIANGLE CONGRUENCE CRITERIA (ASA): Given two triangles $\triangle ABC$ and $\triangle A'B'C'$. If $m\angle CAB = m\angle C'A'B'$ (Angle), $AB = A'B'$ (Side), and $m\angle CBA = m\angle C'B'A'$ (Angle), then the triangles are congruent.

PROOF:

We do not begin at the very beginning of this proof. Revisit your notes on the SAS proof, and recall that there are three cases to consider when comparing two triangles. In the most general case, when comparing two distinct triangles, we translate one vertex to another (choose congruent corresponding angles). A rotation brings congruent, corresponding sides together. Since the ASA criteria allows for these steps, we begin here.

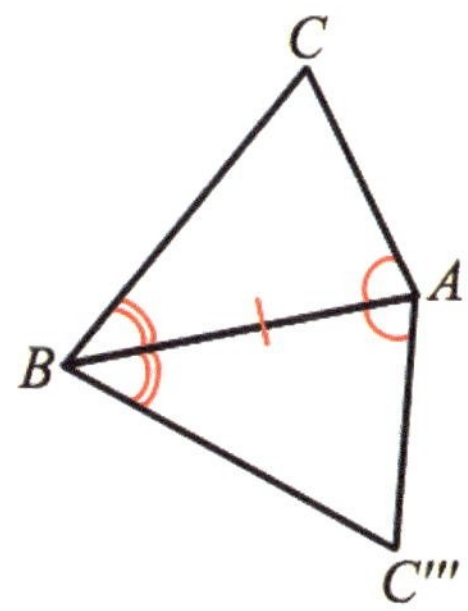

In order to map $\triangle ABC'''$ to $\triangle ABC$, we apply a reflection r across the line AB. A reflection maps A to A and B to B, since they are on line AB. However, we say that $r(C''') = C^*$. Though we know that $r(C''')$ is now in the same half-plane of line AB as C, we cannot assume that C''' maps to C. So we have $r(\triangle ABC''') = \triangle ABC^*$. To prove the theorem, we need to verify that C^* is C.

By hypothesis, we know that $\angle CAB \cong \angle C'''AB$ (recall that $\angle C'''AB$ is the result of two rigid motions of $\angle C'A'B'$, so must have the same angle measure as $\angle C'A'B'$). Similarly, $\angle CBA \cong \angle C'''BA$. Since $\angle CAB \cong r(\angle C'''AB) \cong \angle C^*AB$, and C and C^* are in the same half-plane of line AB, we conclude that $\overrightarrow{AC}$ and $\overrightarrow{AC^*}$ must actually be the same ray. Because the points A and C^* define the same ray as $\overrightarrow{AC}$, the point C^* must be a point somewhere on $\overrightarrow{AC}$. Using the second equality of angles, $\angle CBA \cong r(\angle C'''BA) \cong \angle C^*BA$, we can also conclude that $\overrightarrow{BC}$ and $\overrightarrow{BC^*}$ must be the same ray. Therefore, the point C^* must also be on $\overrightarrow{BC}$. Since C^* is on both $\overrightarrow{AC}$ and $\overrightarrow{BC}$, and the two rays only have one point in common, namely C, we conclude that $C = C^*$.

We have now used a series of rigid motions to map two triangles onto one another that meet the ASA criteria.

SIDE-SIDE-SIDE TRIANGLE CONGRUENCE CRITERIA (SSS): Given two triangles $\triangle ABC$ and $\triangle A'B'C'$, if $AB = A'B'$ (Side), $AC = A'C'$ (Side), and $BC = B'C'$ (Side), then the triangles are congruent.

PROOF:

Again, we do not start at the beginning of this proof, but assume there is a congruence that brings a pair of corresponding sides together, namely the longest side of each triangle.

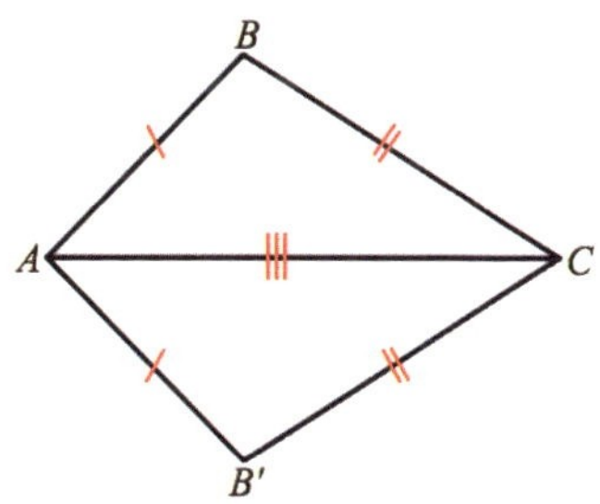

Without any information about the angles of the triangles, we cannot perform a reflection as we have in the proofs for SAS and ASA. What can we do? First we add a construction: Draw an auxiliary line from B to B', and label the angles created by the auxiliary line as r, s, t, and u.

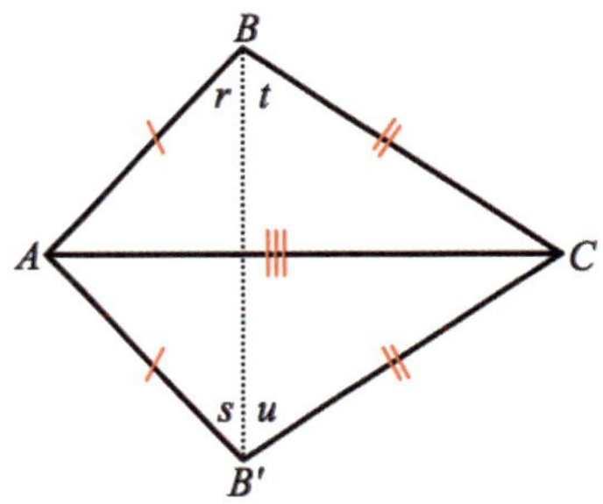

Since $AB = AB'$ and $CB = CB'$, $\triangle ABB'$ and $\triangle CBB'$ are both isosceles triangles respectively by definition. Therefore, $r = s$ because they are base angles of an isosceles triangle ABB'. Similarly, $m\angle t = m\angle u$ because they are base angles of $\triangle CBB'$. Hence, $\angle ABC = m\angle r + m\angle t = m\angle s + m\angle u = \angle AB'C$. Since $m\angle ABC = m\angle AB'C$, we say that $\triangle ABC \cong \triangle AB'C$ by SAS.

We have now used a series of rigid motions and a construction to map two triangles that meet the SSS criteria onto one another. Note that when using the Side-Side-Side triangle congruence criteria as a reason in a proof, you need only state the congruence and *SSS*. Similarly, when using the Angle-Side-Angle congruence as a reason in a proof, you need only state the congruence and *ASA*.

Now we have three triangle congruence criteria at our disposal: SAS, ASA, and SSS. We use these criteria to determine whether or not pairs of triangles are congruent.

Exercises (6 minutes)

These exercises involve applying the newly developed congruence criteria to a variety of diagrams.

Exercises

Based on the information provided, determine whether a congruence exists between triangles. If a congruence exists between triangles or if multiple congruencies exist, state the congruencies and the criteria used to determine them.

1. Given: M is the midpoint of $\overline{HP}$, $m\angle H = m\angle P$

   $\triangle GHM \cong \triangle RPM$, *ASA*

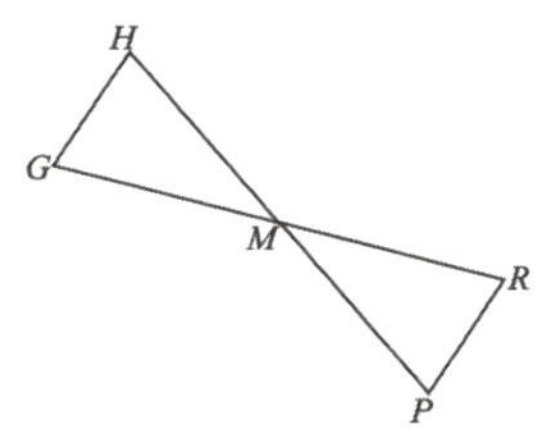

2. Given: Rectangle $JKLM$ with diagonal $\overline{KM}$

   $\triangle JKM \cong \triangle LMK$, *SSS/SAS/ASA*

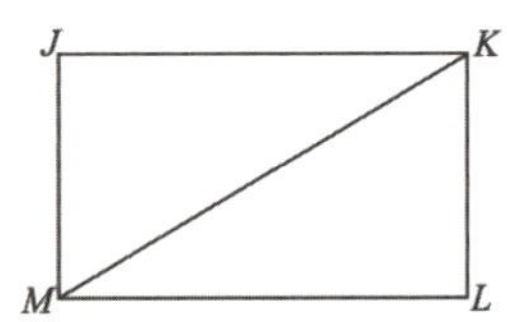

3. Given: $RY = RB$, $AR = XR$

   $\triangle ARY \cong \triangle XRB$, *SAS*

   $\triangle ABY \cong \triangle XYB$, *SAS*

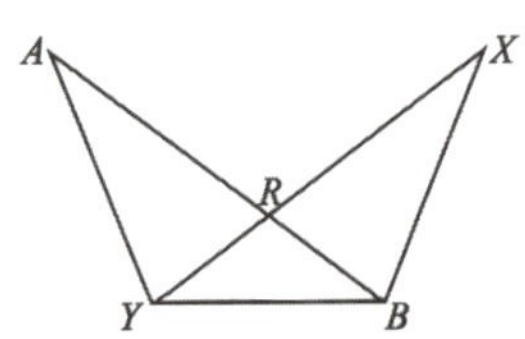

4. Given: $m\angle A = m\angle D, AE = DE$

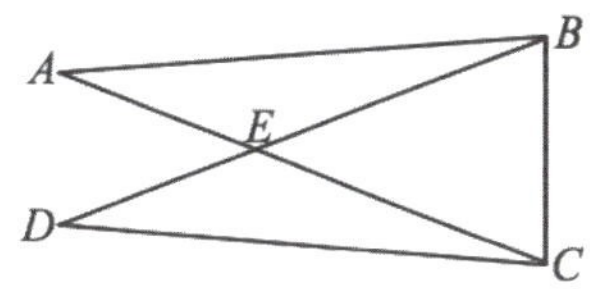

$\triangle AEB \cong \triangle DEC$, ASA

$\triangle DBC \cong \triangle ACB$, SAS/ASA

5. Given: $AB = AC, BD = \frac{1}{4}AB, CE = \frac{1}{4}AC.$

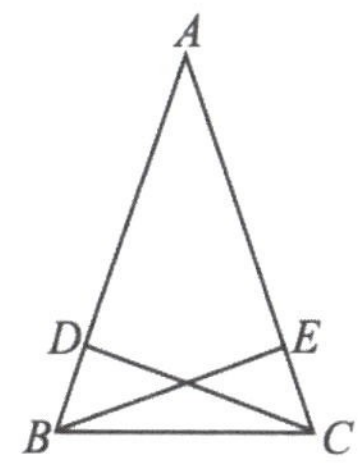

$\triangle ABE \cong \triangle ACD$, SAS

Closing (1 minute)

- **ANGLE-SIDE-ANGLE TRIANGLE CONGRUENCE CRITERIA (ASA):** Given two triangles $\triangle ABC$ and $\triangle A'B'C'$. If $m\angle CAB = m\angle C'A'B'$ (Angle), $AB = A'B'$ (Side), and $m\angle CBA = m\angle C'B'A'$ (Angle), then the triangles are congruent.
- **SIDE-SIDE-SIDE TRIANGLE CONGRUENCE CRITERIA (SSS):** Given two triangles $\triangle ABC$ and $\triangle A'B'C'$. If $AB = A'B'$ (Side), $AC = A'C'$ (Side), and $BC = B'C'$ (Side), then the triangles are congruent.
- The ASA and SSS criteria implies the existence of a congruence that maps one triangle onto the other.

Exit Ticket (7 minutes)

Name ______________________________ Date ______________

Lesson 24: Congruence Criteria for Triangles—ASA and SSS

Exit Ticket

Based on the information provided, determine whether a congruence exists between triangles. If a congruence exists between triangles or if multiple congruencies exist, state the congruencies and the criteria used to determine them.

Given: $BD = CD$, E is the midpoint of $\overline{BC}$

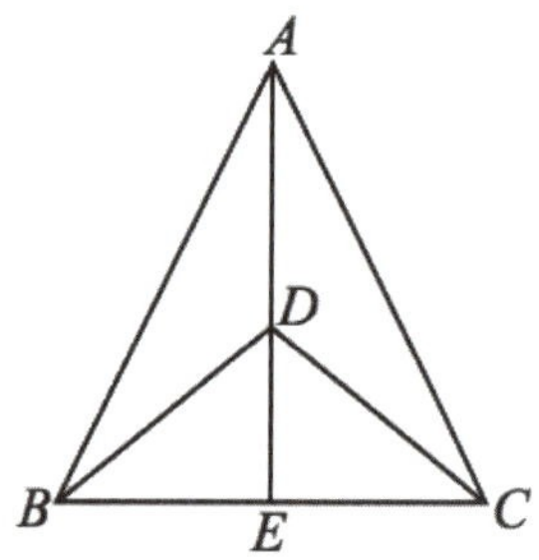

Exit Ticket Sample Solutions

Based on the information provided, determine whether a congruence exists between triangles. If a congruence exists between triangles or if multiple congruencies exist, state the congruencies and the criteria used to determine them.

Given: $BD = CD$, E is the midpoint of $\overline{BC}$

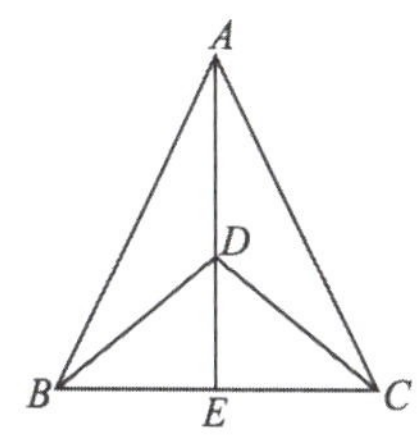

$\triangle BDE \cong \triangle CDE$, *SSS*

$\triangle ABE \cong \triangle ACE$, *SSS or SAS*

$\triangle ABD \cong \triangle ACD$, *SSS or SAS*

Problem Set Sample Solutions

Use your knowledge of triangle congruence criteria to write proofs for each of the following problems.

1. Given: Circles with centers A and B intersect at C and D

   Prove: $\angle CAB \cong \angle DAB$

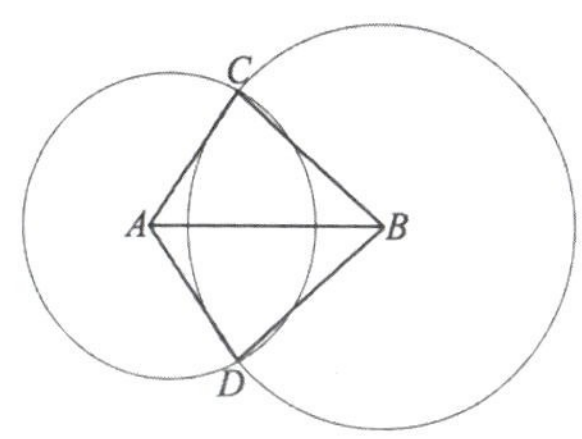

$CA = DA$	*Radius of circle*
$CB = DB$	*Radius of circle*
$AB = AB$	*Reflexive property*
$\triangle CAB \cong \triangle DAB$	*SSS*
$\angle CAB \cong \angle DAB$	*Corresponding angles of congruent triangles are congruent.*

2. Given: $m\angle J = m\angle M$, $JA = MB$, $JK = KL = LM$

   Prove: $\overline{KR} \cong \overline{LR}$

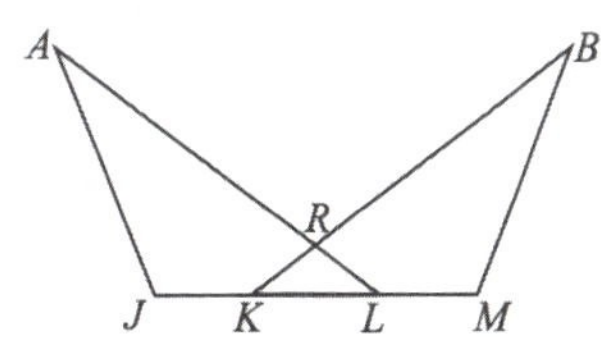

$m\angle J = m\angle M$	*Given*
$JA = MB$	*Given*
$JK = KL = LM$	*Given*
$JL = JK + KL$	*Partition property or segments add*
$KM = KL + LM$	*Partition property or segments add*
$KL = KL$	*Reflexive property*
$JK + KL = KL + LM$	*Addition property of equality*
$JL = KM$	*Substitution property of equality*
$\triangle AJL \cong \triangle BMK$	*SAS*
$\angle RKL \cong \angle RLK$	*Corresponding angles of congruent triangles are congruent.*
$\overline{KR} \cong \overline{LR}$	*If two angles of a triangle are congruent, the sides opposite those angles are congruent.*

3. **Given:** $m\angle w = m\angle x$, and $m\angle y = m\angle z$

Prove: (1) $\triangle ABE \cong \triangle ACE$

(2) $AB = AC$ and $\overline{AD} \perp \overline{BC}$

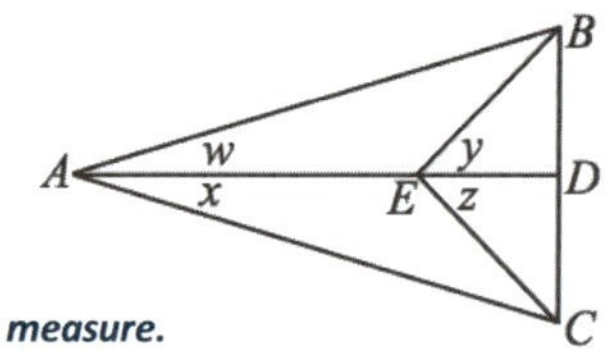

$m\angle y = m\angle z$	*Given*
$m\angle AEB = m\angle AEC$	*Supplements of equal angles are equal in measure.*
$AE = AE$	*Reflexive property*
$m\angle w = m\angle x$	*Given*
$\therefore \triangle ABE \cong \triangle ACE$	*ASA*
$\therefore \overline{AB} \cong \overline{AC}$	*Corresponding sides of congruent triangles are congruent.*
$AD = AD$	*Reflexive property*
$\triangle CAD \cong \triangle BAD$	*SAS*
$m\angle ADC = m\angle ADB$	*Corresponding angles of congruent triangles are equal in measure.*
$m\angle ADC + m\angle ADB = 180°$	*Linear pairs form supplementary angles.*
$2(m\angle ADC) = 180°$	*Substitution property of equality*
$m\angle ADC = 90°$	*Division property of equality*
$\overline{AD} \perp \overline{BC}$	*Definition of perpendicular lines*

4. After completing the last exercise, Jeanne said, "We also could have been given that $\angle w \cong \angle x$ and $\angle y \cong \angle z$. This would also have allowed us to prove that $\triangle ABE \cong \triangle ACE$." Do you agree? Why or why not?

Yes; any time angles are equal in measure, we can also say that they are congruent. This is because rigid motions preserve angle measure; therefore, any time angles are equal in measure, there exists a sequence of rigid motions that carries one onto the other.

Lesson 25: Congruence Criteria for Triangles—AAS and HL

Student Outcomes

- Students learn why any two triangles that satisfy the AAS or HL congruence criteria must be congruent.
- Students learn why any two triangles that meet the AAA or SSA criteria are not necessarily congruent.

Classwork

Opening Exercise (7 minutes)

Opening Exercise

Write a proof for the following question. Once done, compare your proof with a neighbor's.

Given: $DE = DG, EF = GF$

Prove: $\overline{DF}$ is the angle bisector of $\angle EDG$

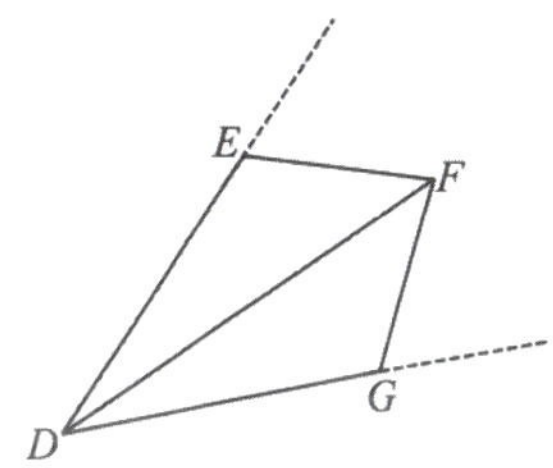

$DE = DG$	*Given*
$EF = GF$	*Given*
$DF = DF$	*Reflexive property*
$\triangle DEF \cong \triangle DGF$	*SSS*
$\angle EDF \cong \angle GDF$	*Corresponding angles of congruent triangles are congruent.*
$\overline{DF}$ *is the angle bisector of* $\angle EDG$.	*Definition of an angle bisector*

Exploratory Challenge (23 minutes)

The included proofs of AAS and HL are not transformational; rather, they follow from ASA and SSS, already proved.

Exploratory Challenge

Today we are going to examine three possible triangle congruence criteria, Angle-Angle-Side (AAS), Side-Side-Angle (SSA), and Angle-Angle-Angle (AAA). Ultimately, only one of the three possible criteria ensures congruence.

ANGLE-ANGLE-SIDE TRIANGLE CONGRUENCE CRITERIA (AAS): Given two triangles $\triangle ABC$ and $\triangle A'B'C'$. If $AB = A'B'$ (Side), $m\angle B = m\angle B'$ (Angle), and $m\angle C = m\angle C'$ (Angle), then the triangles are congruent.

PROOF:

Consider a pair of triangles that meet the AAS criteria. If you knew that two angles of one triangle corresponded to and were equal in measure to two angles of the other triangle, what conclusions can you draw about the third angle of each triangle?

Since the first two angles are equal in measure, the third angles must also be equal in measure.

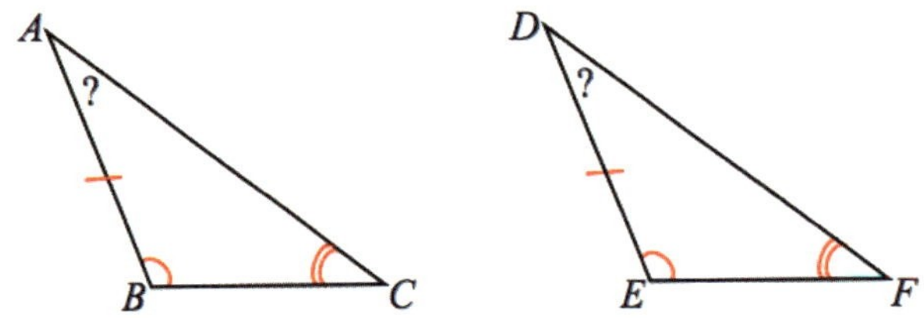

Given this conclusion, which formerly learned triangle congruence criteria can we use to determine if the pair of triangles are congruent?

ASA

Therefore, the AAS criterion is actually an extension of the ___*ASA*___ triangle congruence criterion.

Note that when using the Angle-Angle-Side triangle congruence criteria as a reason in a proof, you need only to state the congruence and *AAS*.

HYPOTENUSE-LEG TRIANGLE CONGRUENCE CRITERIA (HL): Given two right triangles $\triangle ABC$ and $\triangle A'B'C'$ with right angles B and B'. If $AB = A'B'$ (Leg) and $AC = A'C'$ (Hypotenuse), then the triangles are congruent.

PROOF:

As with some of our other proofs, we do not start at the very beginning, but imagine that a congruence exists so that triangles have been brought together such that $A = A'$ and $C = C'$; the hypotenuse acts as a common side to the transformed triangles.

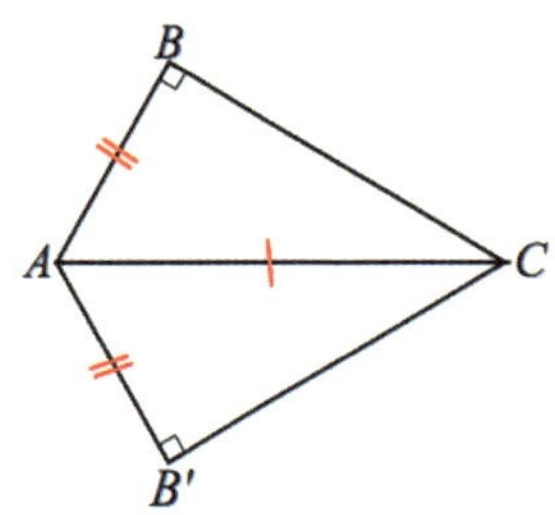

Similar to the proof for SSS, we add a construction and draw $\overline{BB'}$.

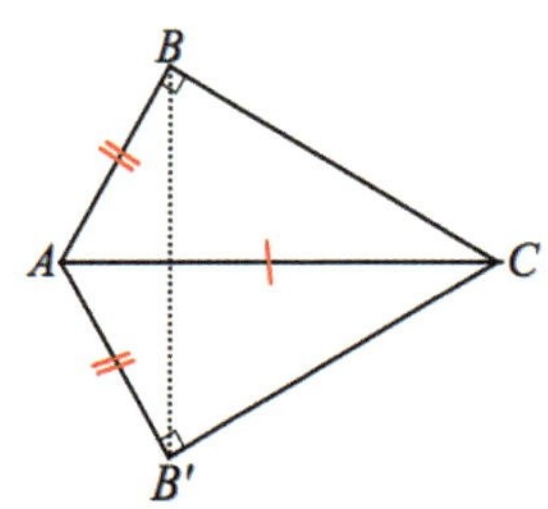

$\triangle ABB'$ is isosceles by definition, and we can conclude that base angles $m\angle ABB' = m\angle AB'B$. Since $\angle CBB'$ and $\angle CB'B$ are both the complements of equal angle measures ($\angle ABB'$ and $\angle AB'B$), they too are equal in measure. Furthermore, since $m\angle CBB' = m\angle CB'B$, the sides of $\triangle CBB'$ opposite them are equal in measure: $BC = B'C'$.

Then, by SSS, we can conclude $\triangle ABC \cong \triangle A'B'C'$. Note that when using the Hypotenuse-Leg triangle congruence criteria as a reason in a proof, you need only to state the congruence and *HL*.

Criteria that *do not* determine two triangles as congruent: SSA and AAA

SIDE-SIDE-ANGLE (SSA): Observe the diagrams below. Each triangle has a set of adjacent sides of measures 11 and 9, as well as the non-included angle of $23°$. Yet, the triangles are not congruent.

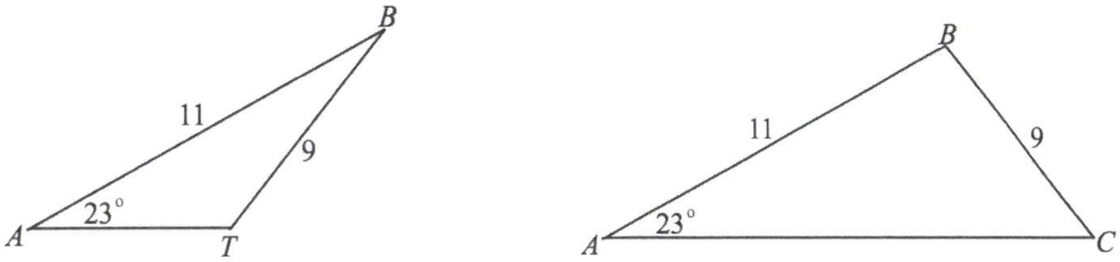

Examine the composite made of both triangles. The sides of length 9 each have been dashed to show their possible locations.

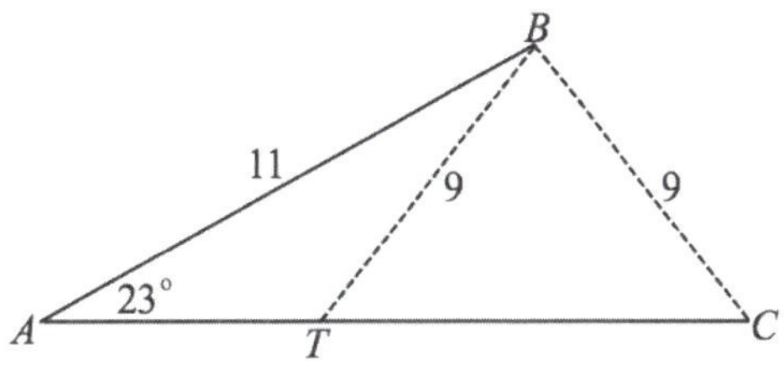

The triangles that satisfy the conditions of SSA cannot *guarantee* congruence criteria. In other words, two triangles under SSA criteria may or may not be congruent; therefore, we cannot categorize SSA as congruence criterion.

ANGLE-ANGLE-ANGLE (AAA): A correspondence exists between $\triangle ABC$ and $\triangle DEF$. Trace $\triangle ABC$ onto patty paper, and line up corresponding vertices.

Based on your observations, why isn't AAA categorized as congruence criteria? Is there any situation in which AAA does guarantee congruence?

Even though the angle measures may be the same, the sides can be proportionally larger; you can have similar triangles in addition to a congruent triangle.

List all the triangle congruence criteria here:

SSS, SAS, ASA, AAS, HL

List the criteria that do not determine congruence here:

SSA, AAA

Examples (8 minutes)

Examples

1. **Given:** $\overline{BC} \perp \overline{CD}$, $\overline{AB} \perp \overline{AD}$, $m\angle 1 = m\angle 2$
   Prove: $\triangle BCD \cong \triangle BAD$

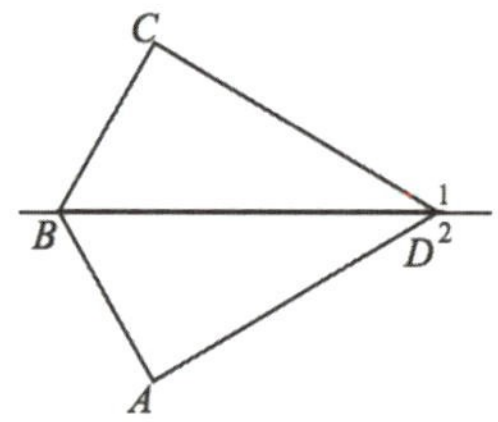

$m\angle 1 = m\angle 2$	*Given*
$\overline{AB} \perp \overline{AD}$	*Given*
$\overline{BC} \perp \overline{CD}$	*Given*
$BD = BD$	*Reflexive property*
$m\angle 1 + m\angle CDB = 180°$	*Linear pairs form supplementary angles.*
$m\angle 2 + m\angle ADB = 180°$	*Linear pairs form supplementary angles.*
$m\angle CDB = m\angle ADB$	*If two angles are equal in measure, then their supplements are equal in measure.*
$m\angle BCD = m\angle BAD = 90°$	*Definition of perpendicular line segments*
$\triangle BCD \cong \triangle BAD$	*AAS*

2. **Given:** $\overline{AD} \perp \overline{BD}$, $\overline{BD} \perp \overline{BC}$, $AB = CD$
   Prove: $\triangle ABD \cong \triangle CDB$

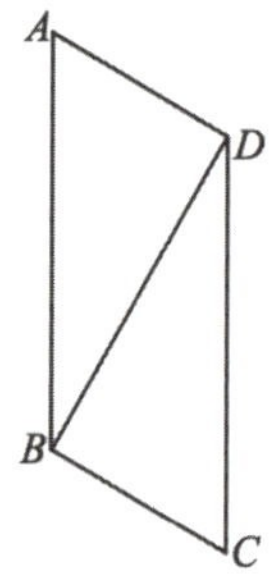

$\overline{AD} \perp \overline{BD}$	*Given*
$\overline{BD} \perp \overline{BC}$	*Given*
$\triangle ABD$ is a right triangle.	*Definition of perpendicular line segments*
$\triangle CDB$ is a right triangle.	*Definition of perpendicular line segments*
$AB = CD$	*Given*
$BD = DB$	*Reflexive property*
$\triangle ABD \cong \triangle CDB$	*HL*

Closing (2 minutes)

- **ANGLE-ANGLE-SIDE TRIANGLE CONGRUENCE CRITERIA (AAS):** Given two triangles $\triangle ABC$ and $\triangle A'B'C'$. If $AB = A'B'$ (Side), $m\angle B = m\angle B'$ (Angle), and $m\angle C = m\angle C'$ (Angle), then the triangles are congruent.
- **HYPOTENUSE-LEG TRIANGLE CONGRUENCE CRITERIA (HL):** Given two right triangles $\triangle ABC$ and $\triangle A'B'C'$ with right angles B and B'. If $AB = A'B'$ (Leg) and $AC = A'C'$ (Hypotenuse), then the triangles are congruent.
- The AAS and HL criteria implies the existence of a congruence that maps one triangle onto the other.
- Triangles that meet either the Angle-Angle-Angle or the Side-Side-Angle criteria do not guarantee congruence.

Exit Ticket (5 minutes)

Name ______________________________ Date ____________

Lesson 25: Congruence Criteria for Triangles—AAS and HL

Exit Ticket

1. Sketch an example of two triangles that meet the AAA criteria but are not congruent.

2. Sketch an example of two triangles that meet the SSA criteria that are not congruent.

Exit Ticket Sample Solutions

1. Sketch an example of two triangles that meet the AAA criteria but are not congruent.

   Responses should look something like the example below.

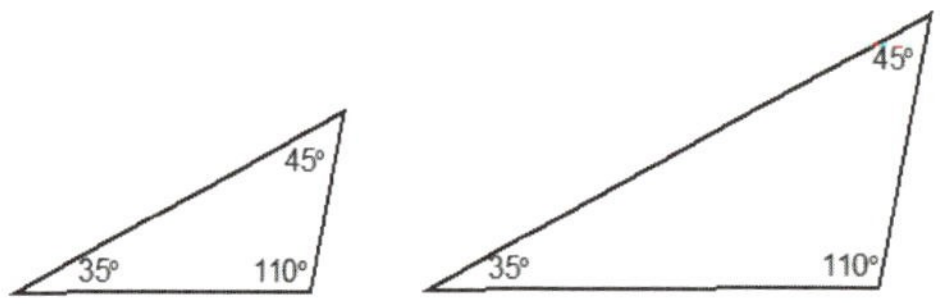

2. Sketch an example of two triangles that meet the SSA criteria that are not congruent.

   Responses should look something like the example below.

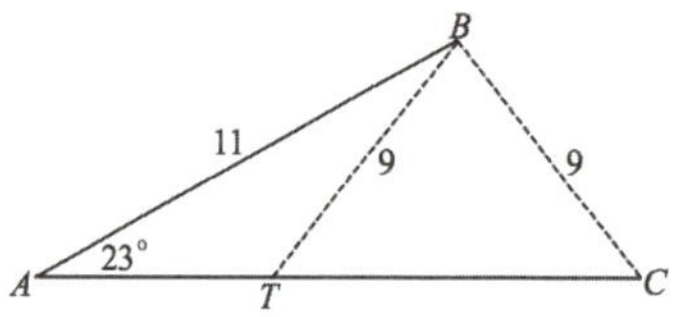

Problem Set Sample Solutions

Use your knowledge of triangle congruence criteria to write proofs for each of the following problems.

1. Given: $\overline{AB} \perp \overline{BC}$, $\overline{DE} \perp \overline{EF}$, $\overline{BC} \parallel \overline{EF}$, $AF = DC$

   Prove: $\triangle ABC \cong \triangle DEF$

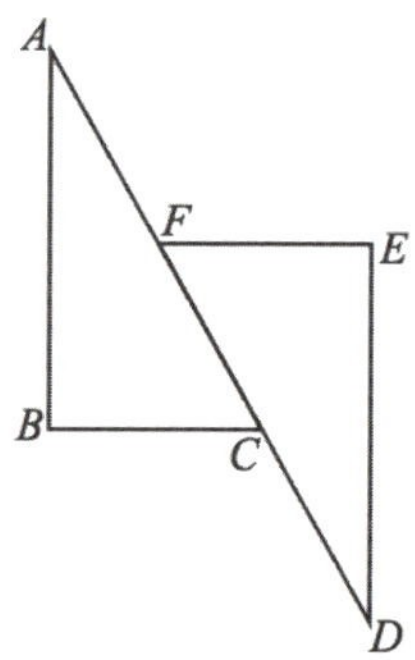

$\overline{AB} \perp \overline{BC}$	*Given*
$\overline{DE} \perp \overline{EF}$	*Given*
$\overline{BC} \parallel \overline{EF}$	*Given*
$AF = DC$	*Given*
$m\angle B = m\angle E = 90^\circ$	*Definition of perpendicular lines*
$m\angle C = m\angle F$	*When two parallel lines are cut by a transversal, the alternate interior angles are equal in measure.*
$FC = FC$	*Reflexive property*
$AF + FC = FC + CD$	*Addition property of equality*
$AC = DF$	*Segment addition*
$\triangle ABC \cong \triangle DEF$	*AAS*

2. In the figure, $\overline{PA} \perp \overline{AR}$ and $\overline{PB} \perp \overline{RB}$ and R is equidistant from $\overleftrightarrow{PA}$ and $\overleftrightarrow{PB}$. Prove that $\overline{PR}$ bisects $\angle APB$.

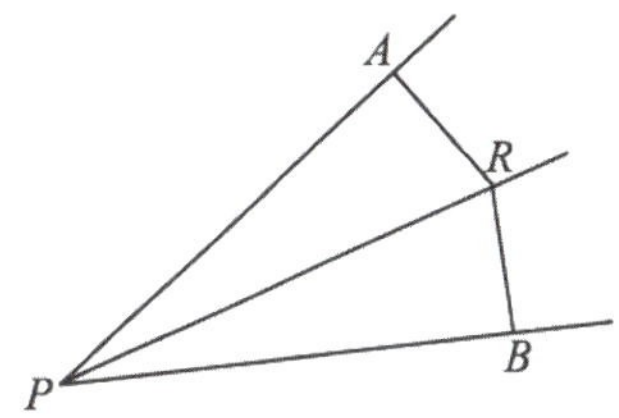

$\overline{PA} \perp \overline{AR}$	*Given*
$\overline{PB} \perp \overline{BR}$	*Given*
$RA = RB$	*Given*
$m\angle A = m\angle R = 90°$	*Definition of perpendicular lines*
$\triangle PAR$, $\triangle PBR$ *are right triangles.*	*Definition of right triangle*
$PR = PR$	*Reflexive property*
$\triangle PAR \cong \triangle PBR$	*HL*
$\angle APR \cong \angle RPB$	*Corresponding angles of congruent triangles are congruent.*
$\overline{PR}$ *bisects* $\angle APB$.	*Definition of an angle bisector*

3. Given: $\angle A \cong \angle P$, $\angle B \cong \angle R$, W is the midpoint of $\overline{AP}$

Prove: $\overline{RW} \cong \overline{BW}$

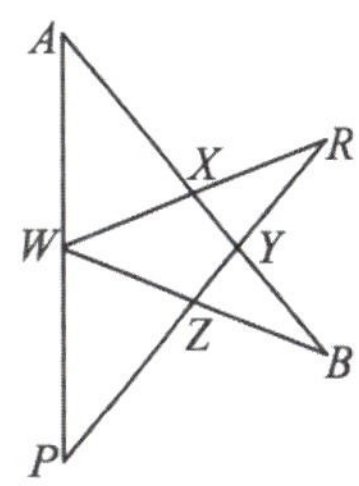

$\angle A \cong \angle P$	*Given*
$\angle B \cong \angle R$	*Given*
W *is the midpoint of* $\overline{AP}$.	*Given*
$AW = PW$	*Definition of midpoint*
$\triangle AWB \cong \triangle PWR$	*AAS*
$\overline{RW} \cong \overline{BW}$	*Corresponding sides of congruent triangles are congruent.*

4. Given: $BR = CU$, rectangle $RSTU$

Prove: $\triangle ARU$ is isosceles

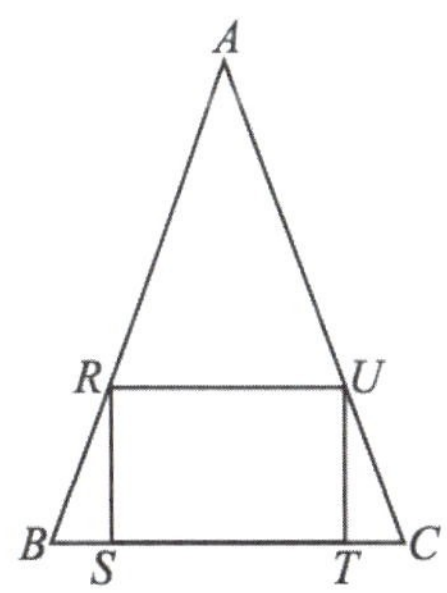

$BR = CU$	*Given*
Rectangle $RSTU$	*Given*
$\overline{BC} \parallel \overline{RU}$	*Definition of a rectangle*
$m\angle RBS = m\angle ARU$	*When two para. lines are cut by a trans., the corr. angles are equal in measure.*
$m\angle UCT = m\angle AUR$	*When two para. lines are cut by a trans., the corr. angles are equal in measure.*
$m\angle RST = 90°$, $m\angle UTS = 90°$	*Definition of a rectangle*
$m\angle RSB + m\angle RST = 180$	*Linear pairs form supplementary angles.*
$m\angle UTC + m\angle UTS = 180$	*Linear pairs form supplementary angles.*
$m\angle RSB = 90°$, $m\angle UTC = 90°$	*Subtraction property of equality*
$\triangle BRS$ *and* $\triangle TUC$ *are right triangles.*	*Definition of right triangle*
$RS = UT$	*Definition of a rectangle*
$\triangle BRS \cong \triangle TUC$	*HL*
$m\angle RBS = m\angle UCT$	*Corresponding angles of congruent triangles are equal in measure.*
$m\angle ARU = m\angle AUR$	*Substitution property of equality*
$\triangle ARU$ *is isosceles.*	*If two angles in a triangle are equal in measure, then it is isosceles.*

Lesson 26: Triangle Congruency Proofs

Student Outcomes

- Students complete proofs requiring a synthesis of the skills learned in the last four lessons.

Classwork

Exercises (39 minutes)

Exercises

1. Given: $\overline{AB} \perp \overline{BC}$, $\overline{BC} \perp \overline{DC}$

   $\overline{DB}$ bisects $\angle ABC$, $\overline{AC}$ bisects $\angle DCB$

   $EB = EC$

   Prove: $\triangle BEA \cong \triangle CED$

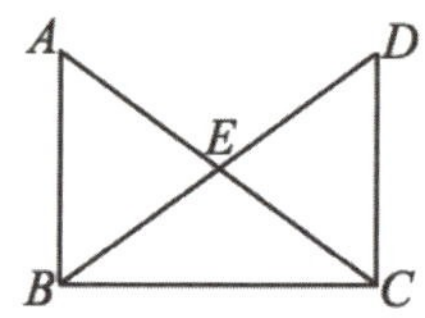

$\overline{AB} \perp \overline{BC}, \overline{BC} \perp \overline{DC}$	*Given*
$m\angle ABC = 90°, m\angle DCB = 90°$	*Definition of perpendicular lines*
$m\angle ABC = m\angle DCB$	*Transitive property*
$\overline{DB}$ *bisects* $\angle ABC$, $\overline{AC}$ *bisects* $\angle DCB$	*Given*
$m\angle ABE = 45°, m\angle DCE = 45°$	*Definition of an angle bisector*
$EB = EC$	*Given*
$m\angle AEB = m\angle DEC$	*Vertical angles are equal in measure.*
$\triangle BEA \cong \triangle CED$	*ASA*

2. Given: $\overline{BF} \perp \overline{AC}$, $\overline{CE} \perp \overline{AB}$

   $AE = AF$

   Prove: $\triangle ACE \cong \triangle ABF$

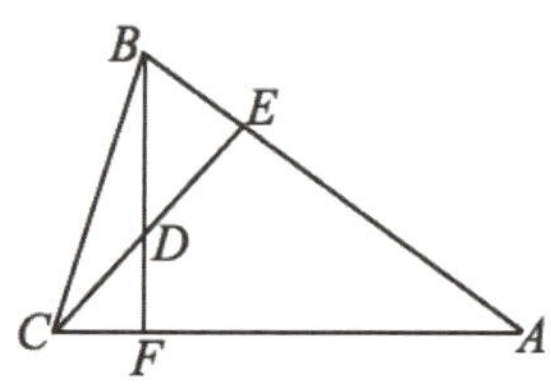

$\overline{BF} \perp \overline{AC}, \overline{CE} \perp \overline{AB}$	*Given*
$m\angle BFA = 90°, m\angle CEA = 90°$	*Definition of perpendicular*
$AE = AF$	*Given*
$m\angle A = m\angle A$	*Reflexive property*
$\triangle ACE \cong \triangle ABF$	*ASA*

EUREKA MATH™

3. Given: $XJ = YK$, $PX = PY$, $\angle ZXJ \cong \angle ZYK$
Prove: $JY = KX$

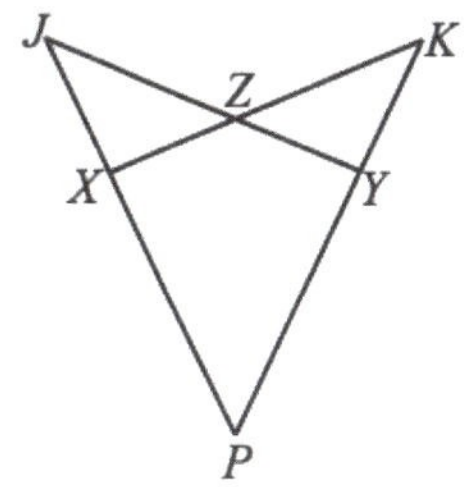

$XJ = YK$, $PX = PY$, $\angle ZXJ \cong \angle ZYK$	*Given*
$\overline{JP} \cong \overline{KP}$	*Segment addition*
$m\angle JZX = m\angle KZY$	*Vertical angles are equal in measure.*
$\triangle JZX \cong \triangle KZY$	*AAS*
$\angle J \cong \angle K$	*Corresponding angles of congruent triangles are congruent.*
$\angle P \cong \angle P$	*Reflexive property*
$\triangle PJY \cong \triangle PKX$	*AAS*
$\overline{JY} \cong \overline{KX}$	*Corresponding sides of congruent triangles are congruent.*
$JY = KX$	*Definition of congruent segments*

4. Given: $JK = JL, \overline{JK} \parallel \overline{XY}$
Prove: $XY = XL$

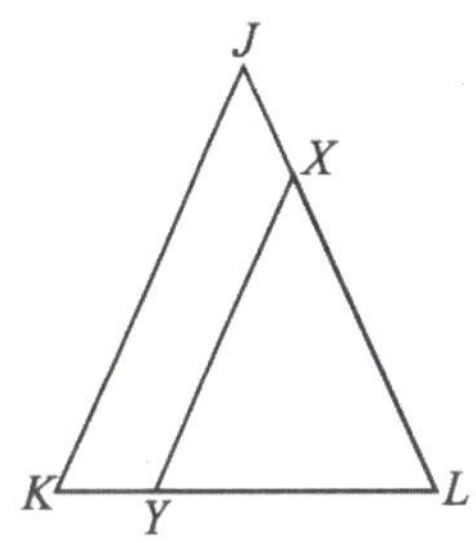

$JK = JL$	*Given*
$m\angle K = m\angle L$	*Base angles of an isosceles triangle are equal in measure.*
$\overline{JK} \parallel \overline{XY}$	*Given*
$m\angle K = m\angle XYL$	*When two parallel lines are cut by a transversal, corresponding angles are equal in measure.*
$m\angle XYL = m\angle L$	*Transitive property*
$XY = XL$	*If two angles of a triangle are congruent, then the sides opposite the angles are equal in length.*

5. Given: $\angle 1 \cong \angle 2, \angle 3 \cong \angle 4$
Prove: $\overline{AC} \cong \overline{BD}$

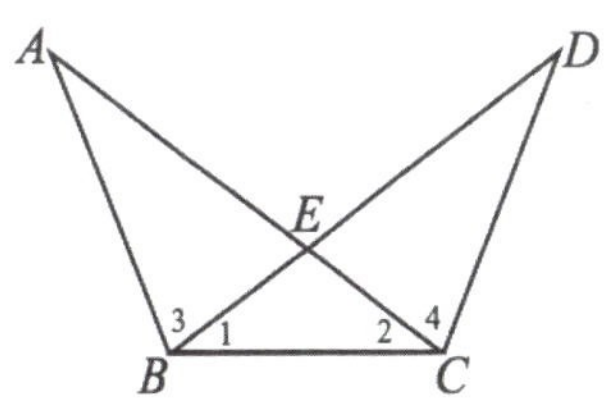

$\angle 1 \cong \angle 2$	*Given*
$\overline{BE} \cong \overline{CE}$	*When two angles of a triangle are congruent, it is an isosceles triangle.*
$\angle 3 \cong \angle 4$	*Given*
$\angle AEB \cong \angle DEC$	*Vertical angles are congruent.*
$\triangle ABE \cong \triangle DCE$	*ASA*
$\angle A \cong \angle D$	*Corresponding angles of congruent triangles are congruent.*
$m\angle ABC = m\angle 1 + m\angle 3$; $m\angle DCB = m\angle 2 + m\angle 4$	*Angle addition*
$m\angle ABC = m\angle DCB$	*Substitution*
$\overline{BC} \cong \overline{BC}$	*Reflexive property*
$\triangle ABC \cong \triangle DCB$	*AAS*
$\overline{AC} \cong \overline{BD}$	*Corresponding sides of congruent triangles are congruent*

6. Given: $m\angle 1 = m\angle 2, m\angle 3 = m\angle 4, AB = AC$

Prove: (a) $\triangle ABD \cong \triangle ACD$

(b) $\angle 5 \cong \angle 6$

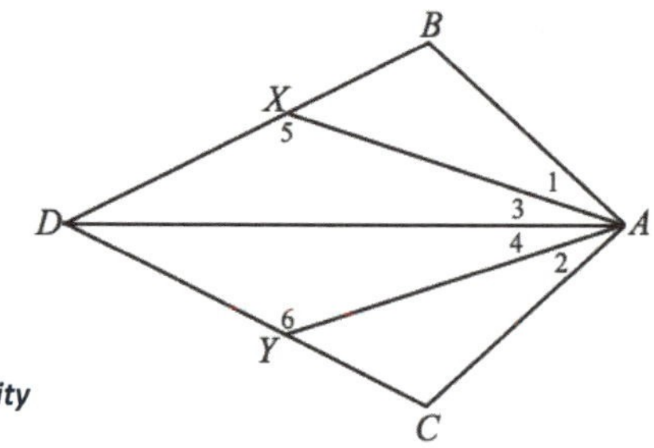

$m\angle 1 = m\angle 2, m\angle 3 = m\angle 4,$ $AB = AC$	*Given*
$m\angle 1 + m\angle 3 = m\angle DAB,$	
$m\angle 2 + m\angle 4 = m\angle DAC$	*Angle addition postulate*
$m\angle DAB = m\angle DAC$	*Substitution property of equality*
$AD = AD$	*Reflexive property*
$\triangle ABD \cong \triangle ACD$	*SAS*
$\angle BDA \cong \angle CDA$	*Corresponding angles of congruent triangles are congruent.*
$\triangle DXA \cong \triangle DYA$	*ASA*
$\angle 5 \cong \angle 6$	*Corresponding angles of congruent triangles are congruent.*

Closing (1 minute)

- Triangle congruency proofs require an application of known facts, definitions, axioms, and geometric properties.

Exit Ticket (5 minutes)

Name ______________________________ Date ______________

Lesson 26: Triangle Congruency Proofs

Exit Ticket

Identify the two triangle congruence criteria that do NOT guarantee congruence. Explain why they do not guarantee congruence, and provide illustrations that support your reasoning.

Exit Ticket Sample Solutions

Identify the two triangle congruence criteria that do NOT guarantee congruence. Explain why they do not guarantee congruence, and provide illustrations that support your reasoning.

Students should identify AAA and SSA as the two types of criteria that do not guarantee congruence. Appropriate illustrations should be included with their justifications.

Problem Set Sample Solutions

Use your knowledge of triangle congruence criteria to write a proof for the following:

In the figure, $\overline{RX}$ and $\overline{RY}$ are the perpendicular bisectors of $\overline{AB}$ and $\overline{AC}$, respectively.

Prove: (a) $\triangle RAX \cong \triangle RBX$

(b) $\overline{RA} \cong \overline{RB} \cong \overline{RC}$

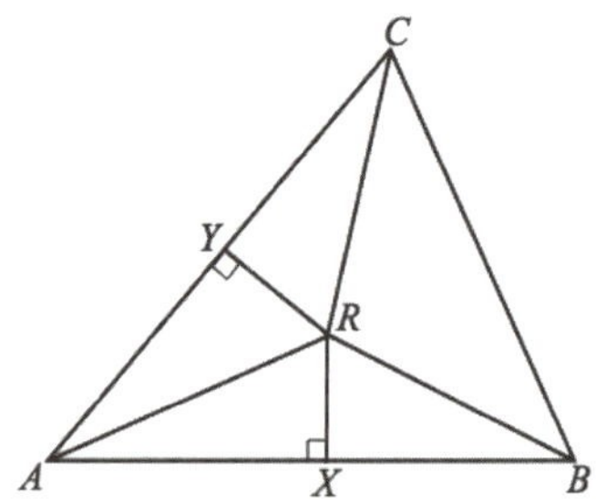

$\overline{RX}$ *is the perpendicular bisector of* $\overline{AB}$.	*Given*
$m\angle RXA = 90°, m\angle RXB = 90°$	Definition of perpendicular bisector
$\overline{AX} \cong \overline{XB}$	Definition of perpendicular bisector
$\overline{RX} \cong \overline{RX}$	*Reflexive property*
$\triangle RAX \cong \triangle RBX$	*SAS*
$\overline{RY}$ *is the perpendicular bisector of* $\overline{AC}$	*Given*
$m\angle RYA = 90°, m\angle RYC = 90°$	Definition of perpendicular bisector
$\overline{AY} \cong \overline{YC}$	Definition of perpendicular bisector
$\overline{RY} \cong \overline{RY}$	*Reflexive property*
$\triangle RAY \cong \triangle RCY$	*SAS*
$\overline{RA} \cong \overline{RB}, \overline{RA} \cong \overline{RC}$	*Corresponding sides of congruent triangles are congruent.*
$\overline{RA} \cong \overline{RB} \cong \overline{RC}$	*Transitive property*

Lesson 27: Triangle Congruency Proofs

Student Outcomes

- Students complete proofs requiring a synthesis of the skills learned in the last four lessons.

Classwork

Exercises (39 minutes)

Exercises

1. Given: $AB = AC$, $RB = RC$

   Prove: $SB = SC$

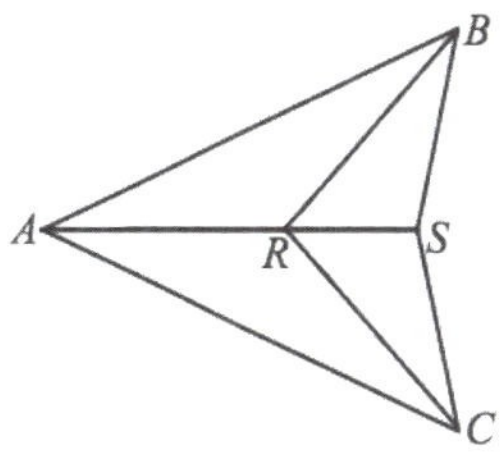

$AB = AC$, $RB = RC$	*Given*
$AR = AR$	*Reflexive property*
$\triangle ARC \cong \triangle ARB$	*SSS*
$m\angle ARC = m\angle ARB$	*Corresponding angles of congruent triangles are equal in measure.*
$m\angle ARC + m\angle SRC = 180°$, $m\angle ARB + m\angle SRB = 180°$	*Linear pairs form supplementary angles.*
$m\angle SRC = m\angle SRB$	*Angles supplementary to either the same angle or to congruent angles are equal in measure.*
$SR = RS$	*Reflexive property*
$\triangle SRB \cong \triangle SRC$	*SAS*
$SB = SC$	*Corresponding sides of congruent angles are equal in length.*

2. Given: Square $ABCS \cong$ Square $EFGS$,

   $\overleftrightarrow{RAB}$, $\overleftrightarrow{REF}$

   Prove: $\triangle ASR \cong \triangle ESR$

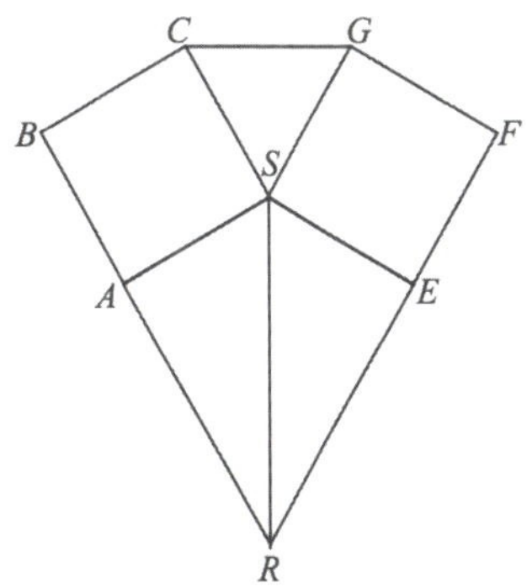

Square $ABCS \cong$ Square $EFGS$	*Given*
$AS = ES$	*Corresponding sides of congruent squares are equal in length.*
$SR = RS$	*Reflexive property*
$\angle BAS$ and $\angle FES$ are right angles.	*Definition of square*
$\angle BAS$ and $\angle SAR$ form a linear pair.	*Definition of linear pair*
$\angle FES$ and $\angle SER$ form a linear pair.	*Definition of linear pair*
$\angle SAR$ and $\angle SER$ are right angles.	*Two angles that are supplementary and congruent each measure $90°$ and are, therefore, right angles.*
$\triangle ASR$ and $\triangle ESR$ are right triangles.	*Definition of right triangle*
$\triangle ASR \cong \triangle ESR$	*HL*

3. **Given:** $JK = JL, JX = JY$

   Prove: $KX = LY$

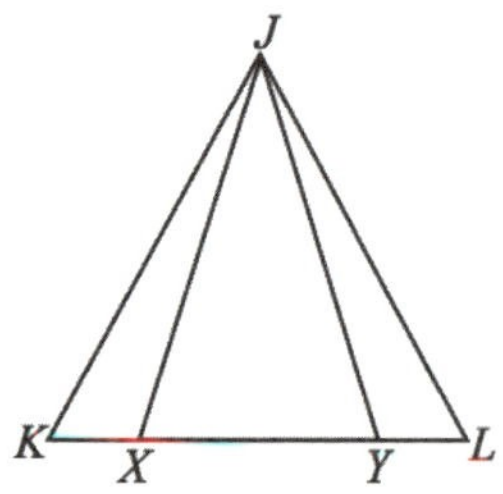

Statement	Reason
$JX = JY$	*Given*
$m\angle JXY = m\angle JYX$	*Base angles of an isosceles triangle are equal in measure.*
$m\angle JXK + m\angle JXY = 180°$, $m\angle JYL + m\angle JYX = 180°$	*Linear pairs form supplementary angles.*
$m\angle JXK + m\angle JXY = m\angle JYL + m\angle JYX$	*Substitution property of equality*
$m\angle JXK + m\angle JXY = m\angle JYL + m\angle JXY$	*Substitution property of equality*
$m\angle JXK = m\angle JYL$	*Angles supplementary to either the same angle or congruent angles are equal in measure.*
$JK = JL$	*Given*
$m\angle K = m\angle L$	*Base angles of an isosceles triangle are equal in measure.*
$\triangle JXK \cong \triangle JYL$	*AAS*
$KX = LY$	*Corresponding sides of congruent triangles are equal in length.*

4. **Given:** $\overline{AD} \perp \overline{DR}, \overline{AB} \perp \overline{BR}$, $\overline{AD} \cong \overline{AB}$

   Prove: $\angle DCR \cong \angle BCR$

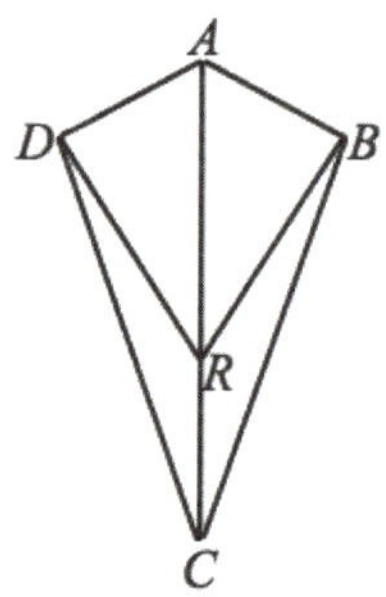

Statement	Reason
Given	
$\triangle ADR$ *and* $\triangle ABR$ *are right triangles.*	*Definition of right triangle*
$\overline{AD} \cong \overline{AB}$	*Given*
$\overline{AR} \cong \overline{AR}$	*Reflexive property*
$\triangle ADR \cong \triangle ABR$	*HL*
$\angle ARD \cong ARB$	*Corresponding angles of congruent triangles are congruent.*
$m\angle ARD + m\angle DRC = 180°$, $m\angle ARB + m\angle BRC = 180°$	*Linear pairs form supplementary angles.*
$m\angle ARD + m\angle DRC = m\angle ARB + m\angle BRC$	*Transitive property*
$m\angle DRC = m\angle BRC$	*Angles supplementary to either the same angle or congruent angles are equal in measure.*
$\overline{DR} \cong \overline{BR}$	*Corresponding sides of congruent triangles are congruent.*
$\overline{RC} \cong \overline{RC}$	*Reflexive property*
$\triangle DRC \cong \triangle BRC$	*SAS*
$\angle DRC \cong \angle BRC$	*Corresponding angles of congruent triangles are congruent.*

5. Given: $AR = AS$, $BR = CS$, $\overline{RX} \perp \overline{AB}$, $\overline{SY} \perp \overline{AC}$

Prove: $BX = CY$

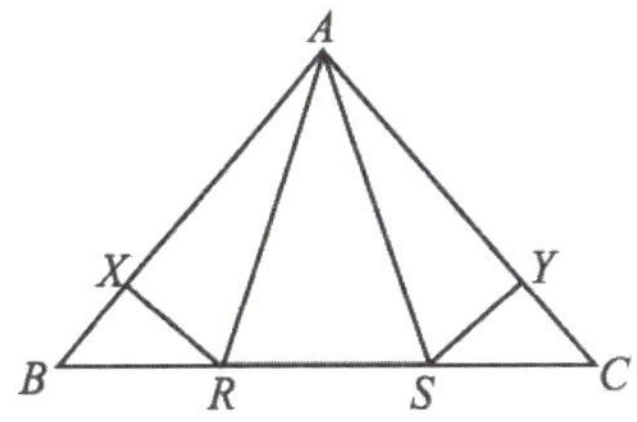

$AR = AS$	*Given*
$m\angle ARS = m\angle ASR$	*Base angles of an isosceles triangle are equal in measure.*
$m\angle ARS + m\angle ARB = 180°$, $m\angle ASR + m\angle ASC = 180°$	*Linear pairs form supplementary angles.*
$m\angle ARS + m\angle ARB = m\angle ASR + m\angle ASC$	*Transitive property*
$m\angle ARB = m\angle ASC$	*Subtraction*
$BR = CS$	*Given*
$\triangle ARB \cong \triangle ASC$	*SAS*
$\angle ABR \cong \angle ACS$	*Corresponding angles of congruent triangles are congruent.*
$\overline{RX} \perp \overline{AB}$, $\overline{SY} \perp \overline{AC}$	*Given*
$m\angle RXB = 90° = m\angle SYC$	*Definition of perpendicular line segments*
$\triangle BRX \cong \triangle SYC$	*AAS*
$BX = CY$	*Corresponding sides of congruent triangles are equal in length.*

6. Given: $AX = BX$, $m\angle AMB = m\angle AYZ = 90°$

Prove: $NY = NM$

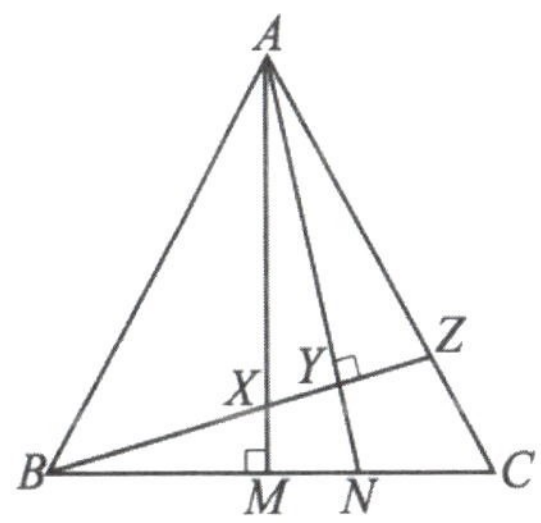

$AX = BX$	*Given*
$m\angle AMB = m\angle AYZ = 90°$	*Given*
$m\angle BXM = m\angle AXY$	*Vertical angles are equal in measure.*
$\triangle BXM \cong \triangle AXY$	*AAS*
$BX + XY = BY$, $AX + XM = AM$	*Segment addition property*
$XM = XY$	*Corresponding sides of congruent triangles are equal in length.*
$BY = AM$	*Substitution property of equality*
$m\angle BYN = 90°$	*Vertical angles are equal in measure.*
$m\angle AMB + m\angle AMN = 180°$	*Linear pairs form supplementary angles.*
$m\angle AMN = 90°$	*Subtraction property of equality*
$m\angle MNY = m\angle MNY$	*Reflexive property*
$\triangle BYN \cong \triangle AMN$	*AAS*
$NY = NM$	*Corresponding sides of congruent triangles are equal in length.*

Closing (1 minute)

- Triangle congruency proofs require an application of known facts, definitions, axioms, and geometric properties.

Exit Ticket (5 minutes)

Name ______________________________ Date ______________

Lesson 27: Triangle Congruency Proofs

Exit Ticket

Given: M is the midpoint of $\overline{GR}$, $\angle G \cong \angle R$

Prove: $\triangle GHM \cong \triangle RPM$

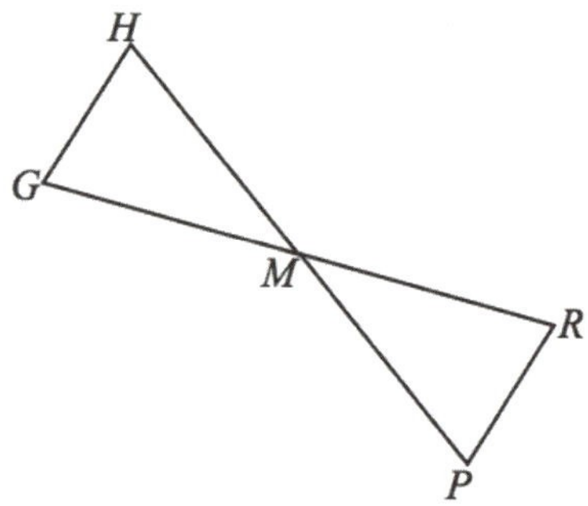

Exit Ticket Sample Solutions

Given: M is the midpoint of $\overline{GR}$, $\angle G \cong \angle R$

Prove: $\triangle GHM \cong \triangle RPM$

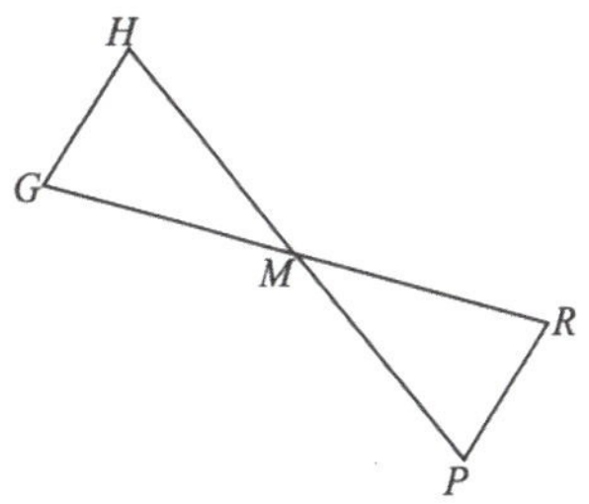

M is the midpoint of $\overline{GR}$.	*Given*
$\angle G \cong \angle R$	*Given*
$GM = RM$	*Definition of midpoint*
$\angle HMG \cong \angle PMR$	*Vertical angles are congruent.*
$\triangle GHM \cong \triangle RPM$	*ASA*

Problem Set Sample Solutions

Use your knowledge of triangle congruence criteria to write a proof for the following:

In the figure $\overline{BE} \cong \overline{CE}$, $\overline{DC} \perp \overline{AB}$, and $\overline{BE} \perp \overline{AC}$; prove $\overline{AE} \cong \overline{RE}$.

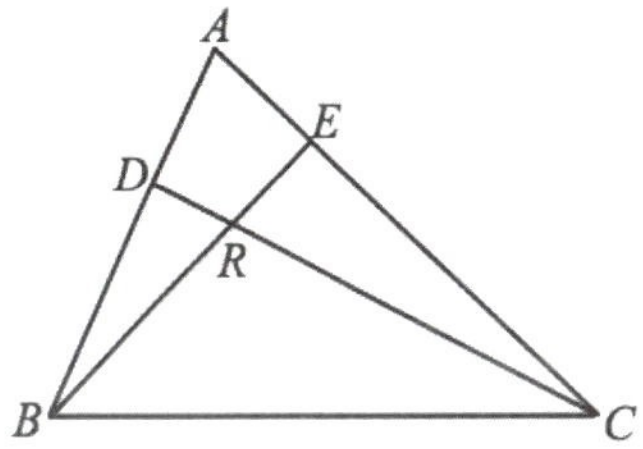

$m\angle ERC = m\angle BRD$	*Vertical angles are equal in measure.*
$\overline{DC} \perp \overline{AB}, \overline{BE} \perp \overline{AC}$	*Given*
$m\angle BDR = 90°$, $m\angle REC = 90°$	*Definition of perpendicular lines*
$m\angle ABE = m\angle RCE$	*Sum of the angle measures in a triangle is $180°$.*
$m\angle BAE = m\angle BRD$	*Sum of the angle measures in a triangle is $180°$.*
$m\angle BAE = m\angle ERC$	*Substitution property of equality*
$\overline{BE} \cong \overline{CE}$	*Given*
$\triangle BAE \cong \triangle CRE$	*AAS*
$\overline{AE} \cong \overline{RE}$	*Corresponding sides of congruent triangles are congruent.*

Topic E

Proving Properties of Geometric Figures

G-CO.C.9, G-CO.C.10, G-CO.C.11

Focus Standards:	G-CO.C.9	Prove theorems about lines and angles. *Theorems include: vertical angles are congruent; when a transversal crosses parallel lines, alternate interior angles are congruent and corresponding angles are congruent; points on a perpendicular bisector of a line segment are exactly those equidistant from the segment's endpoints.*
	G-CO.C.10	Prove theorems about triangles. *Theorems include: measures of interior angles of a triangle sum to* $180°$*; base angles of isosceles triangles are congruent; the segment joining midpoints of two sides of a triangle is parallel to the third side and half the length; the medians of a triangle meet at a point.*
	G-CO.C.11	Prove theorems about parallelograms. *Theorems include: opposite sides are congruent, opposite angles are congruent, the diagonals of a parallelogram bisect each other, and conversely, rectangles are parallelograms with congruent diagonals.*
Instructional Days:	3	
Lesson 28:	Properties of Parallelograms (P)[1]	
Lessons 29–30:	Special Lines in Triangles (P, P)	

In Topic E, students extend their work on rigid motions and proofs to establish properties of triangles and parallelograms. In Lesson 28, students apply their recent experience with triangle congruence to prove problems involving parallelograms. In Lessons 29 and 30, students examine special lines in triangles, namely, midsegments and medians. Students prove why a midsegment is parallel to and half the length of the side of the opposite triangle. In Lesson 30, students prove why the medians are concurrent.

[1]Lesson Structure Key: **P**-Problem Set Lesson, **M**-Modeling Cycle Lesson, **E**-Exploration Lesson, **S**-Socratic Lesson

Lesson 28: Properties of Parallelograms

Student Outcomes

- Students complete proofs that incorporate properties of parallelograms.

Lesson Notes

Throughout this module, we have seen the theme of building new facts with the use of established ones. We see this again in Lesson 28, where triangle congruence criteria are used to demonstrate why certain properties of parallelograms hold true. We begin establishing new facts using only the definition of a parallelogram and the properties we have assumed when proving statements. Students combine the basic definition of a parallelogram with triangle congruence criteria to yield properties taken for granted in earlier grades, such as opposite sides of a parallelogram are parallel.

Classwork

Opening Exercise (5 minutes)

Opening Exercise

a. If the triangles are congruent, state the congruence.

$\triangle AGT \cong \triangle MYJ$

b. Which triangle congruence criterion guarantees part 1?

AAS

c. $\overline{TG}$ corresponds with

$\overline{JY}$

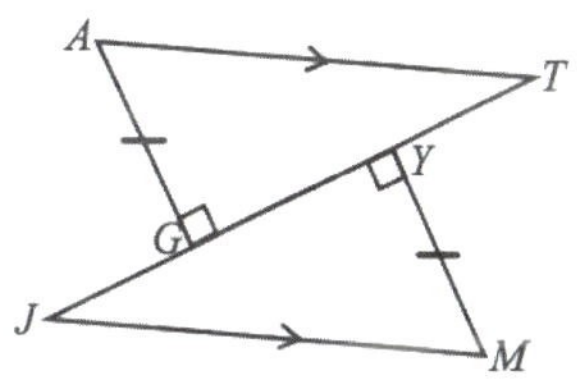

Discussion/Examples 1–7 (35 minutes)

Discussion

How can we use our knowledge of triangle congruence criteria to establish other geometry facts? For instance, what can we now prove about the properties of parallelograms?

To date, we have defined a parallelogram to be a quadrilateral in which both pairs of opposite sides are parallel. However, we have assumed other details about parallelograms to be true, too. We assume that:

- Opposite sides are congruent.
- Opposite angles are congruent.
- Diagonals bisect each other.

Let us examine why each of these properties is true.

Example 1

If a quadrilateral is a parallelogram, then its opposite sides and angles are equal in measure. Complete the diagram, and develop an appropriate *Given* and *Prove* for this case. Use triangle congruence criteria to demonstrate why opposite sides and angles of a parallelogram are congruent.

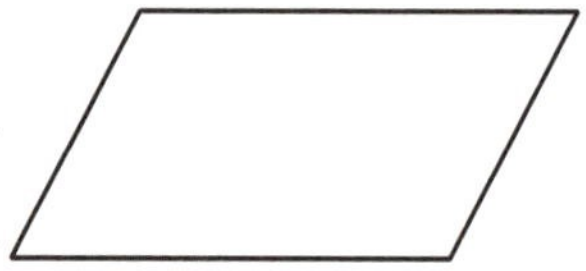

Given: *Parallelogram $ABCD$ ($\overline{AB} \parallel \overline{CD}$, $\overline{AD} \parallel \overline{CB}$)*

Prove: $AD = CB,\ AB = CD,\ m\angle A = m\angle C,\ m\angle B = m\angle D$

Construction: Label the quadrilateral $ABCD$, and mark opposite sides as parallel. Draw diagonal $\overline{BD}$.

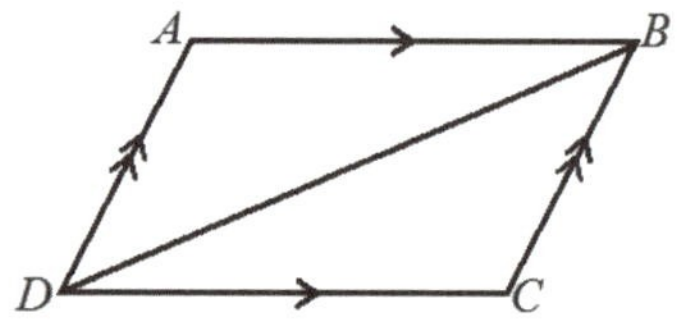

PROOF:

Parallelogram $ABCD$	*Given*
$m\angle ABD = m\angle CDB$	*If parallel lines are cut by a transversal, then alternate interior angles are equal in measure.*
$BD = DB$	*Reflexive property*
$m\angle CBD = m\angle ADB$	*If parallel lines are cut by a transversal, then alternate interior angles are equal in measure.*
$\triangle ABD \cong \triangle CDB$	*ASA*
$AD = CB,\ AB = CD$	*Corresponding sides of congruent triangles are equal in length.*
$m\angle A = m\angle C$	*Corresponding angles of congruent triangles are equal in measure.*
$m\angle ABD + m\angle CBD = m\angle ABC,$	
$m\angle CDB + m\angle ADB = m\angle ADC$	*Angle addition postulate*
$m\angle ABD + m\angle CBD = m\angle CDB + m\angle ADB$	*Addition property of equality*
$m\angle B = m\angle D$	*Substitution property of equality*

Example 2

If a quadrilateral is a parallelogram, then the diagonals bisect each other. Complete the diagram, and develop an appropriate *Given* and *Prove* for this case. Use triangle congruence criteria to demonstrate why diagonals of a parallelogram bisect each other. Remember, now that we have proved opposite sides and angles of a parallelogram to be congruent, we are free to use these facts as needed (i.e., $AD = CB$, $AB = CD$, $\angle A \cong \angle C$, $\angle B \cong \angle D$).

Given: *Parallelogram $ABCD$*

Prove: *Diagonals bisect each other, $AE = CE, DE = BE$*

Construction: Label the quadrilateral $ABCD$. Mark opposite sides as parallel. Draw diagonals $\overline{AC}$ and $\overline{BD}$.

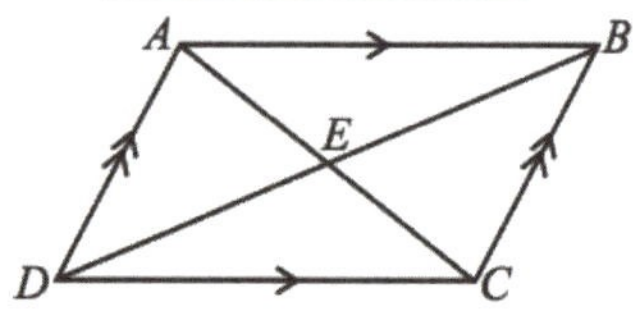

PROOF:

Parallelogram $ABCD$	*Given*
$m\angle BAC = m\angle DCA$	*If parallel lines are cut by a transversal, then alternate interior angles are equal in measure.*
$m\angle AEB = m\angle CED$	*Vertical angles are equal in measure.*
$AB = CD$	*Opposite sides of a parallelogram are equal in length.*
$\triangle AEB \cong \triangle CED$	*AAS*
$AE = CE,\ DE = BE$	*Corresponding sides of congruent triangles are equal in length.*

EUREKA MATH™

Now we have established why the properties of parallelograms that we have assumed to be true are in fact true. By extension, these facts hold for any type of parallelogram, including rectangles, squares, and rhombuses. Let us look at one last fact concerning rectangles. We established that the diagonals of general parallelograms bisect each other. Let us now demonstrate that a rectangle has congruent diagonals.

Students may need a reminder that a rectangle is a parallelogram with four right angles.

Example 3

If the parallelogram is a rectangle, then the diagonals are equal in length. Complete the diagram, and develop an appropriate *Given* and *Prove* for this case. Use triangle congruence criteria to demonstrate why diagonals of a rectangle are congruent. As in the last proof, remember to use any already proven facts as needed.

Given: *Rectangle $GHIJ$*

Prove: *Diagonals are equal in length, $GI = HJ$*

Construction: Label the rectangle $GHIJ$. Mark opposite sides as parallel, and add small squares at the vertices to indicate 90° angles. Draw diagonals $\overline{GI}$ and $\overline{HJ}$.

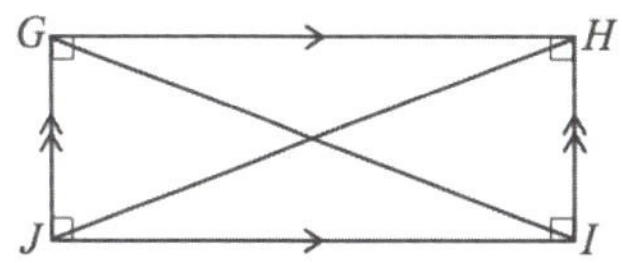

PROOF:

Rectangle $GHIJ$	*Given*
$GJ = IH$	*Opposite sides of a parallelogram are equal in length.*
$GH = HG$	*Reflexive property*
$\angle JGH$, $\angle IHG$ *are right angles.*	*Definition of a rectangle*
$\triangle GHJ \cong \triangle HGI$	*SAS*
$GI = HJ$	*Corresponding sides of congruent triangles are equal in length.*

Converse Properties: Now we examine the converse of each of the properties we proved. Begin with the property, and prove that the quadrilateral is in fact a parallelogram.

Example 4

If both pairs of opposite angles of a quadrilateral are equal, then the quadrilateral is a parallelogram. Draw an appropriate diagram, and provide the relevant *Given* and *Prove* for this case.

Given: *Quadrilateral $ABCD$ with $m\angle A = m\angle C$, $m\angle B = m\angle D$*

Prove: *Quadrilateral $ABCD$ is a parallelogram.*

Construction: Label the quadrilateral $ABCD$. Mark opposite angles as congruent. Draw diagonal $\overline{BD}$. Label the measures of $\angle A$ and $\angle C$ as $x°$. Label the measures of the four angles created by $\overline{BD}$ as $r°$, $s°$, $t°$, and $u°$.

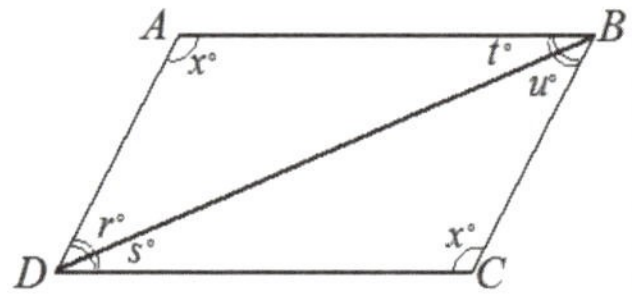

PROOF:

Quadrilateral $ABCD$ with $m\angle A = m\angle C$, $m\angle B = m\angle D$	*Given*
$m\angle D = r + s$, $m\angle B = t + u$	*Angle addition*
$r + s = t + u$	*Substitution*
$x + r + t = 180$, $x + s + u = 180$	*Angles in a triangle add up to $180°$.*
$r + t = s + u$	*Subtraction property of equality, substitution*
$r + t - (r + s) = s + u - (t + u)$	*Subtraction property of equality*
$t - s = s - t$	*Additive inverse property*
$t - s + (s - t) = s - t + (s - t)$	*Addition property of equality*
$0 = 2(s - t)$	*Addition and subtraction properties of equality*
$0 = s - t$	*Division property of equality*
$s = t$	*Addition property of equality*
$s = t \Rightarrow r = u$	*Substitution and subtraction properties of equality*
$\overline{AB} \parallel \overline{CD}$, $\overline{AD} \parallel \overline{BC}$	*If two lines are cut by a transversal such that a pair of alternate interior angles are equal in measure, then the lines are parallel.*
Quadrilateral $ABCD$ is a parallelogram.	*Definition of a parallelogram*

Example 5

If the opposite sides of a quadrilateral are equal, then the quadrilateral is a parallelogram. Draw an appropriate diagram, and provide the relevant *Given* and *Prove* for this case.

Given: *Quadrilateral $ABCD$ with $AB = CD$, $AD = BC$*

Prove: *Quadrilateral $ABCD$ is a parallelogram.*

Construction: Label the quadrilateral $ABCD$, and mark opposite sides as equal. Draw diagonal $\overline{BD}$.

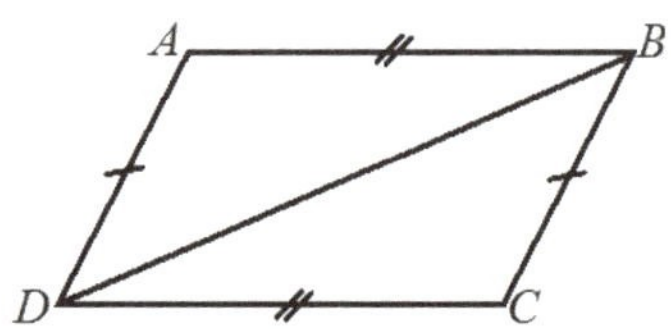

PROOF:

Quadrilateral $ABCD$ with $AB = CD$, $AD = CB$	*Given*
$BD = DB$	*Reflexive property*
$\triangle ABD \cong \triangle CDB$	*SSS*
$\angle ABD \cong \angle CDB$, $\angle ADB \cong \angle CBD$	*Corresponding angles of congruent triangles are congruent.*
$\overline{AB} \parallel \overline{CD}$, $\overline{AD} \parallel \overline{CB}$	*If two lines are cut by a transversal and the alternate interior angles are congruent, then the lines are parallel.*
Quadrilateral $ABCD$ is a parallelogram.	*Definition of a parallelogram*

Example 6

If the diagonals of a quadrilateral bisect each other, then the quadrilateral is a parallelogram. Draw an appropriate diagram, and provide the relevant *Given* and *Prove* for this case. Use triangle congruence criteria to demonstrate why the quadrilateral is a parallelogram.

Given: *Quadrilateral $ABCD$, diagonals $\overline{AC}$ and $\overline{BD}$ bisect each other.*

Prove: *Quadrilateral $ABCD$ is a parallelogram.*

Construction: Label the quadrilateral $ABCD$, and mark opposite sides as equal. Draw diagonals $\overline{AC}$ and $\overline{BD}$.

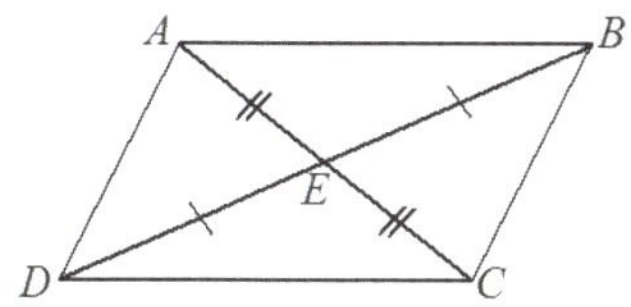

PROOF:

Quadrilateral $ABCD$, diagonals $\overline{AC}$ and $\overline{BD}$ bisect each other.	*Given*
$AE = CE$, $DE = BE$	*Definition of a segment bisector*
$m\angle DEC = m\angle BEA$, $m\angle AED = m\angle CEB$	*Vertical angles are equal in measure.*
$\triangle DEC \cong \triangle BEA$, $\triangle AED \cong \triangle CEB$	*SAS*
$\angle ABD \cong \angle CDB$, $\angle ADB \cong \angle CBD$	*Corresponding angles of congruent triangles are congruent.*
$\overline{AB} \parallel \overline{CD}$, $\overline{AD} \parallel \overline{CB}$	*If two lines are cut by a transversal such that a pair of alternate interior angles are congruent, then the lines are parallel.*
Quadrilateral $ABCD$ is a parallelogram.	*Definition of a parallelogram*

Example 7

If the diagonals of a parallelogram are equal in length, then the parallelogram is a rectangle. Complete the diagram, and develop an appropriate *Given* and *Prove* for this case.

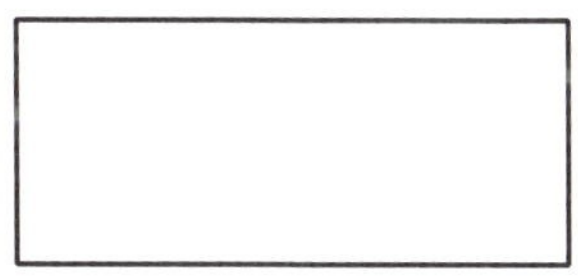

Given: *Parallelogram $GHIJ$ with diagonals of equal length, $GI = HJ$*

Prove: *$GHIJ$ is a rectangle.*

Construction: Label the quadrilateral $GHIJ$. Draw diagonals $\overline{GI}$ and $\overline{HJ}$.

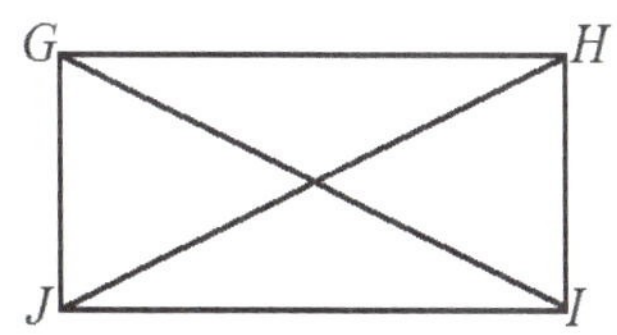

PROOF:

Parallelogram $GHIJ$ with diagonals of equal length, $GI = HJ$	*Given*
$GJ = JG$, $HI = IH$	*Reflexive property*
$GH = IJ$	*Opposite sides of a parallelogram are congruent.*
$\triangle HJG \cong \triangle IGJ$, $\triangle GHI \cong \triangle JIH$	*SSS*
$m\angle G = m\angle J$, $m\angle H = m\angle I$	*Corresponding angles of congruent triangles are equal in measure.*
$m\angle G + m\angle J = 180°$, $m\angle H + m\angle I = 180°$	*If parallel lines are cut by a transversal, then interior angles on the same side are supplementary.*
$2(m\angle G) = 180°$, $2(m\angle H) = 180°$	*Substitution property of equality*
$m\angle G = 90°$, $m\angle H = 90°$	*Division property of equality*
$m\angle G = m\angle J = m\angle H = m\angle I = 90°$	*Substitution property of equality*
$GHIJ$ is a rectangle.	*Definition of a rectangle*

Closing (1 minute)

- A parallelogram is a quadrilateral whose opposite sides are parallel. We have proved the following properties of a parallelogram to be true: Opposite sides are congruent, opposite angles are congruent, diagonals bisect each other.

Exit Ticket (5 minutes)

Name ____________________________________ Date____________________

Lesson 28: Properties of Parallelograms

Exit Ticket

Given: Equilateral parallelogram $ABCD$ (i.e., a rhombus) with diagonals $\overline{AC}$ and $\overline{BD}$

Prove: Diagonals intersect perpendicularly.

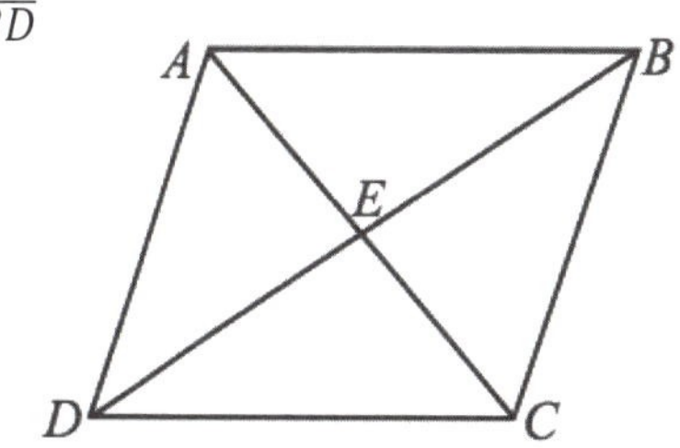

Exit Ticket Sample Solutions

Given: Equilateral parallelogram $ABCD$ (i.e., a rhombus) with diagonals $\overline{AC}$ and $\overline{BD}$

Prove: Diagonals intersect perpendicularly.

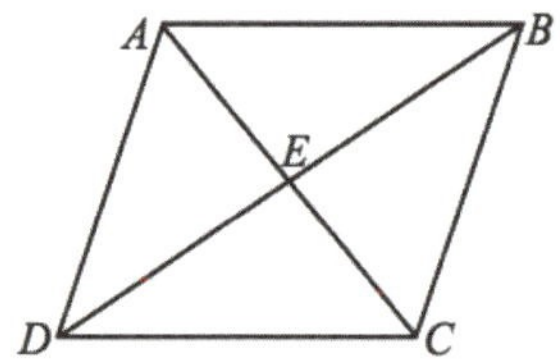

Rhombus $ABCD$	*Given*
$AE = CE,\ DE = BE$	*Diagonals of a parallelogram bisect each other.*
$AB = BC = CD = DA$	*Definition of equilateral parallelogram*
$\triangle AED \cong \triangle AEB \cong \triangle CEB \cong \triangle CED$	*SSS*
$m\angle AED = m\angle AEB = m\angle BEC = m\angle CED$	*Corr. angles of congruent triangles are equal in measure.*
$m\angle AED = m\angle AEB = m\angle BEC = m\angle CED = 90°$	*Angles at a point sum to $360°$. Since all four angles are congruent, each angle measures $90°$.*

Problem Set Sample Solutions

Use the facts you have established to complete exercises involving different types of parallelograms.

1. Given: $\overline{AB} \parallel \overline{CD}$, $AD = AB$, $CD = CB$

   Prove: $ABCD$ is a rhombus.

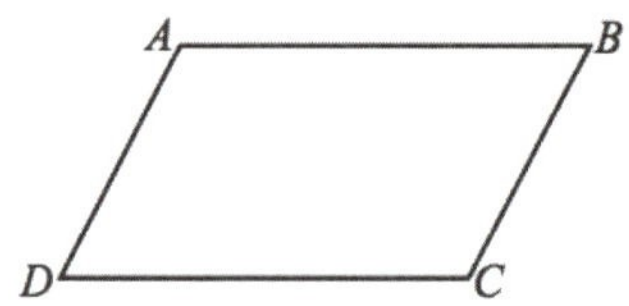

Construction: Draw diagonal $\overline{AC}$.

$AD = AB,\ CD = CB$	*Given*
$AC = CA$	*Reflexive property*
$\triangle ADC \cong \triangle CBA$	*SSS*
$AD = CB,\ AB = CD$	*Corresponding sides of congruent triangles are equal in length.*
$AB = BC = CD = AD$	*Transitive property*
$ABCD$ is a rhombus.	*Definition of a rhombus*

2. Given: Rectangle $RSTU$, M is the midpoint of $\overline{RS}$.

   Prove: $\triangle UMT$ is isosceles.

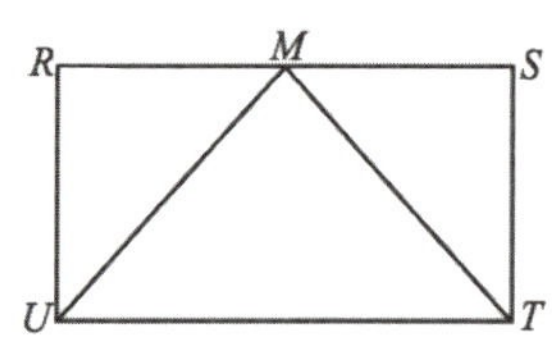

Rectangle $RSTU$	*Given*
$RU = ST$	*Opposite sides of a rectangle are congruent.*
$\angle R$, $\angle S$ are right angles.	*Definition of a rectangle*
M is the midpoint of $\overline{RS}$.	*Given*
$RM = SM$	*Definition of a midpoint*
$\triangle RMU \cong \triangle SMT$	*SAS*
$\overline{UM} \cong \overline{TM}$	*Corresponding sides of congruent triangles are congruent.*
$\triangle UMT$ is isosceles.	*Definition of an isosceles triangle*

3. Given: $ABCD$ is a parallelogram, $\overline{RD}$ bisects $\angle ADC$, $\overline{SB}$ bisects $\angle CBA$.
Prove: $DRBS$ is a parallelogram.

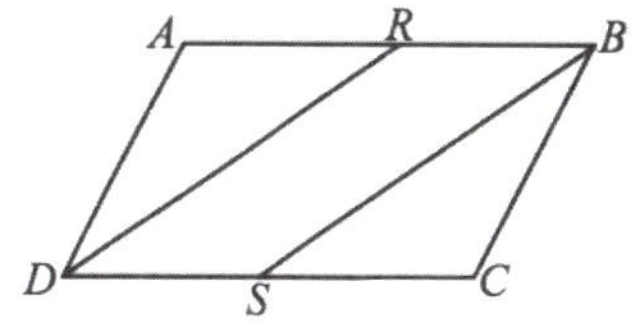

$ABCD$ is a parallelogram; *$\overline{RD}$ bisects $\angle ADC$, $\overline{SB}$ bisects $\angle CBA$.*	*Given*
$AD = CB$	*Opposite sides of a parallelogram are congruent.*
$\angle A \cong \angle C$, $\angle B \cong \angle D$	*Opposite angles of a parallelogram are congruent.*
$\angle RDA \cong \angle RDS$, $\angle SBC \cong \angle SBR$	*Definition of angle bisector*
$m\angle RDA + m\angle RDS = m\angle D, m\angle SBC + m\angle SBR = m\angle B$	*Angle addition*
$m\angle RDA + m\angle RDA = m\angle D, m\angle SBC + m\angle SBC = m\angle B$	*Substitution*
$2(m\angle RDA) = m\angle D$, $2(m\angle SBC) = m\angle B$	*Addition*
$m\angle RDA = \frac{1}{2}m\angle D$, $m\angle SBC = \frac{1}{2}m\angle B$	*Division*
$\angle RDA \cong \angle SBC$	*Substitution*
$\triangle DAR \cong \triangle BCS$	*ASA*
$\angle DRA \cong \angle BSC$	*Corresponding angles of congruent triangles are congruent.*
$\angle DRB \cong \angle BSD$	*Supplements of congruent angles are congruent.*
$DRBS$ is a parallelogram.	*Opposite angles of quadrilateral $DRBS$ are congruent.*

4. Given: $DEFG$ is a rectangle, $WE = YG$, $WX = YZ$
Prove: $WXYZ$ is a parallelogram.

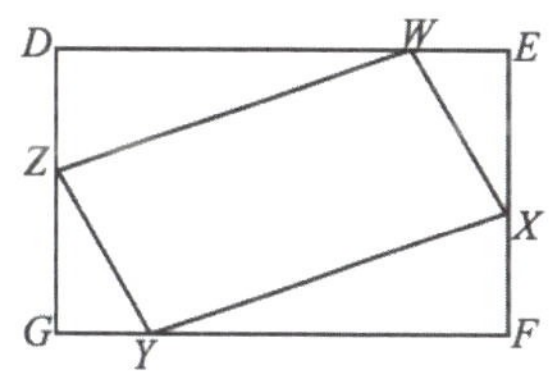

$DE = FG$, $DG = FE$	*Opposite sides of a rectangle are congruent.*
$DEFG$ is a rectangle; $WE = YG$, $WX = YZ$	*Given*
$DE = DW + WE$, $FG = YG + FY$	*Segment addition*
$DW + WE = YG + FY$	*Substitution*
$DW + YG = YG + FY$	*Substitution*
$DW = FY$	*Subtraction*
$m\angle D = m\angle E = m\angle F = m\angle G = 90°$	*Definition of a rectangle*
$\triangle ZGY$, $\triangle XEW$ are right triangles.	*Definition of right triangle*
$\triangle ZGY \cong \triangle XEW$	*HL*
$ZG = XE$	*Corresponding sides of congruent triangles are congruent.*
$DG = ZG + DZ$; $FE = XE + FX$	*Partition property or segment addition*
$DZ = FX$	*Subtraction property of equality*
$\triangle DZW \cong \triangle FXY$	*SAS*
$ZW = XY$	*Corresponding sides of congruent triangles are congruent.*
$WXYZ$ is a parallelogram.	*Both pairs of opposite sides of a parallelogram are congruent.*

5. Given: Parallelogram $ABFE$, $CR = DS$, $\overline{ABC}$ and $\overline{DEF}$ are segments.

Prove: $BR = SE$

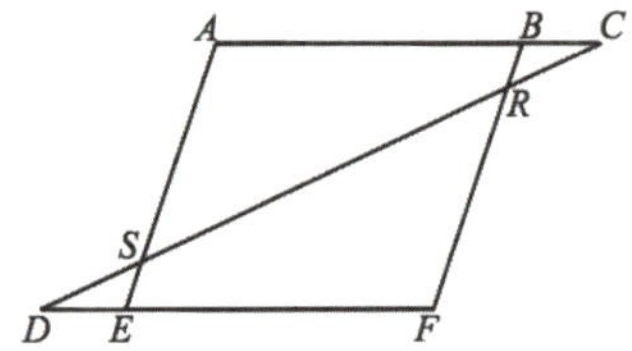

$\overline{AB} \parallel \overline{FE}$	*Opposite sides of a parallelogram are parallel.*
$m\angle BCR = m\angle EDS$	*If parallel lines are cut by a transversal, then alternate interior angles are equal in measure.*
$\angle ABF \cong \angle FEA$	*Opposite angles of a parallelogram are congruent.*
$\angle CBR \cong \angle DES$	*Supplements of congruent angles are congruent.*
$CR = DS$	*Given*
$\triangle CBR \cong \triangle DES$	*AAS*
$BR = SE$	*Corresponding sides of congruent triangles are equal in length.*

Lesson 29: Special Lines in Triangles

Student Outcomes

- Students examine the relationships created by special lines in triangles, namely, midsegments.

Classwork

Opening Exercise (7 minutes)

Opening Exercise

Construct the midsegment of the triangle below. A midsegment is a line segment that joins the midpoints of two sides of a triangle or trapezoid. For the moment, we will work with a triangle.

a. Use your compass and straightedge to determine the midpoints of $\overline{AB}$ and $\overline{AC}$ as X and Y, respectively.

b. Draw midsegment $\overline{XY}$.

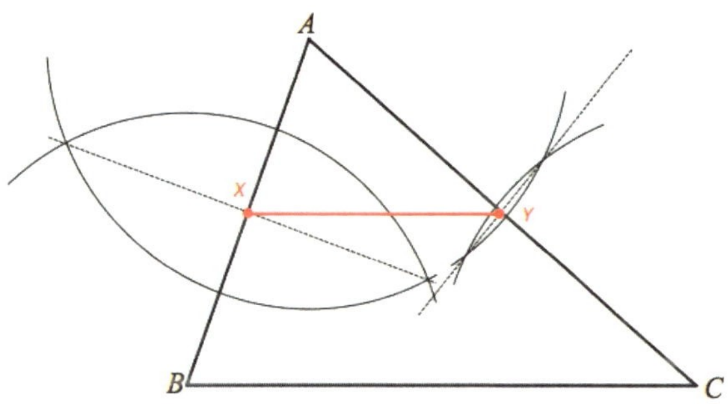

Compare $\angle AXY$ and $\angle ABC$; compare $\angle AYX$ and $\angle ACB$. Without using a protractor, what would you guess is the relationship between these two pairs of angles? What are the implications of this relationship?

$m\angle AXY = m\angle ABC$, $m\angle AYX = m\angle ACB$; $\overline{XY} \parallel \overline{BC}$

Discussion (15 minutes)

Discussion

Note that though we chose to determine the midsegment of $\overline{AB}$ and $\overline{AC}$, we could have chosen any two sides to work with. Let us now focus on the properties associated with a midsegment.

The midsegment of a triangle is parallel to the third side of the triangle and half the length of the third side of the triangle.

We can prove these properties to be true. Continue to work with the figure from the Opening Exercise.

Given: $\overline{XY}$ is a midsegment of $\triangle ABC$.

Prove: $\overline{XY} \parallel \overline{BC}$ and $XY = \frac{1}{2}BC$

Construct the following: In the Opening Exercise figure, draw $\triangle YGC$ according to the following steps. Extend $\overline{XY}$ to point G so that $YG = XY$. Draw $\overline{GC}$.

(1) What is the relationship between XY and YG? Explain why.

Equal, by construction

(2) What is the relationship between $m\angle AYX$ and $m\angle GYC$? Explain why.

Equal, vertical angles are equal in measure.

(3) What is the relationship between AY and YC? Explain why.

Equal, Y is the midpoint of $\overline{AC}$.

(4) What is the relationship between $\triangle AXY$ and $\triangle CGY$? Explain why.

Congruent, SAS

(5) What is the relationship between GC and AX? Explain why.

Equal, corresponding sides of congruent triangles are equal in length.

(6) Since $AX = BX$, what other conclusion can be drawn? Explain why.

$GC = BX$, substitution

(7) What is the relationship between $m\angle AXY$ and $m\angle CGY$? Explain why.

Equal, corresponding angles of congruent triangles are equal in measure.

(8) Based on (7), what other conclusion can be drawn about $\overline{AB}$ and $\overline{GC}$? Explain why.

$\overline{AB} \parallel \overline{GC}$

(9) What conclusion can be drawn about $BXGC$ based on (7) and (8)? Explain why.

$BXGC$ is a parallelogram; one pair of opposite sides is equal and parallel. Also, $\overline{XY} \parallel \overline{BC}$.

(10) Based on (9), what is the relationship between XG and BC?

$XG = BC$, opposite sides of parallelogram are equal.

(11) Since $YG = XY$, $XG =$ ___2___ XY. Explain why.

Substitution

(12) This means $BC =$ ___2___ XY. Explain why.

Substitution

(13) Or by division, $XY =$ ___$\frac{1}{2}$___ B.

Note that Steps (9) and (13) demonstrate our *Prove* statement.

Exercises 1–4 (12 minutes)

Exercises 1–4

Apply what you know about the properties of midsegments to solve the following exercises.

1. $x =$ <u>15</u>
   Perimeter of $\triangle ABC =$ <u>6</u>

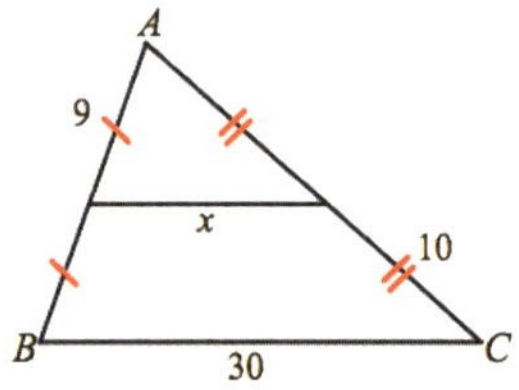

2. $x =$ <u>$50°$</u>
   $y =$ <u>$70°$</u>

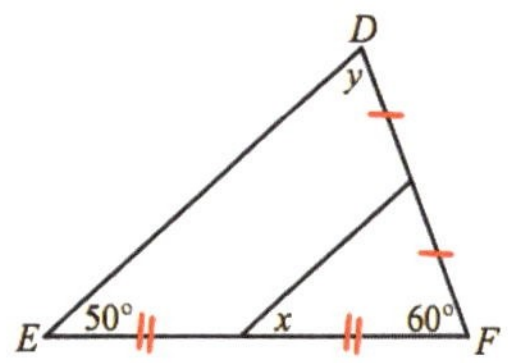

3. In $\triangle RST$, the midpoints of each side have been marked by points X, Y, and Z.
   - Mark the halves of each side divided by the midpoint with a congruency mark. Remember to distinguish congruency marks for each side.
   - Draw midsegments $\overline{XY}$, $\overline{YZ}$, and $\overline{XZ}$. Mark each midsegment with the appropriate congruency mark from the sides of the triangle.

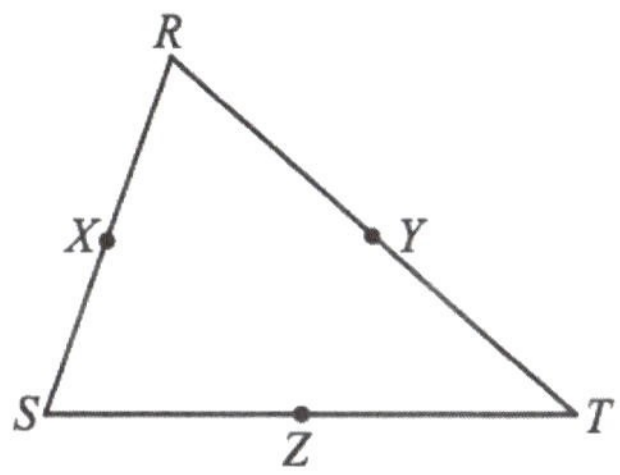

   a. What conclusion can you draw about the four triangles within $\triangle RST$? Explain why.

   All four are congruent. SSS

   b. State the appropriate correspondences among the four triangles within $\triangle RST$.

   $\triangle RXY, \triangle YZT, \triangle XSZ, \triangle ZYX$

   c. State a correspondence between $\triangle RST$ and any one of the four small triangles.

   $\triangle RXY, \triangle RST$

4. Find x.
   $x =$ <u>$9°$</u>

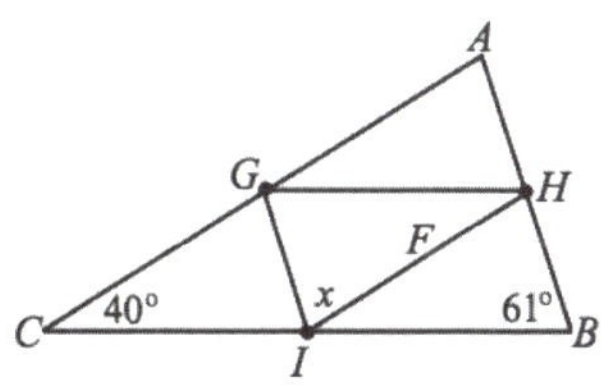

Closing (1 minute)

- What is a midsegment of a triangle, and how does it relate to the other sides of the triangle?
  - *The midsegment of a triangle is parallel to the third side of the triangle and half the length of the third side of the triangle.*

Exit Ticket (5 minutes)

Name ____________________ Date ____________

Lesson 29: Special Lines in Triangles

Exit Ticket

Use the properties of midsegments to solve for the unknown value in each question.

1. R and S are the midpoints of $\overline{XW}$ and $\overline{WY}$, respectively.
   What is the perimeter of $\triangle\, WXY$? ____________

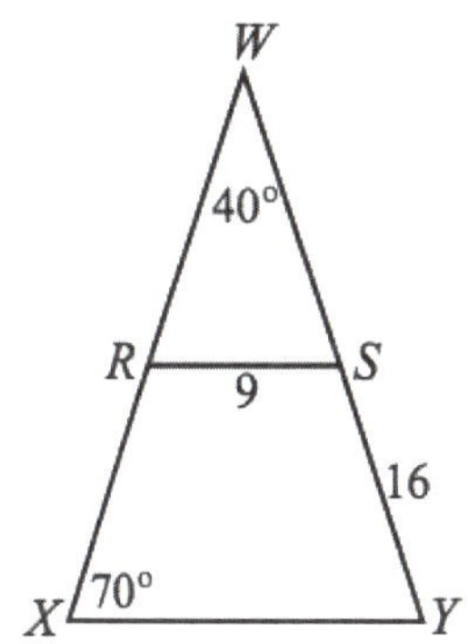

2. What is the perimeter of $\triangle\, EFG$? ____________

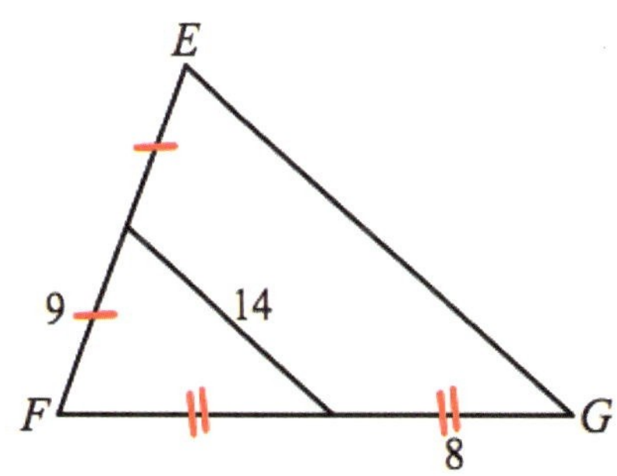

Exit Ticket Sample Solutions

Use the properties of midsegments to solve for the unknown value in each question.

1. R and S are the midpoints of $\overline{XW}$ and $\overline{WY}$, respectively.

   What is the perimeter of $\triangle WXY$? ___82___

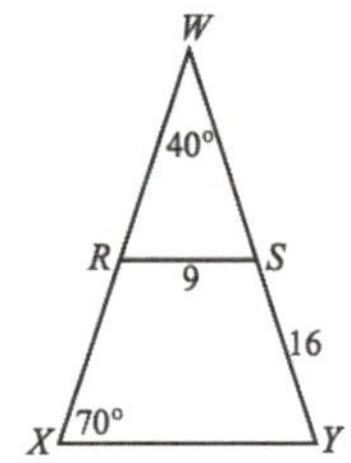

2. What is the perimeter of $\triangle EFG$? ___62___

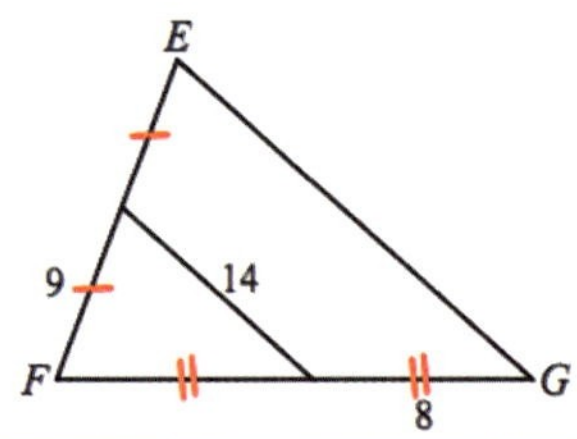

Problem Set Sample Solutions

Use your knowledge of triangle congruence criteria to write proofs for each of the following problems.

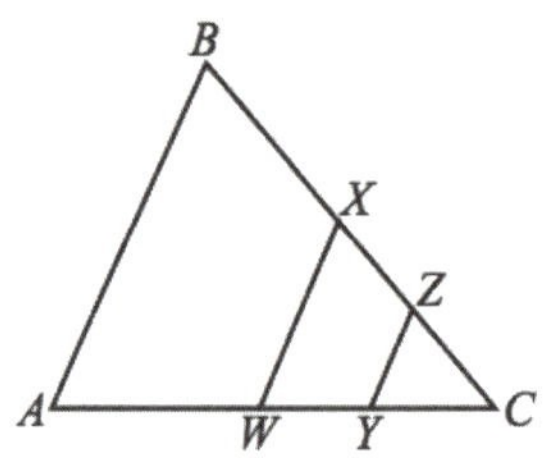

1. $\overline{WX}$ is a midsegment of $\triangle ABC$, and $\overline{YZ}$ is a midsegment of $\triangle CWX$. $BX = AW$

   a. What can you conclude about $\angle A$ and $\angle B$? Explain why.

   $\angle A \cong \angle B$, $BX = AW$, so $CX = CW$; the triangle is isosceles.

   b. What is the relationship of the lengths $\overline{YZ}$ and $\overline{AB}$?

   $YZ = \frac{1}{4}AB$ or $4YZ = AB$

2. W, X, Y, and Z are the midpoints of $\overline{AD}$, $\overline{AB}$, $\overline{BC}$, and $\overline{CD}$, respectively. $AD = 18$, $WZ = 11$, and $BX = 5$. $m\angle WAC = 33°$, $m\angle RYX = 74°$

   a. $m\angle DZW =$ ___74°___

   b. Perimeter of $ABYW =$ ___38___

   c. Perimeter of $ABCD =$ ___56___

   d. $m\angle WAX =$ ___107°___

      $m\angle B =$ ___73°___

      $m\angle YCZ =$ ___107°___

      $m\angle D =$ ___73°___

   e. What kind of quadrilateral is $ABCD$?

   Parallelogram

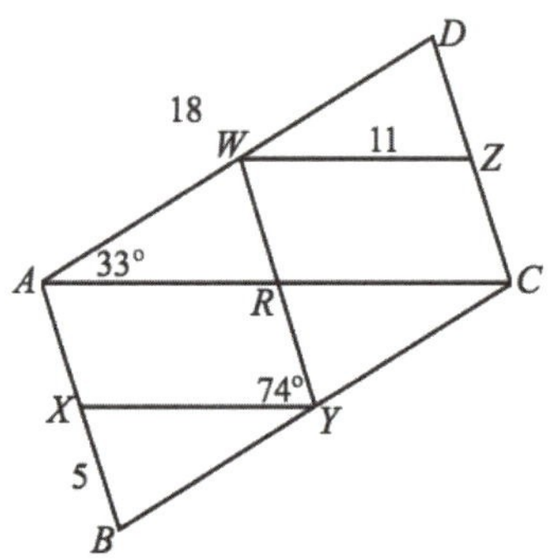

Lesson 30: Special Lines in Triangles

Student Outcomes

- Students examine the relationships created by special lines in triangles, namely, medians.

Lesson Notes

In Lesson 30, we work with a specific set of lines in triangles: the medians. This is an extension of the work we did in Lesson 29, where we proved that a segment joining the midpoints of two sides of a triangle is parallel to and half the length of the third side, among other proofs.

Classwork

Opening Exercise (5 minutes)

Opening Exercise

In $\triangle ABC$ to the right, D is the midpoint of $\overline{AB}$, E is the midpoint of $\overline{BC}$, and F is the midpoint of $\overline{AC}$. Complete each statement below.

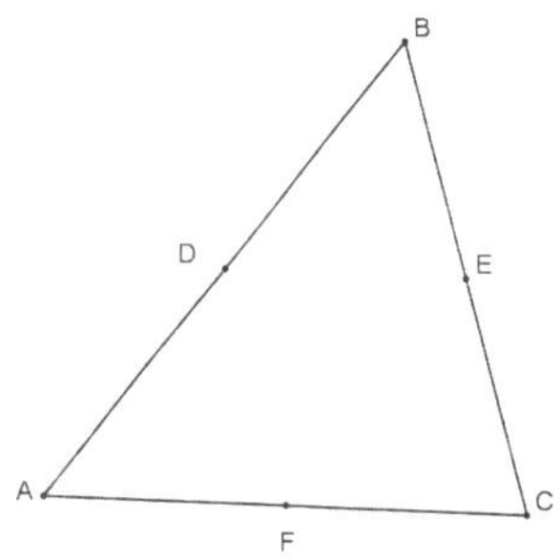

$\overline{DE}$ is parallel to ___$\overline{AC}$___ and measures ___$\frac{1}{2}$___ the length of ___$\overline{AC}$___.

$\overline{DF}$ is parallel to ___$\overline{BC}$___ and measures ___$\frac{1}{2}$___ the length of ___$\overline{BC}$___.

$\overline{EF}$ is parallel to ___$\overline{AB}$___ and measures ___$\frac{1}{2}$___ the length of ___$\overline{AB}$___.

Discussion (5 minutes)

Discussion

In the previous two lessons, we proved that (a) the midsegment of a triangle is parallel to the third side and half the length of the third side and (b) diagonals of a parallelogram bisect each other. We use both of these facts to prove the following assertion:

All medians of a triangle are ___*concurrent*___. **That is, the three medians of a triangle (the segments connecting each vertex to the midpoint of the opposite side) meet at a single point. This point of concurrency is called the** ___*centroid*___**, or the center of gravity, of the triangle. The proof also shows a length relationship for each median: The length from the vertex to the centroid is** ___*twice*___ **the length from the centroid to the midpoint of the side.**

Example 1 (14 minutes)

Example 1

Provide a valid reason for each step in the proof below.

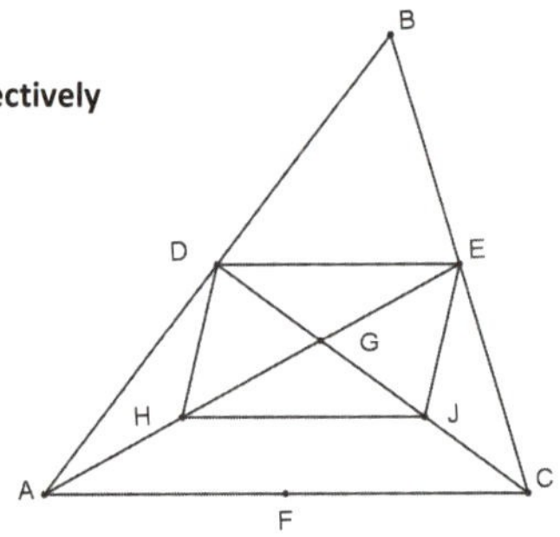

Given: $\triangle ABC$ with D, E, and F the midpoints of sides $\overline{AB}$, $\overline{BC}$, and $\overline{AC}$, respectively

Prove: The three medians of $\triangle ABC$ meet at a single point.

(1) Draw midsegment $\overline{DE}$. Draw $\overline{AE}$ and $\overline{DC}$; label their intersection as point G.

(2) Construct and label the midpoint of $\overline{AG}$ as point H and the midpoint of $\overline{GC}$ as point J.

(3) $\overline{DE} \parallel \overline{AC}$,

$\overline{DE}$ is a midsegment of $\triangle ABC$.

(4) $\overline{HJ} \parallel \overline{AC}$,

$\overline{HJ}$ is a midsegment of $\triangle AGC$.

(5) $\overline{DE} \parallel \overline{HJ}$,

If two segments are parallel to the same segment, then they are parallel to each other.

(6) $DE = \frac{1}{2}AC$ and $HJ = \frac{1}{2}AC$,

Definition of a midsegment

(7) $DEJH$ is a parallelogram.

One pair of sides of a quadrilateral are parallel and equal in length.

(8) $HG = EG$ and $JG = DG$,

Diagonals of a parallelogram bisect each other.

(9) $AH = HG$ and $CJ = JG$,

Definition of a midpoint

(10) $AH = HG = GE$ and $CJ = JG = GD$,

Substitution property of equality

(11) $AG = 2GE$ and $CG = 2GD$,

Partition property or segment addition

(12) We can complete Steps (1)–(11) to include the median from B; the third median, $\overline{BF}$, passes through point G, which divides it into two segments such that the longer part is twice the shorter.

(13) The intersection point of the medians divides each median into two parts with lengths in a ratio of $2:1$; therefore, all medians are concurrent at that point.

The three medians of a triangle are concurrent at the ___*centroid*___, or the center of gravity. This point of concurrency divides the length of each median in a ratio of ___$2:1$___; the length from the vertex to the centroid is ___*twice*___ the length from the centroid to the midpoint of the side.

Example 2 (5 minutes)

Example 2

In $\triangle ABC$, the medians are concurrent at F. $DF = 4$, $BF = 16$, $AG = 30$. Find each of the following measures.

a. $FC =$ ___8___

b. $DC =$ ___12___

c. $AF =$ ___20___

d. $BE =$ ___24___

e. $FG =$ ___10___

f. $EF =$ ___8___

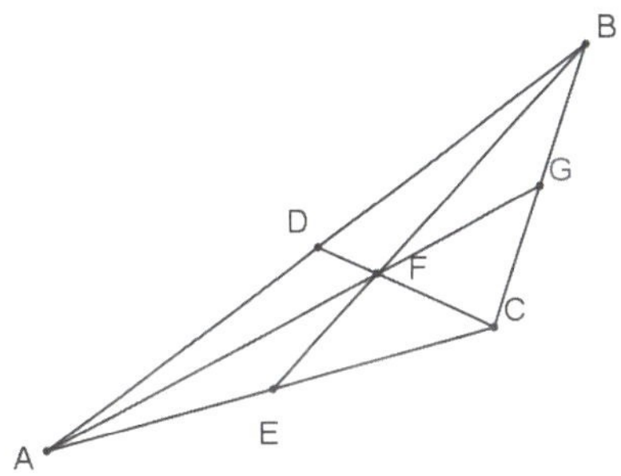

Example 3 (10 minutes)

Example 3

In the figure to the right, $\triangle ABC$ is reflected over $\overline{AB}$ to create $\triangle ABD$. Points P, E, and F are midpoints of $\overline{AB}$, $\overline{BD}$, and $\overline{BC}$, respectively. If $AH = AG$, prove that $PH = GP$.

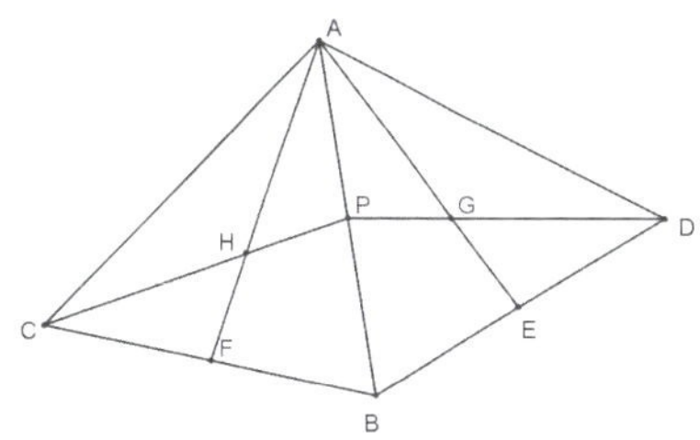

$\triangle ABC$ is reflected over $\overline{AB}$ to create $\triangle ABD$.

$AF = AE$	*Segments preserved, rigid motion*
$\angle FAB \cong \angle EAB$	*Angles preserved, rigid motion*
$AH = AG$	*Given*
$AP = AP$	*Reflexive property*
$\triangle APH \cong \triangle APG$	*SAS*
$PH = GP$	*Corresponding sides of congruent triangles are equal in measure.*

Closing (1 minute)

- What is a centroid, and how does it divide a median?
  - *A centroid is the point of concurrency of the medians of a triangle. It divides a median into two segments such that the distance from the vertex to the centroid is twice that of the distance from the centroid to the midpoint of the side of the triangle.*

Exit Ticket (5 minutes)

Name ______________________________ Date ______________

Lesson 30: Special Lines in Triangles

Exit Ticket

$\overline{DQ}$, $\overline{FP}$, and $\overline{RE}$ are all medians of $\triangle DEF$, and C is the centroid. $DQ = 24$, $FC = 10$, $RC = 7$. Find DC, CQ, FP, and CE.

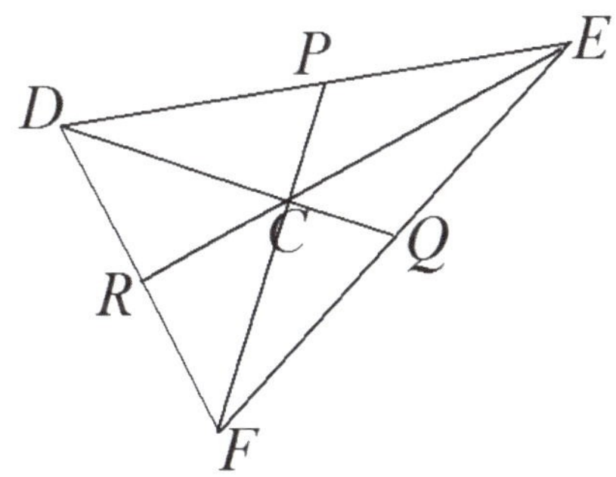

Exit Ticket Sample Solutions

DQ, FP, and RE are all medians of $\triangle DEF$, and C is the centroid. $DQ = 24$, $FC = 10$, $RC = 7$. Find DC, CQ, FP, and CE.

$DC = 16$, $CQ = 8$, $FP = 15$, and $CE = 14$

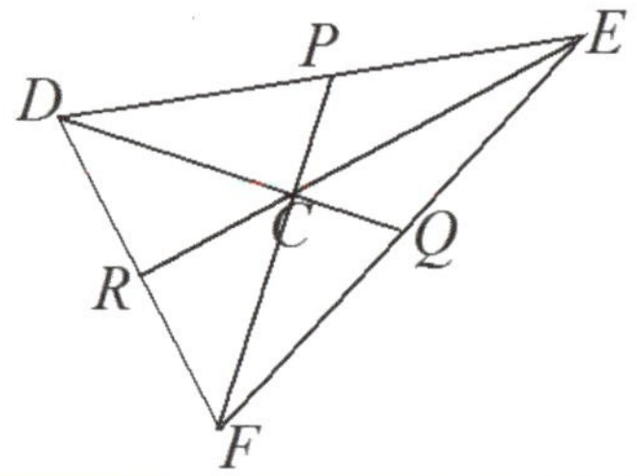

Problem Set Sample Solutions

Ty is building a model of a hang glider using the template below. To place his supports accurately, Ty needs to locate the center of gravity on his model.

1. Use your compass and straightedge to locate the center of gravity on Ty's model.

2. Explain what the center of gravity represents on Ty's model.

   The center of gravity is the centroid.

3. Describe the relationship between the longer and shorter sections of the line segments you drew as you located the center of gravity.

   The centroid divides the length of each median in a ratio of $2:1$.

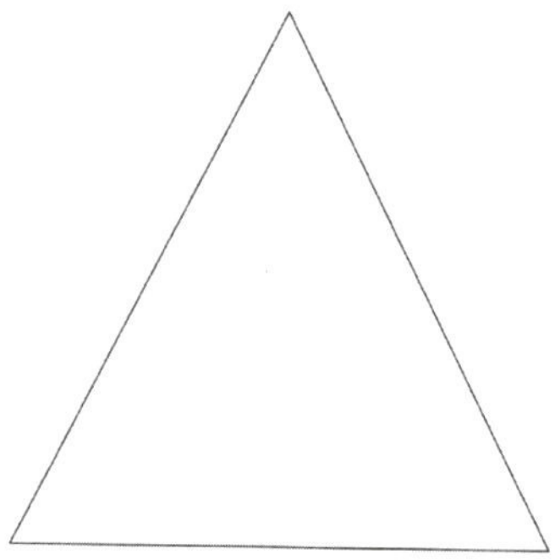

EUREKA MATH

Topic F

Advanced Constructions

G-CO.D.13

Focus Standard:	G-CO.D.13	Construct an equilateral triangle, a square, and a regular hexagon inscribed in a circle.
Instructional Days:	2	
Lesson 31:	Construct a Square and a Nine-Point Circle (E)[1]	
Lesson 32:	Construct a Nine-Point Circle (E)	

In Topic F, Lessons 31 and 32, students are presented with the challenging but interesting construction of a nine-point circle.

[1]Lesson Structure Key: **P**-Problem Set Lesson, **M**-Modeling Cycle Lesson, **E**-Exploration Lesson, **S**-Socratic Lesson

Lesson 31: Construct a Square and a Nine-Point Circle

Student Outcomes

- Students learn to construct a square and begin to construct a nine-point circle.

Lesson Notes

In Lesson 31, students use constructions they already know to construct a square and begin the construction of a nine-point circle. Students articulate the steps needed to do the construction for each. Lessons 31 and 32 are lessons for classes that have been completely successful with all other material. They are also a great opportunity to incorporate technology.

Classwork

Opening Exercise (14 minutes)

Allow students 10 minutes for their attempt, and then share-out steps, or have a student share-out steps. Write their steps on the board so that others actually attempt the instructions as a check.

Opening Exercise

With a partner, use your construction tools and what you learned in Lessons 1–5 to attempt the construction of a square. Once you are satisfied with your construction, write the instructions to perform the construction.

Steps to construct a square:

1. *Extend line segment AB on either side of A and B.*
2. *Construct the perpendicular to $\overline{AB}$ through A; construct the perpendicular to $\overline{AB}$ through B.*
3. *Construct circle A: center A, radius AB; construct circle B: center B, radius BA.*
4. *Select one of the points where circle A meets the perpendicular through A, and call that point D. In the same half-plane as D, select the point where B meets the perpendicular through B, and call that point C.*
5. *Draw segment CD.*

Exploratory Challenge (15 minutes)

In their first attempt at a nine-point circle, ask students to use a triangle that is neither equilateral nor right in the initial step. In the example following the Exploratory challenge, students will examine what happens when an equilateral triangle or right triangle is used.

Exploratory Challenge

Now, we are going to construct a nine-point circle. What is meant by the phrase *nine-point circle*?

A circle that contains a set of nine points

Steps to construct a nine-point circle:

1. Draw a triangle ABC.

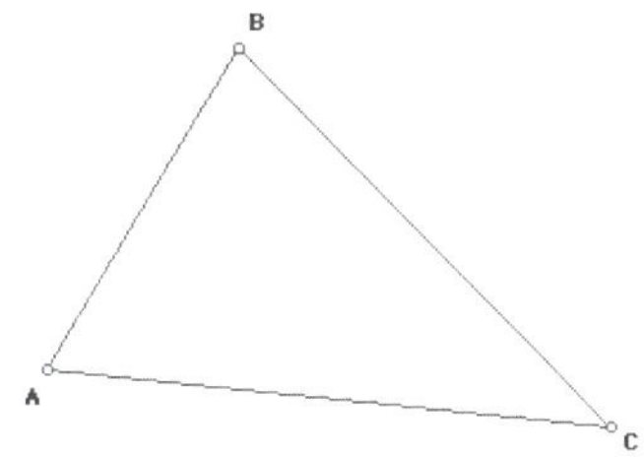

2. Construct the midpoints of the sides $\overline{AB}$, $\overline{BC}$, and $\overline{CA}$, and label them as L, M, and N, respectively.

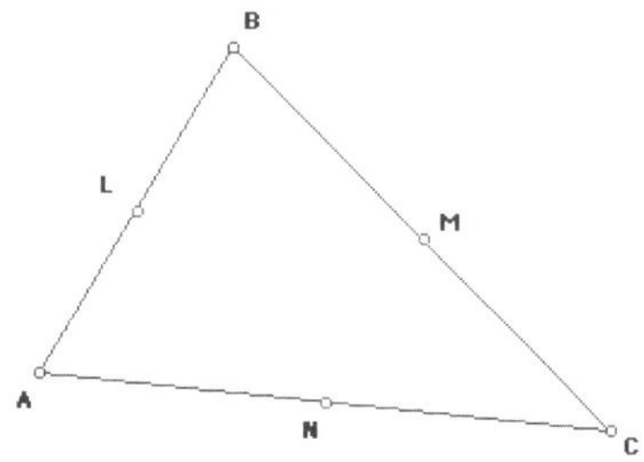

3. Construct the perpendicular from each vertex to the opposite side of the triangle (each is called an *altitude*).

4. Label the intersection of the altitude from C to $\overline{AB}$ as D, the intersection of the altitude from A to $\overline{BC}$ as E, and of the altitude from B to $\overline{CA}$ as F.

5. The altitudes are concurrent at a point; label it H.

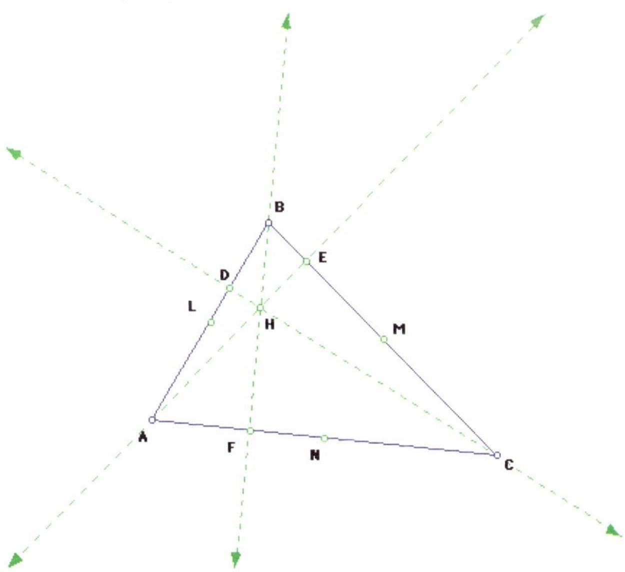

6. Construct the midpoints of $\overline{AH}$, $\overline{BH}$, $\overline{CH}$, and label them X, Y, and Z, respectively.

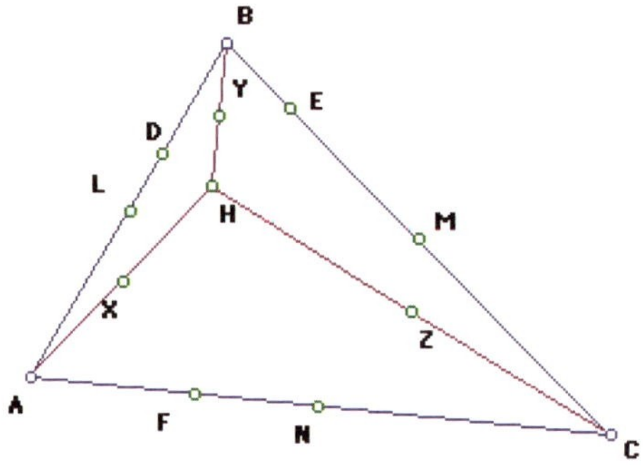

7. The nine points, $L, M, N, D, E, F, X, Y, Z$, are the points that define the nine-point circle.

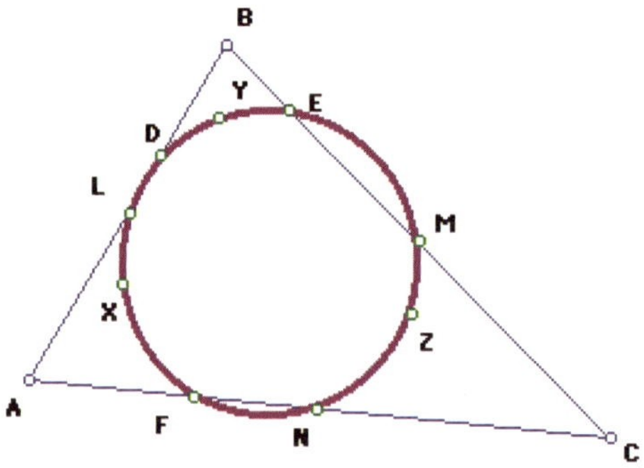

Example (8 minutes)

Example

On a blank white sheet of paper, construct a nine-point circle using an equilateral or right triangle than you used during the notes. What do you notice?

Some of the key points in the nine-point circle construction of a right triangle or equilateral triangle will coincide with each other.

Closing (1 minute)

- The points of a nine-point circle are established in three sets of three: Three are determined by the midpoints of the sides of the triangle, three are determined by the intersections of the altitudes with the sides of the triangle, and the final three are determined by the midpoints of the altitudes.

Exit Ticket (7 minutes)

Name ______________________________ Date ______________

Lesson 31: Construct a Square and a Nine-Point Circle

Exit Ticket

Construct a square $ABCD$ and a square $AXYZ$ so that $\overline{AB}$ contains X and $\overline{AD}$ contains Z.

Exit Ticket Sample Solutions

Construct a square $ABCD$ and a square $AXYZ$ so that $\overline{AB}$ contains X and $\overline{AD}$ contains Z.

Sample construction might look like the following:

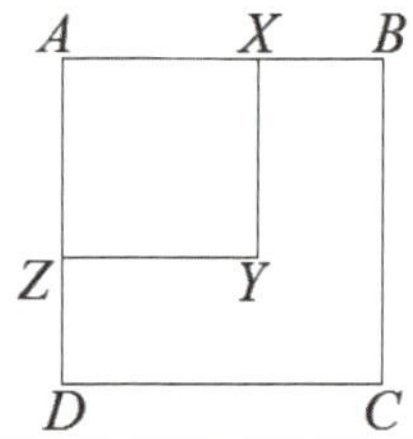

Problem Set Sample Solutions

Construct square $ABCD$ and square $GHIJ$ so that

a. Each side of $GHIJ$ is half the length of each $ABCD$.

b. $\overline{AB}$ contains $\overline{GH}$.

c. The midpoint of $\overline{AB}$ is also the midpoint of $\overline{GH}$.

Sample construction might look like the following:

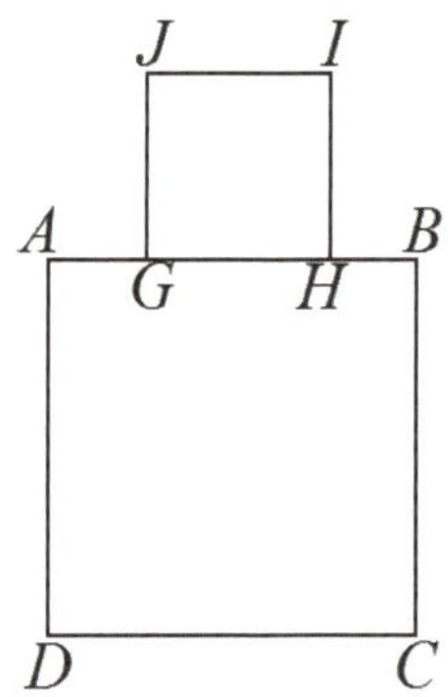

Lesson 32: Construct a Nine-Point Circle

Student Outcomes

- Students complete the construction of a nine-point circle.

Lesson Notes

In Lesson 32, students continue the construction of a nine-point circle. Students articulate the steps needed to do the construction for each.

Note that the Problem Set exercise is challenging and time-consuming. While a good opportunity to build students' ability to persevere, it may be modified to lessen the burden of so many constructions.

Classwork

Opening Exercise (5 minutes)

Opening Exercise

During this unit, we have learned many constructions. Now that you have mastered these constructions, write a list of advice for someone who is about to learn the constructions you have learned for the first time. What did and did not help you? What tips did you wish you had at the beginning that would have made it easier along the way?

Exploratory Challenge 1 (15 minutes)

Exploratory Challenge 1

Yesterday, we began the nine-point circle construction. What did we learn about the triangle that we start our construction with? Where did we stop in the construction?

We continue our construction today.

There are two constructions for finding the center of the nine-point circle. With a partner, work through both constructions.

Construction 1

1. **To find the center of the circle, draw inscribed $\triangle LMN$.**
2. **Find the circumcenter of $\triangle LMN$, and label it as U.**

Recall that the circumcenter of a triangle is the center of the circle that circumscribes the triangle, which, in this case, is the nine-point circle.

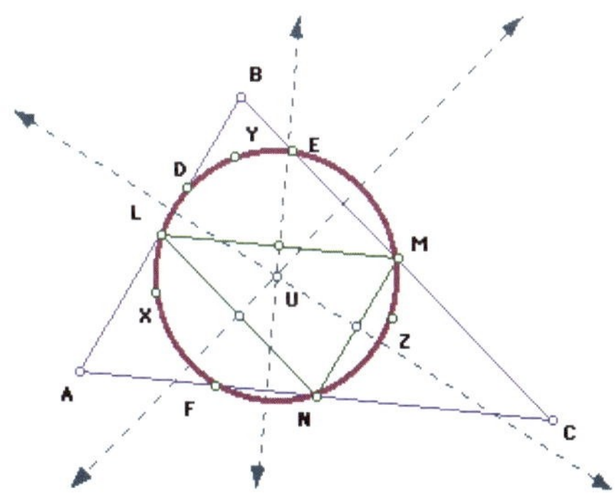

Construction 2

1. Construct the circle that circumscribes $\triangle ABC$.
2. Find the circumcenter of $\triangle ABC$, which is the center of the circle that circumscribes $\triangle ABC$. Label its center CC.
3. Draw the segment that joins point H (the orthocenter from the construction of the nine-point circle in Lesson 31) to the point CC.
4. Find the midpoint of the segment you drew in Step 3, and label that point U.

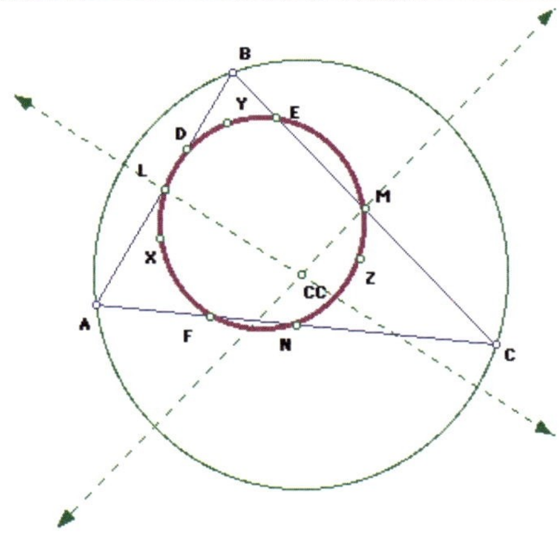

Describe the relationship between the midpoint you found in Step 4 of the second construction and the point U in the first construction.

The center U is the midpoint of the segment from the orthocenter H to CC.

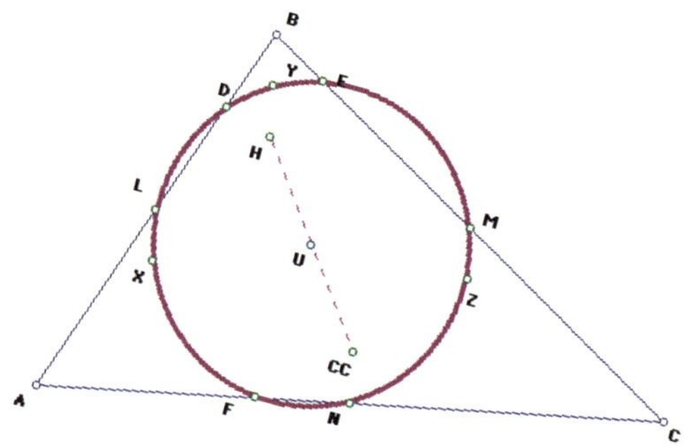

Exploratory Challenge 2 (14 minutes)

Exploratory Challenge 2

Construct a square $ABCD$. Pick a point E between B and C, and draw a segment from point A to a point E. The segment forms a right triangle and a trapezoid out of the square. Construct a nine-point circle using the right triangle.

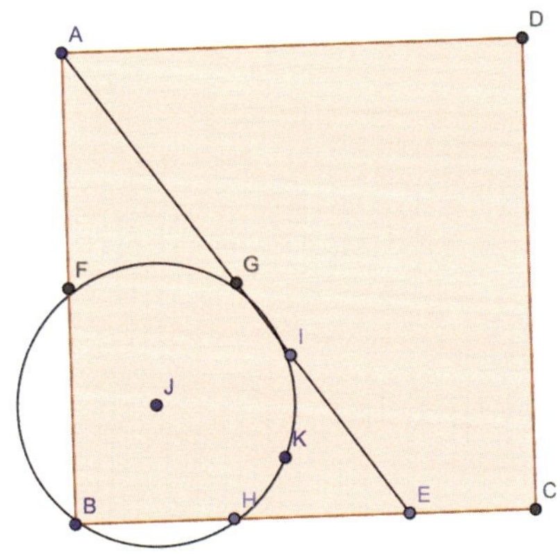

Closing (1 minute)

- The center of a nine-point circle can be found by determining the circumcenter of a triangle inscribed in the nine-point circle.

Exit Ticket (10 minutes)

Name ______________________________ Date ______________

Lesson 32: Construct a Nine-Point Circle

Exit Ticket

Construct a nine-point circle, and then inscribe a square in the circle (so that the vertices of the square are on the circle).

Exit Ticket Sample Solutions

Construct a nine-point circle, and then inscribe a square in the circle (so that the vertices of the square are on the circle).

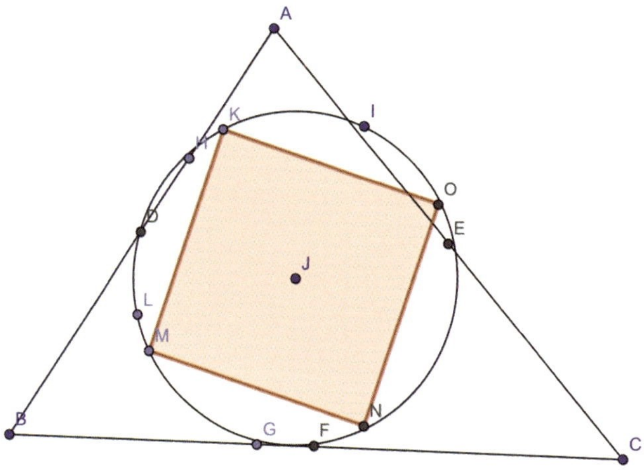

Problem Set Sample Solutions

Take a blank sheet of $8\frac{1}{2}$ inch by 11 inch white paper, and draw a triangle with vertices on the edge of the paper. Construct a nine-point circle within this triangle. Then, draw a triangle with vertices on that nine-point circle, and construct a nine-point circle within that. Continue constructing nine-point circles until you no longer have room inside your constructions.

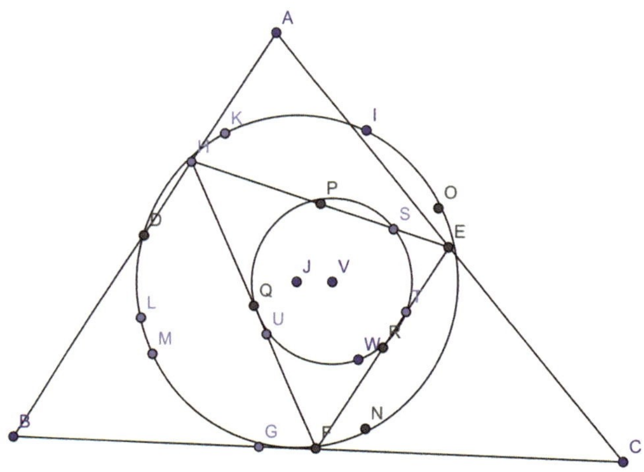

EUREKA MATH™

Topic G

Axiomatic Systems

G-CO.A.1, G-CO.A.2, G-CO.A.3, G-CO.A.4, G-CO.A.5, G-CO.B.6, G-CO.B.7, G-CO.B.8, G-CO.C.9, G-CO.C.10, G-CO.C.11, G-CO.D.12, G-CO.D.13

Focus Standards:		
	G-CO.A.1	Know precise definitions of angle, circle, perpendicular line, parallel line, and line segment, based on the undefined notions of point, line, distance along a line, and distance around a circular arc.
	G-CO.A.2	Represent transformations in the plane using, e.g., transparencies and geometry software; describe transformations as functions that take points in the plane as inputs and give other points as outputs. Compare transformations that preserve distance and angle to those that do not (e.g., translation versus horizontal stretch).
	G-CO.A.3	Given a rectangle, parallelogram, trapezoid, or regular polygon, describe the rotations and reflections that carry it onto itself.
	G-CO.A.4	Develop definitions of rotations, reflections, and translations in terms of angles, circles, perpendicular lines, parallel lines, and line segments.
	G-CO.A.5	Given a geometric figure and a rotation, reflection, or translation, draw the transformed figure using, e.g., graph paper, tracing paper, or geometry software. Specify a sequence of transformations that will carry a given figure onto another.
	G-CO.B.6	Use geometric descriptions of rigid motions to transform figures and to predict the effect of a given rigid motion on a given figure; given two figures, use the definition of congruence in terms of rigid motions to decide if they are congruent.
	G-CO.B.7	Use the definition of congruence in terms of rigid motions to show that two triangles are congruent if and only if corresponding pairs of sides and corresponding pairs of angles are congruent.
	G-CO.B.8	Explain how the criteria for triangle congruence (ASA, SAS, and SSS) follow from the definition of congruence in terms of rigid motions.
	G-CO.C.9	Prove theorems about lines and angles. *Theorems include: vertical angles are congruent; when a transversal crosses parallel lines, alternate interior angles are congruent and corresponding angles are congruent; points on a perpendicular bisector of a line segment are exactly those equidistant from the segment's endpoints.*

	G-CO.C.10	Prove theorems about triangles. *Theorems include: measures of interior angles of a triangle sum to* $180°$*; base angles of isosceles triangles are congruent; the segment joining midpoints of two sides of a triangle is parallel to the third side and half the length; the medians of a triangle meet at a point.*
	G-CO.C.11	Prove theorems about parallelograms. *Theorems include: opposite sides are congruent, opposite angles are congruent, the diagonals of a parallelogram bisect each other, and conversely, rectangles are parallelograms with congruent diagonals.*
	G-CO.D.12	Make formal geometric constructions with a variety of tools and methods (compass and straightedge, string, reflective devices, paper folding, dynamic geometric software, etc.). *Copying a segment; copying an angle; bisecting a segment; bisecting an angle; constructing perpendicular lines, including the perpendicular bisector of a line segment; and constructing a line parallel to a given line through a point not on the line.*
	G-CO.D.13	Construct an equilateral triangle, a square, and a regular hexagon inscribed in a circle.
Instructional Days:	2	
Lessons 33–34:	Review of the Assumptions (P, P)[1]	

In Topic G, students review material covered throughout the module. Additionally, students discuss the structure of geometry as an axiomatic system.

[1]Lesson Structure Key: **P**-Problem Set Lesson, **M**-Modeling Cycle Lesson, **E**-Exploration Lesson, **S**-Socratic Lesson

Lesson 33: Review of the Assumptions

Student Outcomes

- Students examine the basic geometric assumptions from which all other facts can be derived.
- Students review the principles addressed in Module 1.

Classwork

Review Exercises (38 minutes)

Review Exercises

We have covered a great deal of material in Module 1. Our study has included definitions, geometric assumptions, geometric facts, constructions, unknown angle problems and proofs, transformations, and proofs that establish properties we previously took for granted.

In the first list below, we compile all of the geometric assumptions we took for granted as part of our reasoning and proof-writing process. Though these assumptions were only highlights in lessons, these assumptions form the basis from which all other facts can be derived (e.g., the other facts presented in the table). College-level geometry courses often do an in-depth study of the assumptions.

The latter tables review the facts associated with problems covered in Module 1. Abbreviations for the facts are within brackets.

Geometric Assumptions (Mathematicians call these *axioms*.)

1. (Line) Given any two distinct points, there is exactly one line that contains them.
2. (Plane Separation) Given a line contained in the plane, the points of the plane that do not lie on the line form two sets, called *half-planes*, such that
   a. Each of the sets is convex.
   b. If P is a point in one of the sets and Q is a point in the other, then $\overline{PQ}$ intersects the line.
3. (Distance) To every pair of points A and B there corresponds a real number $\text{dist}\,(A, B) \geq 0$, called the *distance* from A to B, so that
   a. $\text{dist}(A, B) = \text{dist}(B, A)$
   b. $\text{dist}(A, B) \geq 0$, and $\text{dist}(A, B) = 0 \Leftrightarrow A$ and B coincide.
4. (Ruler) Every line has a coordinate system.
5. (Plane) Every plane contains at least three noncollinear points.
6. (Basic Rigid Motions) Basic rigid motions (e.g., rotations, reflections, and translations) have the following properties:
   a. Any basic rigid motion preserves lines, rays, and segments. That is, for any basic rigid motion of the plane, the image of a line is a line, the image of a ray is a ray, and the image of a segment is a segment.
   b. Any basic rigid motion preserves lengths of segments and angle measures of angles.
7. ($180°$ Protractor) To every $\angle AOB$, there corresponds a real number $m\angle AOB$, called the *degree* or *measure* of the angle, with the following properties:
   a. $0° < m\angle AOB < 180°$
   b. Let $\overrightarrow{OB}$ be a ray on the edge of the half-plane H. For every r such that $0° < r° < 180°$, there is exactly one ray $\overrightarrow{OA}$ with A in H such that $m\angle AOB = r°$.
   c. If C is a point in the interior of $\angle AOB$, then $m\angle AOC + m\angle COB = m\angle AOB$.
   d. If two angles $\angle BAC$ and $\angle CAD$ form a linear pair, then they are supplementary (e.g., $m\angle BAC + m\angle CAD = 180°$).

8. **(Parallel Postulate) Through a given external point, there is at most one line parallel to a given line.**

Fact/Property	Guiding Questions/Applications	Notes/Solutions
Two angles that form a linear pair are supplementary.	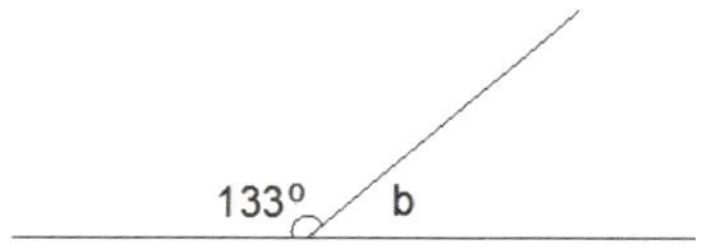	$m\angle b = 47°$
The sum of the measures of all adjacent angles formed by three or more rays with the same vertex is $360°$.	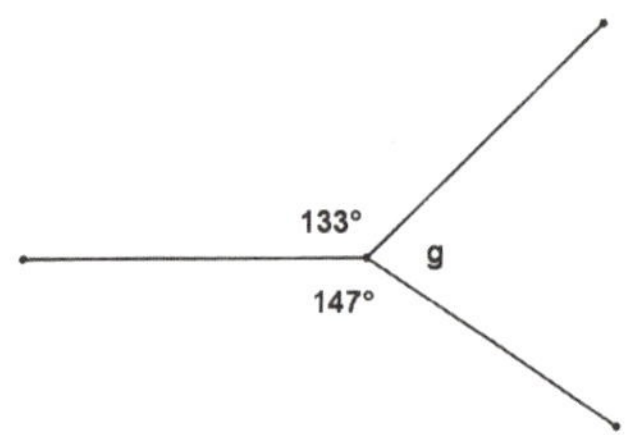	$m\angle g = 80°$
Vertical angles have equal measure.	Use the fact that linear pairs form supplementary angles to prove that vertical angles are equal in measure. 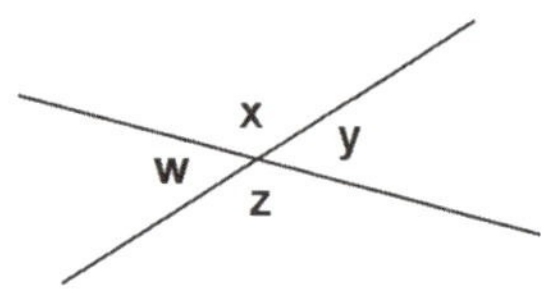	$m\angle w + m\angle x = 180°$ $m\angle y + m\angle x = 180°$ $m\angle w + m\angle x = m\angle y + m\angle x$ $\therefore\ m\angle w = m\angle y$
The bisector of an angle is a ray in the interior of the angle such that the two adjacent angles formed by it have equal measure.	In the diagram below, $\overline{BC}$ is the bisector of $\angle ABD$, which measures $64°$. What is the measure of $\angle ABC$? 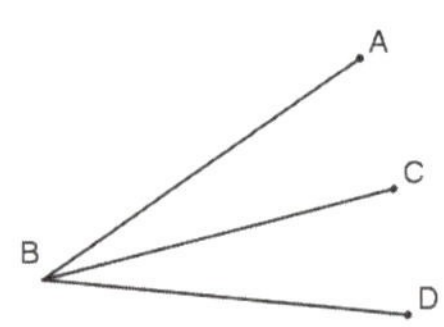	$32°$
The perpendicular bisector of a segment is the line that passes through the midpoint of a line segment and is perpendicular to the line segment.	In the diagram below, $\overline{DC}$ is the perpendicular bisector of $\overline{AB}$, and $\overline{CE}$ is the angle bisector of $\angle ACD$. Find the measures of $\overline{AC}$ and $\angle ECD$. 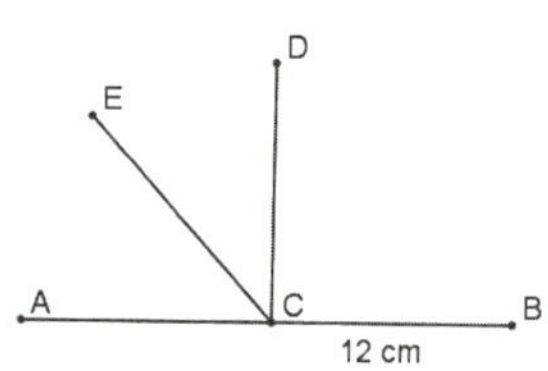	$AC = 12$, $m\angle ECD = 45°$

The sum of the 3 angle measures of any triangle is $180°$.	Given the labeled figure below, find the measures of $\angle DEB$ and $\angle ACE$. Explain your solutions. 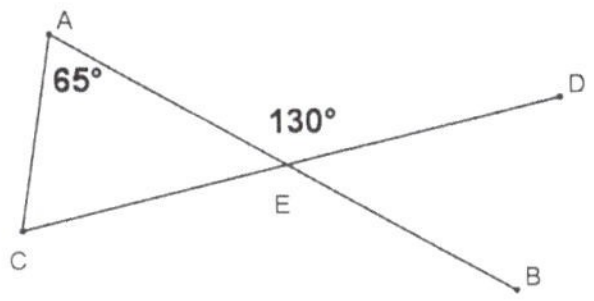	*$m\angle DEB = 50°$, $m\angle ACE = 65°$* *$m\angle DEB + m\angle AED = 180°$ and angle sum of a triangle*
When one angle of a triangle is a right angle, the sum of the measures of the other two angles is $90°$.	This fact follows directly from the preceding one. How is simple arithmetic used to extend the angle sum of a triangle property to justify this property?	*Since a right angle is $90°$ and angles of a triangle sum to $180°$, by arithmetic the other two angles must add up to $90°$.*
An exterior angle of a triangle is equal to the sum of its two opposite interior angles.	In the diagram below, how is the exterior angle of a triangle property proved? 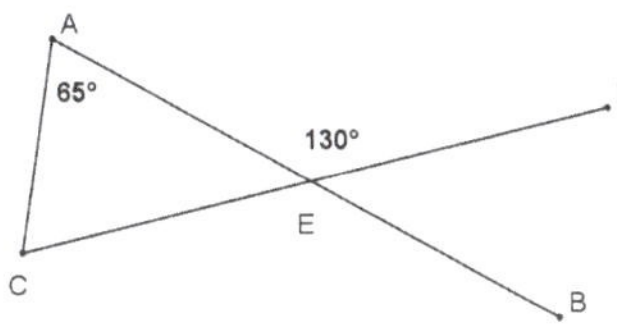	*The sum of two interior opposite angles and the third angle of a triangle is $180°$, which is equal to the angle sum of the third angle and the exterior angle. Thus, the exterior angle of a triangle is equal to the sum of the interior opposite angles.*
Base angles of an isosceles triangle are congruent.	The triangle in the figure above is isosceles. How do we know this?	*The base angles are equal.*
All angles in an equilateral triangle have equal measure.	If the figure above is changed slightly, it can be used to demonstrate the equilateral triangle property. Explain how this can be demonstrated. 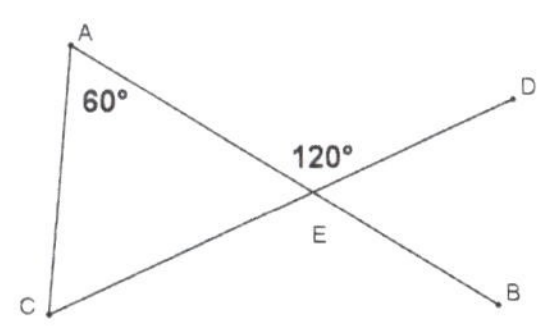	*$m\angle AEC$ is $60°$; angles on a line. $m\angle C$ is also $60°$ by the angle sum of a triangle property. Thus, each interior angle is $60°$.*

The facts and properties in the immediately preceding table relate to angles and triangles. In the table below, we review facts and properties related to parallel lines and transversals.

Fact/Property	Guiding Questions/Applications	Notes/Solutions
If a transversal intersects two parallel lines, then the measures of the corresponding angles are equal.	Why does the property specify *parallel* lines?	*If the lines are not parallel, then the corresponding angles are not congruent.*
If a transversal intersects two lines such that the measures of the corresponding angles are equal, then the lines are parallel.	The converse of a statement turns the relevant property into an *if and only if* relationship. Explain how this is related to the guiding question about corresponding angles.	*The "if and only if" specifies the only case in which corresponding angles are congruent (when two lines are parallel).*
If a transversal intersects two parallel lines, then the interior angles on the same side of the transversal are supplementary.	This property is proved using (in part) the corresponding angles property. Use the diagram below ($\overline{AB} \parallel \overline{CD}$) to prove that $\angle AGH$ and $\angle CHG$ are supplementary. 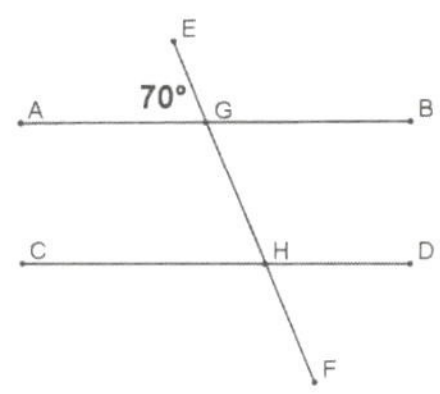	*$m\angle AGH$ is $110°$ because they form a linear pair, and $\angle CHG$ is $70°$ because of corresponding angles. Thus, interior angles on the same side are supplementary.*
If a transversal intersects two lines such that the same-side interior angles are supplementary, then the lines are parallel.	Given the labeled diagram below, prove that $\overline{AB} \parallel \overline{CD}$. 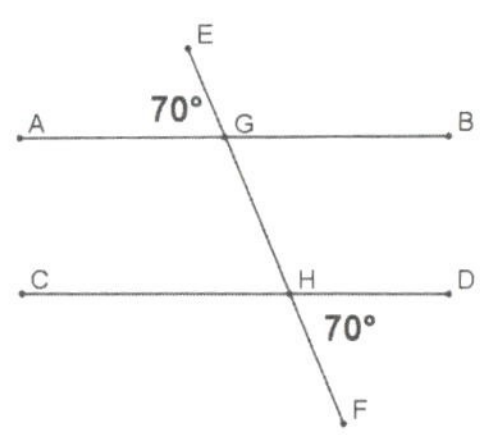	*$m\angle AGH = 110°$ due to a linear pair, and $m\angle GHC = 70°$ due to vertical angles. Then, $\overline{AB} \parallel \overline{CD}$ because the corresponding angles are congruent.*
If a transversal intersects two parallel lines, then the measures of alternate interior angles are equal.	1. Name both pairs of alternate interior angles in the diagram above. 2. How many different angle measures are in the diagram?	*1.* $\angle GHC$, $\angle HGB$ $\angle AGH$, $\angle DHG$ *2.* 2
If a transversal intersects two lines such that measures of the alternate interior angles are equal, then the lines are parallel.	Although not specifically stated here, the property also applies to *alternate exterior angles*. Why is this true?	*The alternate exterior angles are vertical angles to the alternate interior angles.*

EUREKA MATH™

Closing (1 minute)

- The list of eight axioms is the list from which all other geometry facts can be derived.

Exit Ticket (6 minutes)

Name ______________________________ Date ____________

Lesson 33: Review of the Assumptions

Exit Ticket

1. Which assumption(s) must be used to prove that vertical angles are congruent?

2. If two lines are cut by a transversal such that corresponding angles are NOT congruent, what must be true? Justify your response.

Exit Ticket Sample Solutions

1. Which assumption(s) must be used to prove that vertical angles are congruent?

   The "protractor postulate" must be used. If two angles, BAC and CAD, form a linear pair, then they are supplementary (e.g., $m\angle BAC + m\angle CAD = 180°$).

2. If two lines are cut by a transversal such that corresponding angles are NOT congruent, what must be true? Justify your response.

   The lines are not parallel. Corresponding angles are congruent if and only if the lines are parallel. The "if and only if" part of this statement requires that, if the angles are NOT congruent, then the lines are NOT parallel.

Problem Set Sample Solutions

Use any of the assumptions, facts, and/or properties presented in the tables above to find x and y in each figure below. Justify your solutions.

1. $x =$ __$52°$__, $y =$ __$56°$__

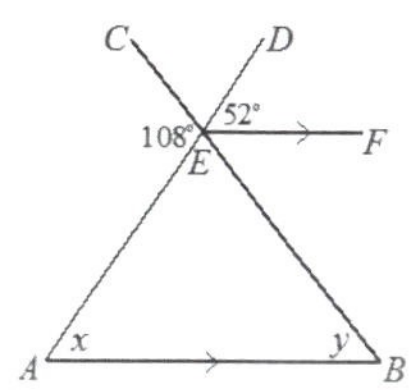

$m\angle AEB$ is $72°$	*Linear pairs form supplementary angles.*
$m\angle FEB$ is $56°$	*Linear pairs form supplementary angles.*
$x = 52°$	*If two parallel lines are cut by a transversal, then the corresponding angles are congruent.*
$y = 56°$	*Angles in a triangle add up to $180°$.*

2. You need to draw an auxiliary line to solve this problem.

   $x =$ __$45°$__, $y =$ __$45°$__

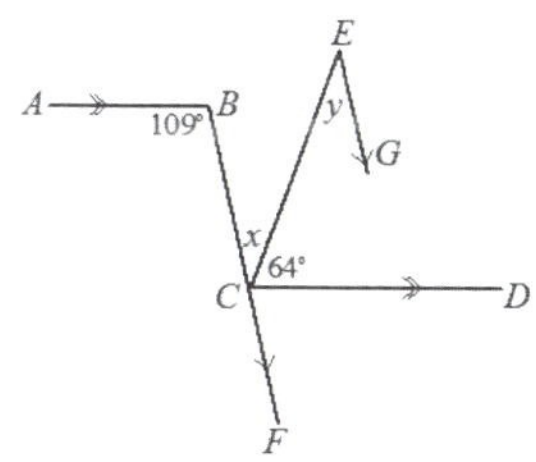

   $\angle ABC$ and $\angle DCB$ are alternate interior angles because $\overline{AB} \parallel \overline{CD}$; $x = 45°$.

   Angles x and y are also alternate interior angles because $\overline{BC} \parallel \overline{EG}$; $y = 45°$.

3. $x =$ __$73°$__, $y =$ __$39°$__

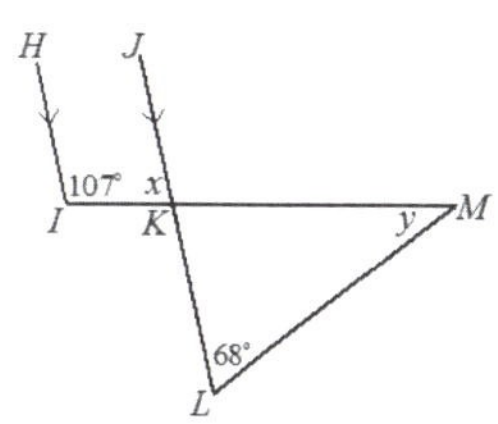

   $\angle HIK$ and $\angle JKI$ are supplementary because they are same side interior angles and $\overline{JK} \parallel \overline{HI}$; therefore, the value of x is $73°$. $\angle MKL$ and $\angle JKI$ are vertical angles. So, using the fact that the sum of angles in a triangle is $180°$, we find that the value of y is $39°$.

4. Given the labeled diagram at the right, prove that $\angle VWX \cong \angle XYZ$.

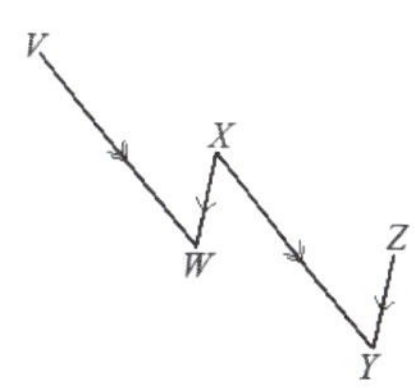

$\angle VWX \cong \angle YXW$	*When two parallel lines are cut by a transversal, the alternate interior angles are congruent.*
$\angle XYZ \cong \angle YXW$	*When two parallel lines are cut by a transversal, the alternate interior angles are congruent.*
$\therefore \angle VWX = \angle XYZ$	*Substitution property of equality*

Lesson 34: Review of the Assumptions

Student Outcomes

- Students review the principles addressed in Module 1.

Lesson Notes

In Lesson 33, we reviewed many of the assumptions, facts, and properties used in this module to derive other facts and properties in geometry. We continue this review process with the table of facts and properties below, beginning with those related to rigid motions.

Classwork

Review Exercises (38 minutes)

Assumption/Fact/Property	Guiding Questions/Applications	Notes/Solutions
Given two triangles ABC and $A'B'C'$ so that $AB = A'B'$ (Side), $m\angle A = m\angle A'$ (Angle), and $AC = A'C'$ (Side), then the triangles are congruent. [SAS]	The figure below is a parallelogram $ABCD$. What parts of the parallelogram satisfy the SAS triangle congruence criteria for $\triangle ABD$ and $\triangle CDB$? Describe a rigid motion(s) that maps one onto the other. (Consider drawing an auxiliary line.) 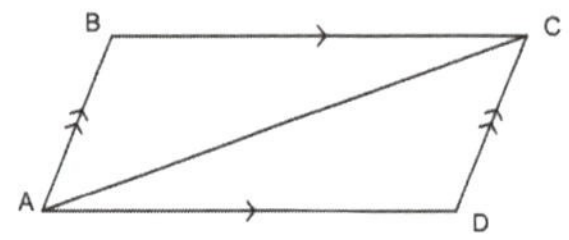 	*$AD = CB$, property of a parallelogram* *$m\angle ABD = m\angle CDB$, alternate interior angles* *$BD = BD$, reflexive property* *$\triangle ABD \cong \triangle CDB$, SAS* *$180^\circ$ rotation about the midpoint of $\overline{BD}$*
Given two triangles ABC and $A'B'C'$, if $m\angle A = m\angle A'$ (Angle), $AB = A'B'$ (Side), and $m\angle B = m\angle B'$ (Angle), then the triangles are congruent. [ASA]	In the figure below, $\triangle CDE$ is the image of the reflection of $\triangle ABE$ across line FG. Which parts of the triangle can be used to satisfy the ASA congruence criteria? 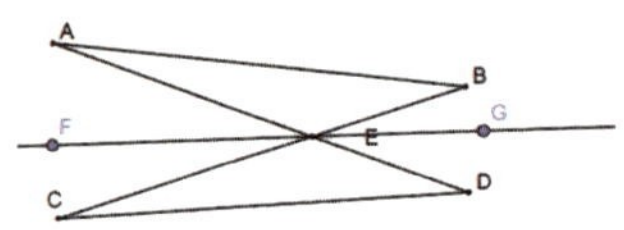 	*$m\angle AEB = m\angle CED$, vertical angles are equal in measure.* *$BE = DE$, reflections map segments onto segments of equal length.* *$\angle B \cong \angle D$, reflections map angles onto angles of equal measure.*

Given two triangles ABC and $A'B'C'$, if $AB = A'B'$ (Side), $AC = A'C'$ (Side), and $BC = B'C'$ (Side), then the triangles are congruent. [SSS]	$\triangle ABC$ and $\triangle ADC$ are formed from the intersections and center points of circles A and C. Prove $\triangle ABC \cong \triangle ADC$ by SSS. 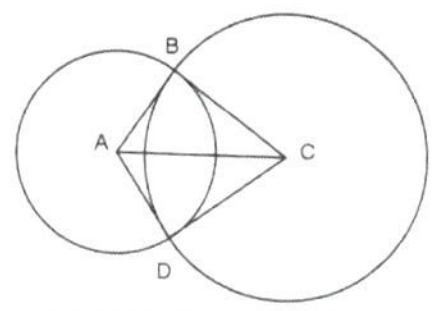	*AC is a common side.* *$AB = AD$, they are both radii of the same circle.* *$BC = DC$, they are both radii of the same circle.* *Thus, $\triangle ABC \cong \triangle ADC$ by SSS.*
Given two triangles ABC and $A'B'C'$, if $AB = A'B'$ (Side), $m\angle B = m\angle B'$ (Angle), and $m\angle C = m\angle C'$ (Angle), then the triangles are congruent. [AAS]	The AAS congruence criterion is essentially the same as the ASA criterion for proving triangles congruent. Why is this true? 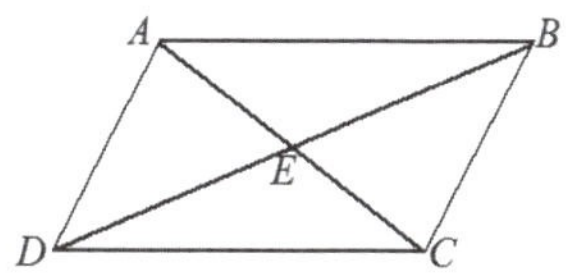	*If two angles of a triangle are congruent to two angles of a second triangle, then the third pair must also be congruent. Therefore, if one pair of corresponding sides is congruent, we treat the given corresponding sides as the included side, and the triangles are congruent by ASA.*
Given two right triangles ABC and $A'B'C'$ with right angles $\angle B$ and $\angle B'$, if $AB = A'B'$ (Leg) and $AC = A'C'$ (Hypotenuse), then the triangles are congruent. [HL]	In the figure below, CD is the perpendicular bisector of AB, and $\triangle ABC$ is isosceles. Name the two congruent triangles appropriately, and describe the necessary steps for proving them congruent using HL. 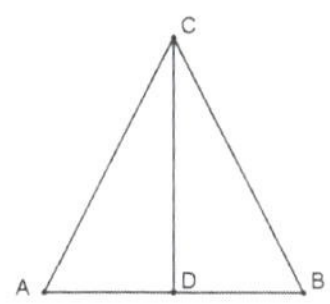	*$\triangle ADC \cong \triangle BDC$* *Given $\overline{CD} \perp \overline{AB}$, both $\triangle ADC$ and $\triangle BDC$ are right triangles. $\overline{CD}$ is a common side. Given $\triangle ABC$ is isosceles, $\overline{AC} \cong \overline{CB}$.*
The opposite sides of a parallelogram are congruent. The opposite angles of a parallelogram are congruent. The diagonals of a parallelogram bisect each other.	In the figure below, $BE \cong DE$, and $\angle CBE \cong \angle ADE$. Prove $ABCD$ is a parallelogram. 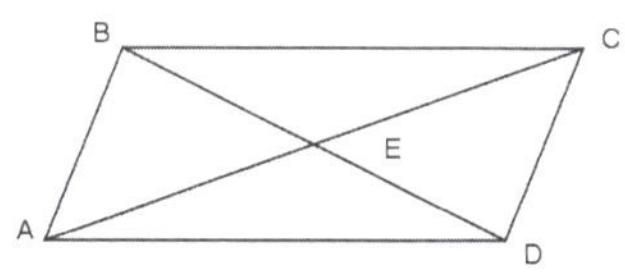	*$\angle BEC \cong \angle AED$, vertical angles are equal in measure.* *$\overline{BE} \cong \overline{DE}$, and $\angle CBE \cong \angle ADE$, given.* *$\triangle BEC \cong \triangle DEA$, ASA.* *By similar reasoning, we can show that $\triangle BEA \cong \triangle DEC$.* *Since $\overline{AB} \cong \overline{DC}$ and $\overline{BC} \cong \overline{DA}$, $ABCD$ is a parallelogram because the opposite sides are congruent (property of parallelogram).*
The midsegment of a triangle is a line segment that connects the midpoints of two sides of a triangle; the midsegment is parallel to the third side of the triangle and is half the length of the third side.	$\overline{DE}$ is the midsegment of $\triangle ABC$. Find the perimeter of $\triangle ABC$, given the labeled segment lengths. 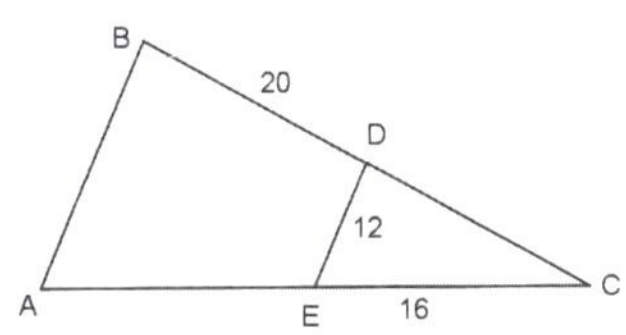	*96*

The three medians of a triangle are concurrent at the centroid; the centroid divides each median into two parts, from vertex to centroid and centroid to midpoint, in a ratio of $2:1$.	If $\overline{AE}$, $\overline{BF}$, and $\overline{CD}$ are medians of $\triangle ABC$, find the length of $\overline{BG}$, $\overline{GE}$, and $\overline{CG}$, given the labeled lengths. 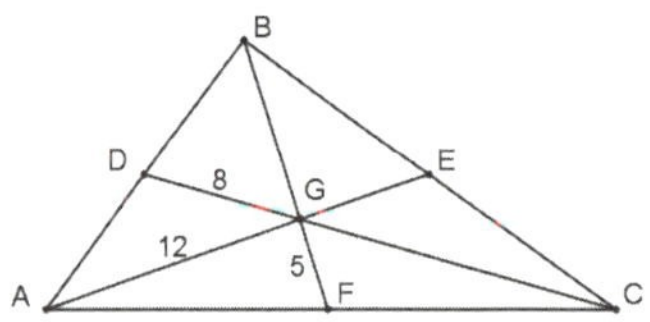	$BG = 10$ $GE = 6$ $CG = 16$

Closing (1 minute)

- In the second half of the module, we studied triangle congruence criteria, we proved assumed properties of parallelograms, and we discovered the significance of special lines in triangles.

Exit Ticket (6 minutes)

Name ______________________________ Date ______________

Lesson 34: Review of the Assumptions

Exit Ticket

The inner parallelogram in the figure is formed from the midsegments of the four triangles created by the outer parallelogram's diagonals. The lengths of the smaller and larger midsegments are as indicated. If the perimeter of the outer parallelogram is 40, find the value of x.

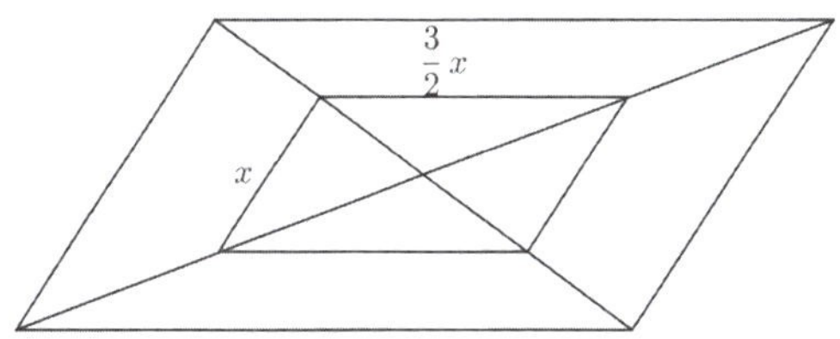

Exit Ticket Sample Solutions

The inner parallelogram in the figure is formed from the midsegments of the four triangles created by the outer parallelogram's diagonals. The lengths of the smaller and larger midsegments are as indicated. If the perimeter of the outer parallelogram is 40, find the value of x.

$x = 4$

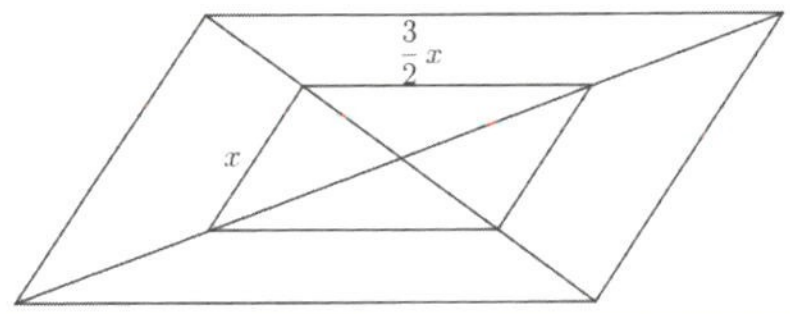

Problem Set Sample Solutions

Use any of the assumptions, facts, and/or properties presented in the tables above to find x and/or y in each figure below. Justify your solutions.

1. Find the perimeter of parallelogram $ABCD$. Justify your solution.

   $100, 15 = x + 4, x = 11$

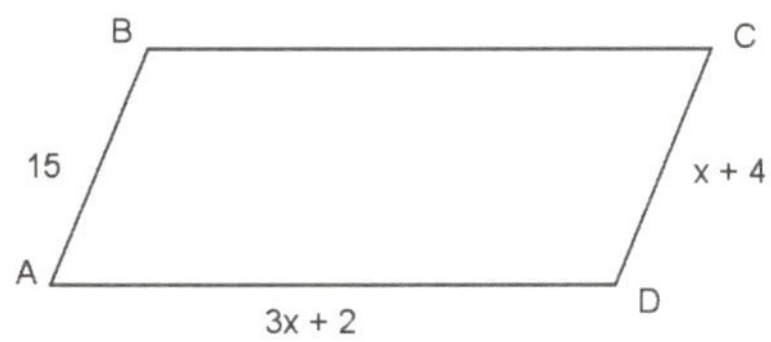

2. $AC = 34$

   $AB = 26$

   $BD = 28$

   Given parallelogram $ABCD$, find the perimeter of $\triangle CED$. Justify your solution.

   57

   $CE = \frac{1}{2}AC; CE = 17$

   $CD = AB; CE = 26$

   $ED = \frac{1}{2}BD; ED = 14$

   Perimeter $= 17 + 26 + 14 = 57$

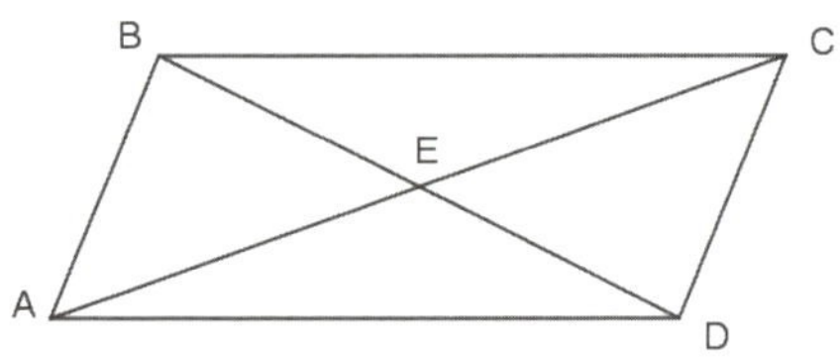

3. $XY = 12$

   $XZ = 20$

   $ZY = 24$

   F, G, and H are midpoints of the sides on which they are located. Find the perimeter of $\triangle FGH$. Justify your solution.

   28

   The midsegment is half the length of the side of the triangle it is parallel to.

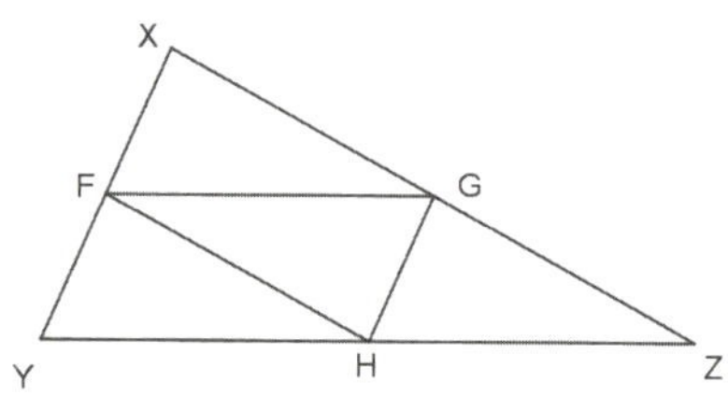

4. $ABCD$ is a parallelogram with $AE = CF$.
Prove that $DEBF$ is a parallelogram.

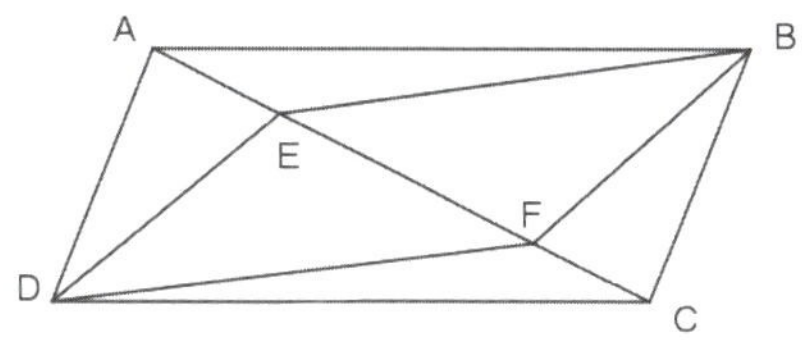

$AE = CF$	*Given*
$AD = BC$	*Property of a parallelogram*
$m\angle DAE = m\angle BCF$	*If parallel lines are cut by a transversal, then alternate interior angles are equal in measure.*
$\triangle ADE \cong \triangle CBF$	*SAS*
$DE = BF$	*Corresponding sides of congruent triangles are congruent.*
$AB = DC$	*Property of a parallelogram*
$m\angle BAE = m\angle DCF$	*If parallel lines are cut by a transversal, then alternate interior angles are equal in measure.*
$\triangle BAE \cong \triangle DCF$	*SAS*
$BE = DF$	*Corresponding sides of congruent triangles are congruent.*
$\therefore$ $ABCD$ *is a parallelogram.*	*If both sets of opposite sides of a quadrilateral are equal in length, then the quadrilateral is a parallelogram.*

5. C is the centroid of $\triangle RST$.
$RC = 16, CL = 10, TJ = 21$

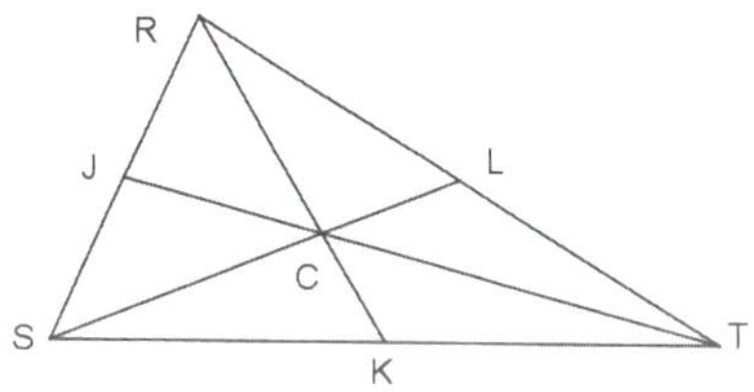

$SC =$ 20

$TC =$ 14

$KC =$ 8

Name ______________________________ Date ______________

1. Each of the illustrations on the next page shows in black a plane figure consisting of the letters F, R, E, and D evenly spaced and arranged in a row. In each illustration, an alteration of the black figure is shown in gray. In some of the illustrations, the gray figure is obtained from the black figure by a geometric transformation consisting of a single rotation. In others, this is not the case.

   a. Which illustrations show a single rotation?

   b. Some of the illustrations are not rotations or even a sequence of rigid transformations. Select one such illustration, and use it to explain why it is not a sequence of rigid transformations.

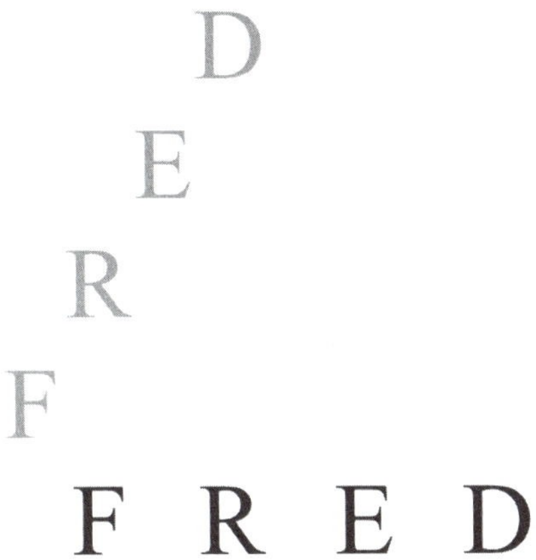

Illustration 1

Illustration 2

Illustration 3

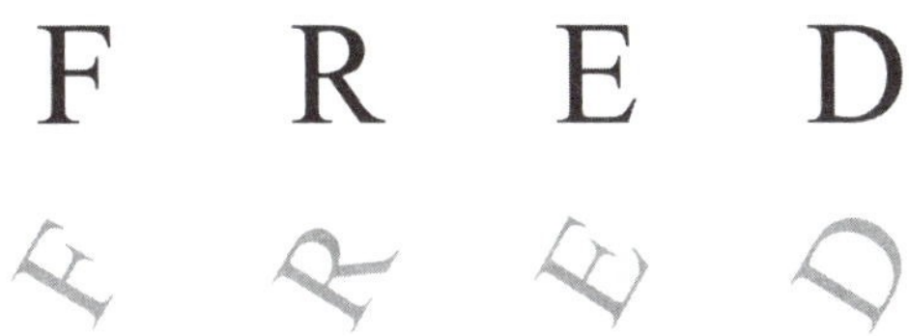

Illustration 4

Illustration 5

Illustration 6

2. In the figure below, $\overline{CD}$ bisects $\angle ACB$, $AB = BC$, $m\angle BEC = 90°$, and $m\angle DCE = 42°$.

   Find the measure of $\angle A$.

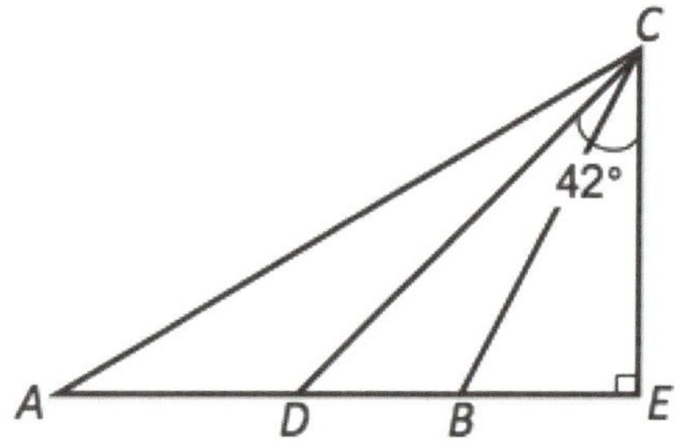

3. In the figure below, $\overline{AD}$ is the angle bisector of $\angle BAC$. $\overline{BAP}$ and $\overline{BDC}$ are straight lines, and $\overline{AD} \parallel \overline{PC}$.

   Prove that $AP = AC$.

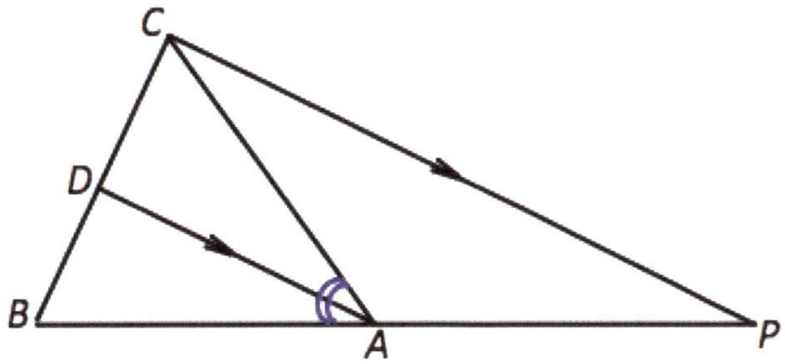

4. $\triangle ABC$ and $\triangle DEF$, in the figure below are such that $\overline{AB} \cong \overline{DE}$, $\overline{AC} \cong \overline{DF}$, and $\angle A \cong \angle D$.

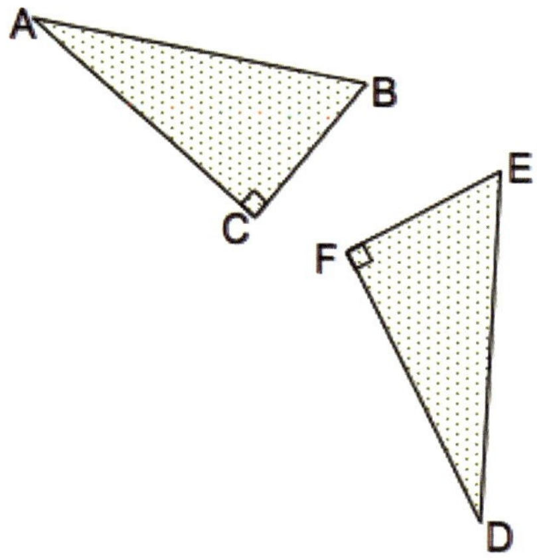

a. Which criteria for triangle congruence (ASA, SAS, SSS) implies that $\triangle ABC \cong \triangle DEF$?

b. Describe a sequence of rigid transformations that shows $\triangle ABC \cong \triangle DEF$.

5.

a. Construct a square $ABCD$ with side $\overline{AB}$. List the steps of the construction.

b. Three rigid motions are to be performed on square $ABCD$. The first rigid motion is the reflection through $\overline{BD}$. The second rigid motion is a $90°$ clockwise rotation around the center of the square.

Describe the third rigid motion that will ultimately map $ABCD$ back to its original position. Label the image of each rigid motion A, B, C, and D in the blanks provided.

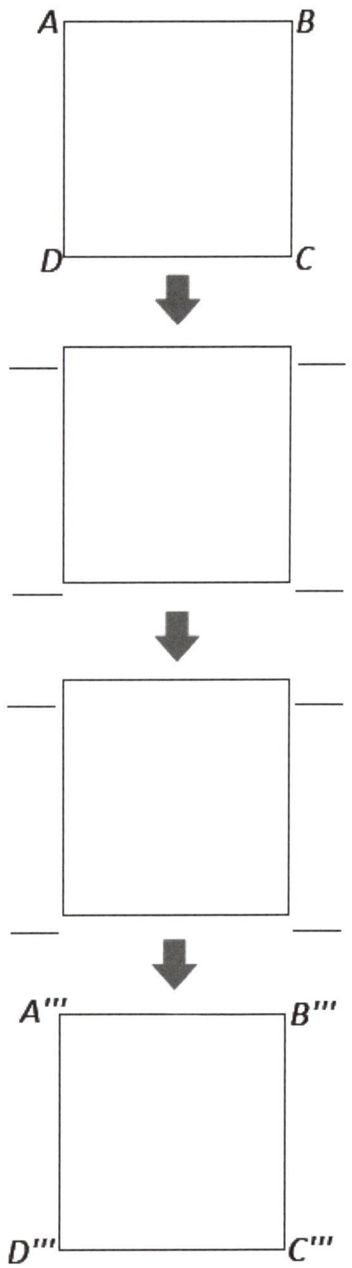

Rigid Motion 1 Description: Reflection through $\overline{BD}$

Rigid Motion 2 Description: $90°$ clockwise rotation around the center of the square.

Rigid Motion 3 Description:

6. Suppose that $ABCD$ is a parallelogram and that M and N are the midpoints of $\overline{AB}$ and $\overline{CD}$, respectively. Prove that $AMCN$ is a parallelogram.

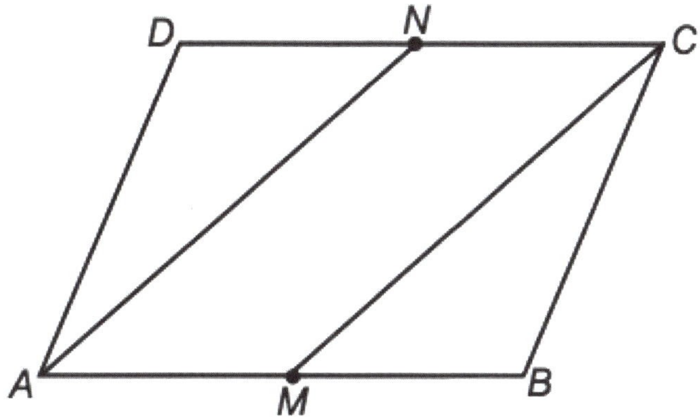

A Progression Toward Mastery					
Assessment Task Item		**STEP 1** **Missing or incorrect answer and little evidence of reasoning or application of mathematics to solve the problem.**	**STEP 2** **Missing or incorrect answer but evidence of some reasoning or application of mathematics to solve the problem.**	**STEP 3** **A correct answer with some evidence of reasoning or application of mathematics to solve the problem, OR an incorrect answer with substantial evidence of solid reasoning or application of mathematics to solve the problem.**	**STEP 4** **A correct answer supported by substantial evidence of solid reasoning or application of mathematics to solve the problem.**
1	**a–b** **G-CO.A.2**	Student identifies illustration 2 or 5 for part (a) and provides a response that shows no understanding of what a sequence of rigid transformations entails in part (b).	Student correctly identifies illustrations 2 and 5 for part (a) and provides a response that shows little understanding of what a sequence of rigid transformations entails in part (b).	Student correctly identifies illustrations 2 and 5 for part (a) and provides a response that shows an understanding of what a sequence of rigid transformations entails but states a less than perfect solution.	Student correctly identifies illustrations 2 and 5 for part (a) and provides a response that correctly reasons why any one of illustrations 1, 3, 4, or 6 is not a sequence of rigid transformations.
2	**G-CO.C.10**	Student provides a response that shows little or no understanding of angle sum properties and no correct answer. OR Student states the correct answer without providing any evidence of the steps to get there.	Student provides a response that shows the appropriate work needed to correctly calculate the measure of angle A but makes one conceptual error and one computational error, two conceptual errors, or two computational errors.	Student provides a response that shows the appropriate work needed to correctly calculate the measure of angle A but makes one conceptual error, such as labeling $m\angle CDE = 132°$, or one computational error, such as finding $m\angle CDE \neq 48°$.	Student provides a response that shows all the appropriate work needed to correctly calculate the measure of angle A.
3	**G-CO.C.10**	Student writes a proof that demonstrates little or no understanding of the method needed to achieve the conclusion.	Student writes a proof that demonstrates an understanding of the method needed to reach the conclusion, but two steps are missing or incorrect.	Student writes a proof that demonstrates an understanding of the method needed to reach the conclusion, but one step is missing or incorrect.	Student writes a complete and correct proof that clearly leads to the conclusion that $AP = AC$.

4	a–b G-CO.B.7 G-CO.B.8	Student provides a response that shows little or no evidence of understanding for parts (a) or (b).	Student provides a response that shows the correct triangle congruence criteria in part (a) and provides a sequence that contains more than one error in part (b).	Student provides a response that shows the correct triangle congruence criteria in part (a) and provides a sequence that contains an error in part (b).	Student provides a response that shows the correct triangle congruence criteria in part (a) and provides any valid sequence of transformations in part (b).
5	a–b G-CO.A.3 G-CO.D.13	Student draws a construction that is not appropriate and provides an underdeveloped list of steps. Student provides a response that contains errors with the vertex labels and the description for Rigid Motion 3 in part (b).	Student draws a construction, but two steps are either missing or incorrect in the construction or list of steps. Student correctly provides vertex labels in the diagram for part (b) but gives an incorrect Rigid Motion 3 description. OR Student correctly describes the Rigid Motion 3 but provides incorrect vertex labels.	Student draws a construction, but one step is missing or incorrect in the construction or in list of steps, such as the marks to indicate the length of side $\overline{AD}$. Student correctly provides vertex labels in the diagram for part (b) but gives an incorrect Rigid Motion 3 description. OR Student correctly describes the Rigid Motion 3 but provides incorrect vertex labels.	Student draws a correct construction showing all appropriate marks and correctly writes out the steps of the construction. Student correctly provides vertex labels in the diagram for part (b) and gives a correct Rigid Motion 3 description.
6	G-CO.C.11	Student writes a proof that demonstrates little or no understanding of the method needed to achieve the conclusion.	Student writes a proof that demonstrates an understanding of the method needed to reach the conclusion, but two steps are missing or incorrect.	Student writes a proof that demonstrates an understanding of the method needed to reach the conclusion, but one step is missing or incorrect.	Student writes a complete and correct proof that clearly leads to the conclusion that $AMCN$ is a parallelogram.

Name ______________________________ Date ______________

1. Each of the illustrations on the next page shows in black a plane figure consisting of the letters F, R, E, and D evenly spaced and arranged in a row. In each illustration, an alteration of the black figure is shown in gray. In some of the illustrations, the gray figure is obtained from the black figure by a geometric transformation consisting of a single rotation. In others, this is not the case.

   a. Which illustrations show a single rotation?

   Illustrations 2 and 5

   b. Some of the illustrations are not rotations or even a sequence of rigid transformations. Select one such illustration, and use it to explain why it is not a sequence of rigid transformations.

   Illustration 1 shows translations of individual letters F, R, E, and D; but each letter is translated a different distance. Since translation requires a shift of the entire plane by the same distance, Illustration 1 does not qualify.

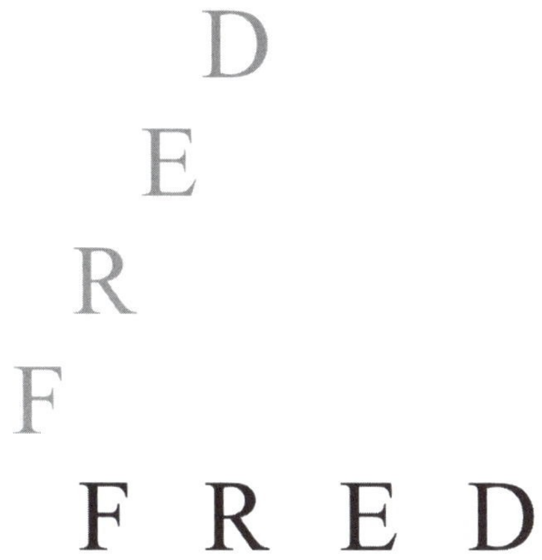

Illustration 1

Illustration 2

Illustration 3

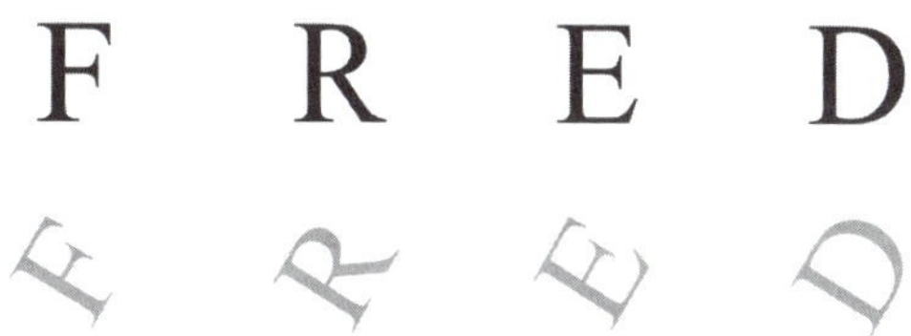

Illustration 4

Illustration 5

Illustration 6

2. In the figure below, $\overline{CD}$ bisects $\angle ACB$, $AB = BC$, $m\angle BEC = 90°$, and $m\angle DCE = 42°$.

Find the measure of $\angle A$.

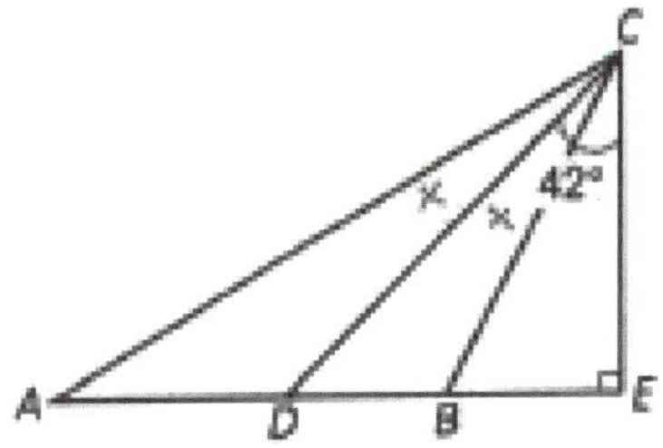

Label the angles as shown.
($\angle ACD \cong \angle DCB$ since $\overline{CD}$ bisects $\angle ACB$)
Since $AB = BC$, ΔABC is isosceles, therefore $2x = a$.

$m\angle A + m\angle ACE + m\angle E = 180°$

$a + (x + 42) + 90 = 180$

$2x + x + 132 = 180$

$x = 16$

Since $a = 2x$, $m\angle A = 32°$

3. In the figure below, $\overline{AD}$ is the angle bisector of $\angle BAC$. $\overline{BAP}$ and $\overline{BDC}$ are straight lines, and $\overline{AD} \parallel \overline{PC}$.

   Prove that $AP = AC$.

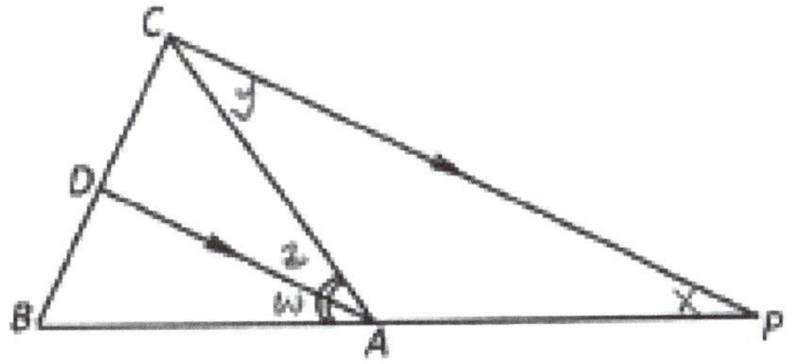

Label w, x, y, and z as shown.

Statements	Reasons
1. $\overline{AD}$ is the angle bisector of ∠BAC	Given
2. $\overline{AD} \parallel \overline{PC}$	Given
3. z = w	Definition of angle bisector
4. z = y	If two parallel lines are cut by a transversal, alt. int. angles are equal in measure.
5. w = x	If two parallel lines are cut by a transversal, corr. angles are equal in measure.
6. x = y	Transitive property
7. △ACP is isosceles	Base angles are congruent
8. AC = AP	Definition of isosceles triangle

4. $\triangle ABC$ and $\triangle DEF$, in the figure below are such that $\overline{AB} \cong \overline{DE}$, $\overline{AC} \cong \overline{DF}$, and $\angle A \cong \angle D$.

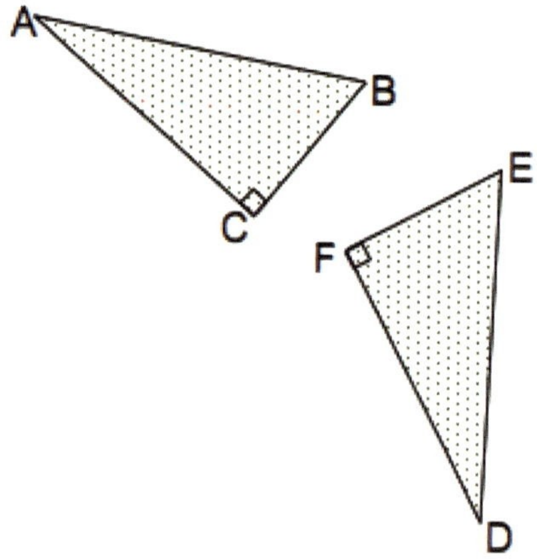

a. Which criteria for triangle congruence (ASA, SAS, SSS) implies that $\triangle ABC \cong \triangle DEF$?

Side-Angle-Side

b. Describe a sequence of rigid transformations that shows $\triangle ABC \cong \triangle DEF$.

1. Translate $\triangle DEF$ so that F is mapped onto C
2. Rotate the image of $\triangle DEF$ about C so that E is mapped onto B
3. Reflect the image of the rotation across $\overleftrightarrow{BC}$

5.

a. Construct a square $ABCD$ with side $\overline{AB}$. List the steps of the construction.

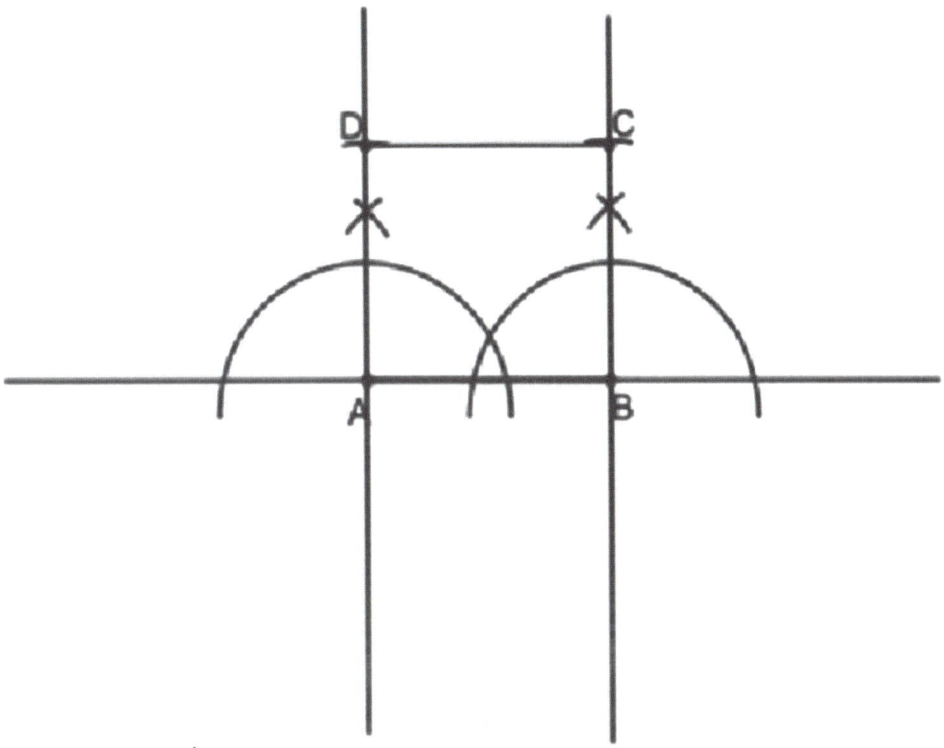

1. Extend $\overline{AB}$ in both directions.
2. Construct a perpendicular bisector to AB through A; construct a perpendicular bisector to AB through B.
3. Construct a circle with center A and radius AB; construct a circle with center B and radius AB.
4. Select a point where circle A meets the perpendicular through A and call that point D. On the same side of AB as D, select the point where circle B meets the perpendicular through B and call that point C.
5. Draw segment CD.

b. Three rigid motions are to be performed on square $ABCD$. The first rigid motion is the reflection through $\overline{BD}$. The second rigid motion is a $90°$ clockwise rotation around the center of the square.

Describe the third rigid motion that will ultimately map $ABCD$ back to its original position. Label the image of each rigid motion A, B, C, and D in the blanks provided.

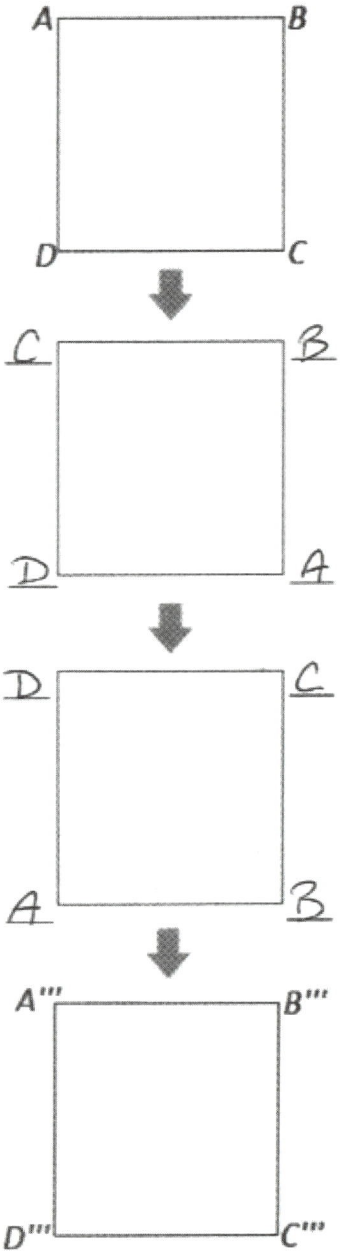

Rigid Motion 1 Description: Reflection through $\overline{BD}$

Rigid Motion 2 Description: $90°$ clockwise rotation around the center of the square.

Rigid Motion 3 Description: Reflection through the line connecting the midpoint of AD and the midpoint of BC.

6. Suppose that $ABCD$ is a parallelogram and that M and N are the midpoints of $\overline{AB}$ and $\overline{CD}$, respectively. Prove that $AMCN$ is a parallelogram.

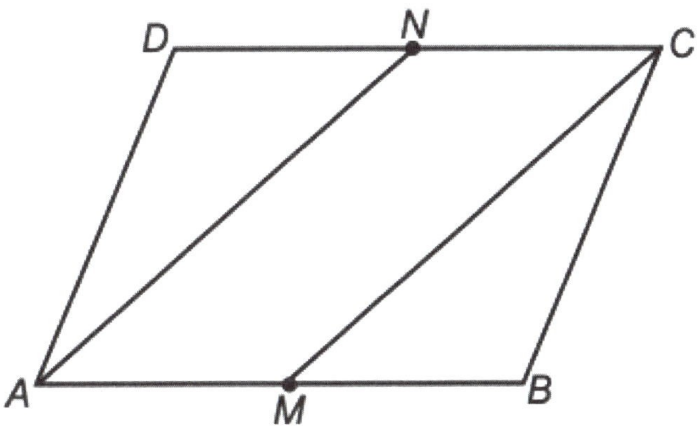

Statements	Reasons
1. M and N are the midpoints of $\overline{AB}$ and $\overline{CD}$	Given
2. ABCD is a parallelogram	Given
3. AB = DC	Opposite sides of a parallelogram are congruent
4. NC = $\frac{1}{2}$DC	N is the midpoint of DC
5. AM = $\frac{1}{2}$AB	M is the midpoint of AB
6. AM = $\frac{1}{2}$DC	Substitution
7. AM = NC	Transitive Property
8. $\overline{AB} \parallel \overline{DC}$	Definition of a parallelogram
9. $\overline{AM} \parallel \overline{NC}$	M is on line AB, N is on line DC
10. AMCN is a parallelogram	opposite sides are equal in length and parallel